Digital Communication Systems Using *SystemVue*™

LIMITED WARRANTY AND DISCLAIMER OF LIABILITY

THE CD-ROM THAT ACCOMPANIES THE BOOK MAY BE USED ON A SINGLE PC ONLY. THE LICENSE DOES NOT PERMIT THE USE ON A NETWORK (OF ANY KIND). YOU FURTHER AGREE THAT THIS LICENSE GRANTS PERMISSION TO USE THE PRODUCTS CONTAINED HEREIN, BUT DOES NOT GIVE YOU RIGHT OF OWNERSHIP TO ANY OF THE CONTENT OR PRODUCT CONTAINED ON THIS CD-ROM. USE OF THIRD-PARTY SOFTWARE CONTAINED ON THIS CD-ROM IS LIMITED TO AND SUBJECT TO LICENSING TERMS FOR THE RESPECTIVE PRODUCTS.

FIREWALL MEDIA (AN IMPRINT OF LAXMI PUBLICATIONS PVT. LTD.) AND/OR ANYONE WHO HAS BEEN INVOLVED IN THE WRITING, CREATION OR PRODUCTION OF THE ACCOMPANYING CODE ("THE SOFTWARE") OR THE THIRD-PARTY PRODUCTS CONTAINED ON THE CD-ROM OR TEXTUAL MATERIAL IN THE BOOK, CANNOT AND DO NOT WARRANT THE PERFORMANCE OR RESULTS THAT MAY BE OBTAINED BY USING THE SOFTWARE OR CONTENTS OF THE BOOK. THE AUTHOR AND PUBLISHER HAVE USED THEIR BEST EFFORTS TO ENSURE THE ACCURACY AND FUNCTIONALITY OF THE TEXTUAL MATERIAL AND PROGRAMS CONTAINED HEREIN. WE, HOWEVER, MAKE NO WARRANTY OF ANY KIND, EXPRESS OR IMPLIED, REGARDING THE PERFORMANCE OF THESE PROGRAMS OR CONTENTS. THE SOFTWARE IS SOLD "AS IS" WITHOUT WARRANTY (EXCEPT FOR DEFECTIVE MATERIALS USED IN MANUFACTURING THE DISK OR DUE TO FAULTY WORKMANSHIP).

THE AUTHOR, THE PUBLISHER, DEVELOPERS OF THIRD-PARTY SOFTWARE, AND ANYONE INVOLVED IN THE PRODUCTION AND MANUFACTURING OF THIS WORK SHALL NOT BE LIABLE FOR DAMAGES OF ANY KIND ARISING OUT OF THE USE OF (OR THE INABILITY TO USE) THE PROGRAMS, SOURCE CODE, OR TEXTUAL MATERIAL CONTAINED IN THIS PUBLICATION. THIS INCLUDES, BUT IS NOT LIMITED TO, LOSS OF REVENUE OR PROFIT, OR OTHER INCIDENTAL OR CONSEQUENTIAL DAMAGES ARISING OUT OF THE USE OF THE PRODUCT.

THE SOLE REMEDY IN THE EVENT OF A CLAIM OF ANY KIND IS EXPRESSLY LIMITED TO REPLACEMENT OF THE BOOK, AND/OR CD-ROM AND ONLY AT THE DISCRETION OF FIREWALL MEDIA (AN IMPRINT OF LAXMI PUBLICATIONS PVT. LTD.).

THE USE OF "IMPLIED WARRANTY" AND CERTAIN "EXCLUSIONS" VARIES FROM STATE TO STATE, AND MAY NOT APPLY TO THE PURCHASER OF THIS PRODUCT.

DIGITAL COMMUNICATION SYSTEMS USING SYSTEMVUE™

DENNIS SILAGE
TEMPLE UNIVERSITY

FIREWALL MEDIA
(An Imprint of Laxmi Publications Pvt. Ltd.)
BANGALORE • CHENNAI • COCHIN • GUWAHATI • HYDERABAD
JALANDHAR • KOLKATA • LUCKNOW • MUMBAI • RANCHI
NEW DELHI • BOSTON, USA

Published by :

FIREWALL MEDIA

(An Imprint of Laxmi Publications Pvt. Ltd.)
113, Golden House, Daryaganj,
New Delhi-110002

Phone : 011-43 53 25 00
Fax : 011-43 53 25 28

www.laxmipublications.com
info@laxmipublications.com

Economy Edition Price : **Rs. 295.00** *Only.* *Edition* : 2009

OFFICES

India

© **Bangalore**	080-26 61 15 61
© **Chennai**	044-24 34 47 26
© **Cochin**	0484-239 70 04
© **Guwahati**	0361-254 36 69, 251 38 81
© **Hyderabad**	040-24 65 23 33
© **Jalandhar**	0181-222 12 72
© **Kolkata**	033-22 27 43 84
© **Lucknow**	0522-220 95 78
© **Mumbai**	022-24 91 54 15, 24 92 78 69
© **Ranchi**	0651-221 47 64

USA

Boston
11, Leavitt Street, Hingham,
MA 02043, USA

FDI-2869-295-DIGITAL COMM SYSTEMS **C—16306/08/08**

Printed at : Sanjeev Offset Printrs, Delhi.

Judgment comes from experience.
Experience comes from poor judgment.

— Robert E. Lee's Truce

To my loving wife Kathleen.

Acknowledgments

The text presented here would not have been possible without the support and encouragement of Agilent EESof and Frank Vincze. Several of my former students have utilized SystemVue for their graduate thesis research and in their professional careers. James Brennan, Michael DeLuca, Robert Esposito, and Soly Pangaribuan provided assistance in the formulation of the topics included in the text.

Contents

Preface

The text *Digital Communication Systems Using* SystemVue is a direct outgrowth of experience in teaching analog and digital communication systems at the undergraduate, professional, and graduate level. The sea of change in this material for the undergraduate and professional student is the introduction of channel *noise* and *nonlinearities* in the analysis of communication systems within the last decade. Prior to this time, analog and digital communication systems were presented by *analytical equations* without noise, and with a hardware laboratory without significant non-linearities.

TESLA by Tesoft is a communication system simulator that was used early in the last decade. TESLA provided a netlist description of a communication system and utilized functions, rather than tokens, for components. SystemView by Elanix, which featured tokens and a graphical user interface, was introduced by the middle of the last decade. SystemVue by Agilent EESof (*http://eesof.tm.agilent.com*) is the current manifestation of this communication system simulator, with advanced capabilities for design and analysis well beyond those presented even in this text.

The motivation for this text is the insight that can be obtained from the implementation of a digital communication system as interconnected tokens in a simulation environment. For example, analytical expressions provide nearly automatic solutions to the spectrum of a modulated signal, but are these spectra really what occurs? There is something rewarding in assembling a digital communication system from tokens, executing a simulation, and then obtaining the spectrum of a temporal signal, all without benefit of the *pro forma* analytical solution or, for that matter, any equations at all.

A SystemVue simulation is a *software brassboard* that allows the virtual construction of a digital communication system to explore the what-ifs of design with channel noise and non-linearities. However, for the undergraduate or graduate student and the professional seeking continuing education, SystemVue can animate the typical block diagrams of a digital communication system. These block diagrams are offered in a conventional text as if their mere appearance will somehow validate the analytical solution.

Digital Communication Systems Using SystemVue is a supplement to a standard text in digital communications. Although not totally devoid of theory, the text relies upon the simulation results in digital communication systems to illustrate and validate the concepts. However, several analyses are presented where it seems that some standard texts are lacking. This includes the correlation receiver for baseband and bandpass signals in the presence of additive white Gaussian noise, source coding to provide a reduction in bit error, and the spectra of baseband line codes.

Audio .wav files are also used as an input to a SystemVue simulation to provide a perceptible assessment of the performance of a digital communication system. For example, μ-law companding (compression and expansion) of a speech signal for pulse code modulation (PCM) is routinely featured in a standard text. However, in this text the μ-law companding PCM system is simulated and the speech processing is audible, as described in Chapter 6, "Sampling and Quantization."

Digital Communication Systems Using SystemVue is intended as a supplementary text for undergraduate students in a contemporary course in communication systems, graduate students in a rigorous first course in digital communications who have not had an exposure to simulation in their undergraduate preparation, and as a reference text for professionals who now seek to acquaint themselves with modern digital communications once having had perhaps a course primarily in analog communications. The concepts presented in this text are also described in standard texts on digital communication systems. The "References" sections at the end of each chapter contain a list of suitable undergraduate and graduate texts and reference books.

Digital Communication Systems Using SystemVue is intended for a broad audience. For the undergraduate student taking a course in digital communication systems, the text provides simulations of these systems and the opportunity to go beyond the lecture course or even the communication hardware laboratory with its traditional experiments. A SystemVue simulation of a digital communication system can augment the hardware laboratory with its complex instruments and techniques, but usually fixed and predictable results.

For the graduate student in a first course in digital communications, the text provides support for the foundations of baseband and bandpass modulation and demodulation, but with the opportunity now to develop investigations and term projects. Finally for the professional, the text facilitates an expansive review and experience with the tenets of digital communication systems.

ON THE CD

The SystemVue Textbook Edition application software and the simulation models described in this text are provided on the accompanying CD-ROM. The SystemVue Textbook Edition software allows the manipulation of parameters of these simulation models to explore the what-ifs of digital communication system design. However, the SystemVue Textbook Edition software is limited in its application by intent. An evaluation version of SystemVue Professional would allow these simulation models to be extended and new designs to be investigated. Additional instructional and reference materials and undergraduate laboratories using SystemVue Professional for educational use are available on the author's Web site: *http://astro.temple.edu/~silage.*

1 Communication Simulation Techniques

In This Chapter

- Capabilities and Limitations of Simulation
- Introduction to SystemVue
- SystemVue Simulation Token Library
- SystemVue Simulation Displays
- Analog Communication Systems Using SystemVue

Communication systems convey information from a source or transmitter over a channel to a sink or receiver. Contemporary communication systems often do so in the presence of additive noise and mild to severe system nonlinearities, which tend to corrupt the transmission. Examining the performance of a communication system as a set of analytical expressions, even if random noise and system nonlinearities can be described adequately, seems to provide little insight or motivation for a deeper understanding.

Communication systems have traditionally been designed by *analysis software* for the analytical equations of the inherent signal processing and by empirical methods using hardware *brassboards* in the laboratory. Because of the increased complexity evident in the modulation and demodulation of signals, *simulation software* is now a *software brassboard* that allows the design and virtual construction of complex communication systems without possessing a complete analytical understanding of the underlying process. Although communication system simulators do exhibit some limitations, their extent can usually be compensated for in the measurement of the performance of the system being modeled.

SystemVue™ by Agilent EESof (*http://eesof.tm.agilent.com*) is one such useful communication system simulation tool. This chapter introduces SystemVue (formerly known as SystemView by Elanix®) in a quick-start manner, with emphasis on the design and analysis environment and the available tokens from the libraries. Examples of the use of SystemVue for the simulation of familiar analog communication systems are provided to introduce the techniques.

The source files for these SystemVue communication system simulation techniques are located in the Chapter 1 folder on the CD-ROM and are identified by the figure number (such as Fig1-55.svu). The appendix provides a description of the complete contents of the CD-ROM.

CAPABILITIES AND LIMITATIONS OF SIMULATION

Communication system simulation software provides a library of modular subroutines, each of which is a complete processing unit or *module*. The modules can be conveniently interconnected either by a primitive text-based *netlist* or a graphical user interface (GUI). The modules are specified by their type and parameters, and they perform signal-processing tasks integral to the simulation of a communication system. These tasks include simple arithmetic addition and multiplication of signals for phase rotation and frequency translation, temporal delays, baseband and bandpass filtering, non-linear processing, deterministic signal and *pseudorandom* data generators, noise generators, and measurement modules, such as temporal and spectral displays [Gardner97].

Modules connected by a GUI are *tokens*. The simulation software allows the collection of GUI tokens as *metasystems* for simplicity of the *top-level* rendering of the communication system model. The simulation of a communication system also requires the visualization of the results. Temporal waveforms, amplitude, phase and power spectra, signal correlations, eye diagrams, signal constellation, and bit error rate (BER) plots render the simulation of the design and the assessment of performance.

However, the simulation of a communication system is not a substitute for the hardware brassboard in the laboratory. Simulation is an approximate model and may not replicate the performance of hardware components exactly. A simulation also executes slower than virtually any brassboard hardware system, and long simulation times are required to assess the performance in the presence of pseudorandom input data and noise. Regardless, simulation provides the means to study the *what-ifs* of communication system design without resorting to the construction of hardware brassboards and the use of complex instrumentation and measurements.

INTRODUCTION TO SYSTEMVUE

SystemVue is an environment for design in *baseband* and *bandpass* communication systems. SystemVue has extensive professional capabilities that support digital signal processing (DSP) in DSP microprocessors and programmable gate arrays (PGAs), analog radio frequency (RF) design, and design to standards such as IEEE 802.11g, Bluetooth and UWB. SystemVue also can be used to simulate the typical *block diagram* representations of the fundamentals of communication systems.

Temporal and spectral signal displays can illustrate both the ideal and degraded performance due to additive noise and system non-linearities. SystemVue also has a sound capability utilizing .wav files for input and output. Sound can be used to extend the analysis to an aural assessment of communication system performance.

The SystemVue simulation environment is presented here in a quick-start manner to begin the investigation of communication system design. However, SystemVue provides an extensive Help Navigator that describes each of the operations and all of the tokens from the various libraries, which can be used as a supplement and reference. SystemVue Application Notes and the Examples from this text are all available from the Help option of the Design window menu.

Design Window

The Design window of the SystemVue simulation environment, as shown in Figure 1.1, consists of a standard Windows menu, a toolbar, horizontal and vertical scroll bars, a Design Area, a Message Area, and a Token Reservoir. The Windows menu provides the usual File operations of opening and printing a SystemVue simulation model, and should be familiar from experience in other Windows applications. The Edit operations enable you to copy all or parts of the Design Area to a word processing application, such as Microsoft Word, to document the work. However, the SystemVue Textbook Edition software provided on the CD-ROM is a limited application and does not allow a simulation model from the text examples to be saved or the tokens to be copied or deleted.

ON THE CD

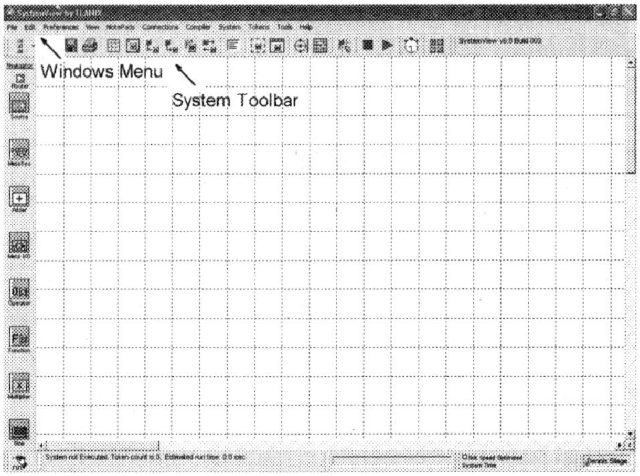

FIGURE 1.1 Design window of the SystemVue simulation environment.

The Preferences...Options menu provides several customization features that can be used to tailor the mode of operation of the Design window. The System Colors feature, as shown in Figure 1.2, allows you to select a background and grid color.

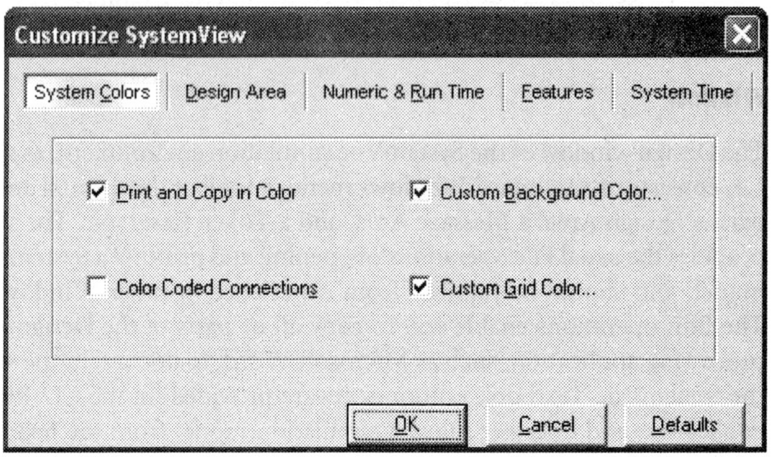

FIGURE 1.2 System Colors feature of the Preferences...Options menu.

The Design Area feature, as shown in Figure 1.3, has a useful Snap to Grid and Drag & Drop Connect mode to easily place SystemVue tokens on the Design window. However, the SystemVue Textbook Edition software provided on the CD-ROM does not allow additional tokens to be placed and connected on the Design window.

ON THE CD

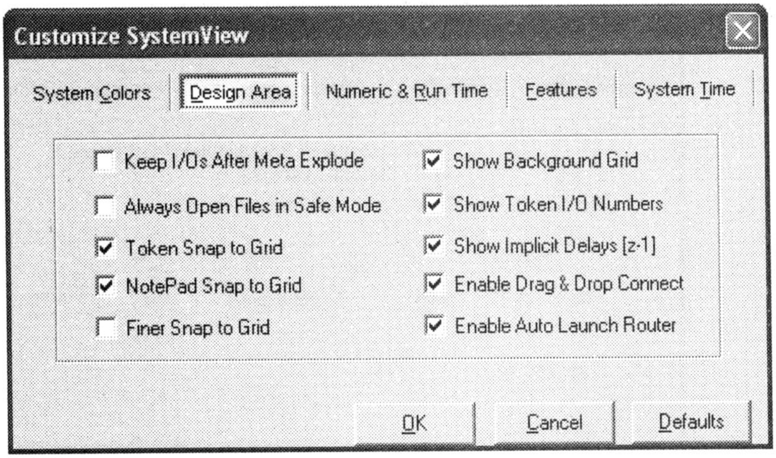

FIGURE 1.3 Design Area feature of the Preferences...Options menu.

The Numeric & Run Time feature, as shown in Figure 1.4, provides a convenient Alert when Simulation Complete alarm that is useful for long simulation times.

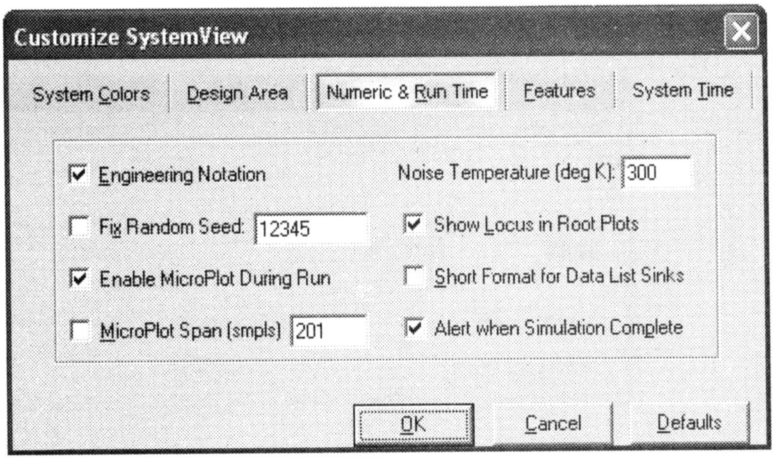

FIGURE 1.4 Numeric & Run Time feature of the Preferences...Options menu.

Among the operations in Features, as shown in Figure 1.5, are audio effects (chimes) and a screensaver. However, since the SystemVue Textbook Edition software provided on the CD-ROM does not allow a simulation model from the examples in the text to be saved, the file backup and insurance features are not provided.

ON THE CD

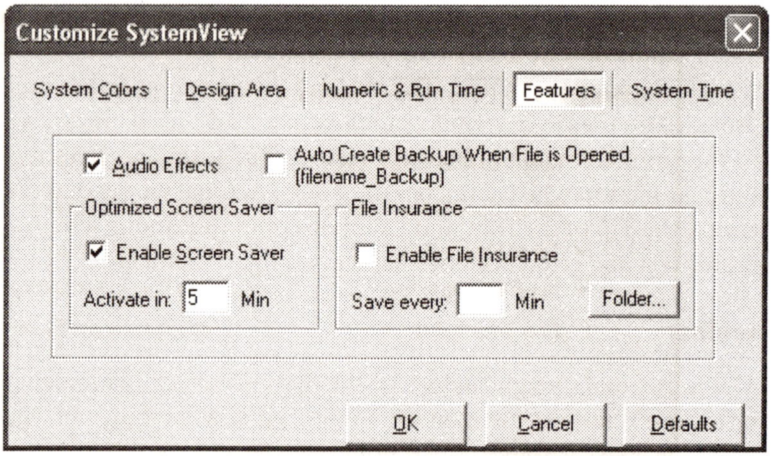

FIGURE 1.5 System Features operation of the Preferences...Options menu.

Although the System Time operation of the Preferences...Options menu (not shown) can set the default simulation sampling rate and the number of samples, the System Time Specification feature of the Design toolbar provides more options, as shown in Figure 1.6.

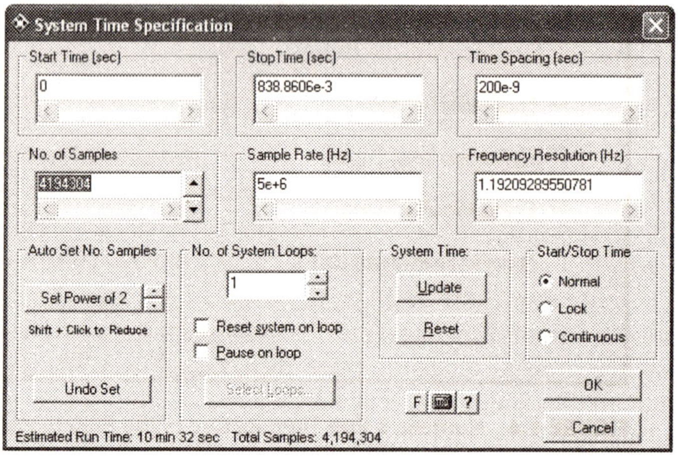

FIGURE 1.6 System Time Specification feature of the Design toolbar.

The View menu (refer to Figure 1.1) provides the capability to zoom the Design window display, resize the Design window, view MetaSystems, and launch the Windows Calculator and Units Converter. The NotePads menu allows you to label the Design window with titles, comments, and parameters. The Connections menu checks the completeness of the communication system simulation by indicating token inputs or outputs that are not connected.

The Compiler, System, Tokens, and Tools menus either duplicate functions provided on the Design window toolbar or are advanced functions that are not used in the simulations described throughout this text. The Help menu includes the Help Navigator, which launches the Adobe Acrobat viewer of the *SystemVue User's Guide*, the Tip of the Day, and Example communications system simulations used in this text and referenced by chapter and figure number.

Design Toolbar

The horizontal Design toolbar of the SystemVue simulation environment is just below the Windows menu in the Design window, as shown in Figure 1.1, and presents the most common operations. The SystemVue Textbook Edition software provided on the CD-ROM is a limited application and does not provide all of the indicated functions. Positioning the Windows cursor over a Design toolbar button provides a *pop-up label* that describes the function.

The first toolbar button is the Change Library Reservoir, which toggles between the Main Libraries and the Optional Libraries of the SystemVue simulation token library. The Open System button opens recently accessed simulation models. Although the Save System button is not available in the SystemVue Textbook Edition software because the simulation model cannot be saved, the Print System button allows the Design window to be printed.

Some Design toolbar buttons are not available in the SystemVue Textbook Edition software. These include the Clear Design Area button, which clears all the system tokens and connections, the Delete Object button, which removes tokens and notepads, the Disconnect Tokens button, which removes a connection, the Connect Token button, which places a connection, and the Duplicate Tokens button, which generates an identical token.

The Reverse Token I/O (input/output) button changes the direction of the input and output of a SystemVue token for display on the Design window. Usually a system is drawn from left to right, with the input on the left of the token and the output on the right, but a feedback loop is often drawn in reverse for clarity.

The New NotePad button of the Design toolbar duplicates one of the operations of the NotePads menu. The Create MetaSystem button *drag-and-drop* selects a group of SystemVue tokens that are displayed as a single token, but is not available in the SystemVue Textbook Edition software. MetaSystem tokens are used for clarity when the simulation becomes complex. However, the View MetaSystem button is available, and selects a MetaSystem token to be expanded and displayed as a separate Design window.

The *View System Root Locus* button and the View System Bode Plot button of the Design toolbar are used for the analysis of a feedback control system and are not used here. The Redraw System button clears and then redisplays the system. If a SystemVue token is moved, its connections are dragged to the new position on the Design window. The Redraw System button conveniently redraws the token connections.

The Cancel Operation button of the Design toolbar halts a SystemVue simulation in progress. The Escape key on the keyboard duplicates this operation. The Run System button initiates a SystemVue simulation and the F5 function key on the keyboard duplicates this operation.

The Define System Time button of the Design toolbar provides the System Time Specification feature (refer to Figure 1.6). System Time Specification is a very important part of the SystemVue communication system simulation, and is set by interactive parameters. Analog signals are simulated in SystemVue by sampled data values that are uniformly spaced in time. These samples are not quantized or rounded off and can be considered to be instantaneous points in time with continuous amplitudes.

A common method for the initial specification of the System Time is to input the Sample Rate (Hz) f_s and the resulting Time Spacing (sec) Δt is then displayed. Alternatively, the Time Spacing is inputted and the resulting Sample Rate is displayed. After the Sample Rate or Time Spacing is set, the Number of Samples N or the Stop Time (sec) is inputted.

Either the Sample Rate or Time Spacing is then displayed. The Start Time (sec) parameter is usually set at its default value of 0 seconds. The Sample Rate f_s or Time Spacing Δt and the Number of Samples N or Stop Time parameters then determine the Frequency Resolution (Hz) Δf (or spectral resolution) of the SystemVue simulation, as given by Equation 1.1.

$$\Delta f = \frac{1}{N\,\Delta t} = \frac{f_s}{N} = \frac{1}{T_{sim}} \tag{1.1}$$

In Equation 1.1, Δf is the frequency resolution of the Fourier spectrum that can be computed from the number of samples N in the simulated temporal signal and Δt is the uniform time spacing between the samples in the SystemVue simulation. The total simulation time T_{sim} sec in Equation 1.1 is the difference between the Stop Time and the Start Time.

An appropriate value for the Sample Rate $f_s = 1/\Delta t$ is necessary for an accurate SystemVue simulation of a communication system, even if the Fourier spectrum is not required. An analog signal, which has a continuous representation in time and amplitude, is simulated by samples that are discrete in time, but continuous in amplitude. An analog sinusoidal signal, also called a single *tone*, is given by Equation 1.2.

$$m(t) = A \sin 2\pi f_m t \tag{1.2}$$

In Equation 1.2, $m(t)$ is the analog signal, A is the peak amplitude in volts (V) and f_m is the sinusoidal frequency in Hz. The SystemVue samples that correspond to the single sinusoidal signal are given by Equation 1.3.

$$m[n] = A \sin 2\pi f_m n \Delta t \tag{1.3}$$

In Equation 1.3, $m[n]$ is the *discrete-time* or *sampled* signal with continuous amplitude at the nth data sample and n ranges from 0 to $T_{sim}/\Delta t = N$.

The Analysis Window button of the Design toolbar, initiates the Analysis window where temporal, spectral, and specialized waveforms are computed and displayed. The Change Library Reservoir button is shown in Figure 1.7. SystemVue provides a set of Main Libraries and Optional Libraries on the vertical Token Library toolbar, as shown in Figure 1.7.

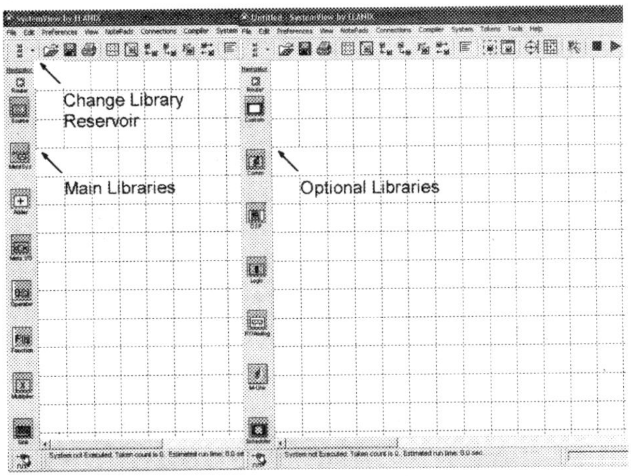

FIGURE 1.7 Main Libraries and Optional Libraries of the Token Library toolbar.

Although the SystemVue Textbook Edition software does not allow tokens to be added to a simulation model or changed to a token from another Library, the type and functional parameters of an existing token can be modified to explore the what-ifs of communication system design. Double-clicking on a token in a simulation model opens the SystemVue token library window, and selecting a token within that Library then opens the parameter window.

SYSTEMVUE SIMULATION TOKEN LIBRARY

The SystemVue library provides a set of tokens that are used to simulate a communication system. The Main Library tokens are primarily divided into the basic classes of sources, sinks, operators, and functions. The Optional Library provides tokens for specialized classes, such as custom, logic, and communication systems (refer to Figure 1.7). The Adder Token and Multiplier Token are not libraries, but are tokens that output the sum and product of any number of input signals.

Not all of the available tokens or libraries will be described in this quick-start introduction to SystemVue. The Custom Library, the DSP Library, the RF/Analog Library, the M-Link (Matlab script) Library, and the Scheduler Library are not used in the SystemVue simulations here. However, the versatility and effectiveness of a SystemVue simulation of a communication system can still be readily appreciated with the tokens employed.

Source Token Library

The Source Token Library is the initial generic token on the vertical Main Libraries toolbar (refer to Figure 1.7). Figure 1.8 shows the Periodic Token feature of the Source Library. The available periodic source tokens include the Swept Frequency token, the Phase Shift Keyed (PSK) Carrier token, the Pulse Train token, the Sawtooth token, and the Sinusoid token. Each of these tokens has a specification window to set the amplitude, frequency, phase, pulse width, offset voltage, and specialized parameters. Communication system examples can illustrate the use of these tokens.

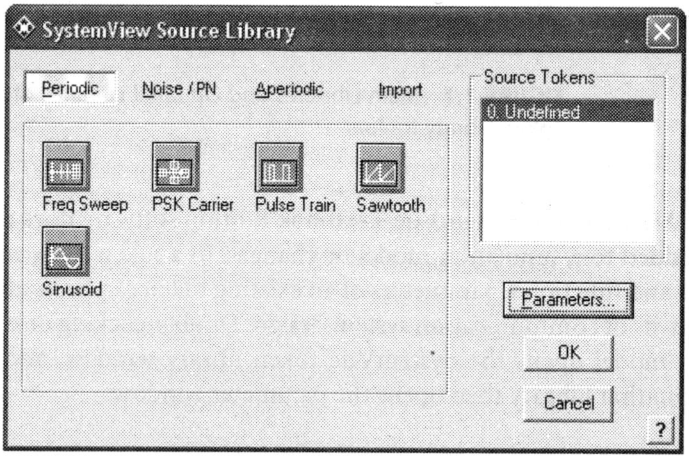

FIGURE 1.8 Periodic Token feature of the Source Library.

The Swept Frequency token can be used to provide a variable frequency, constant amplitude input to a communication system to assess its performance, such as an input to a linear system filter to measure its frequency response. The Phase Shift Keyed Carrier token can provide bandpass modulation of a data source by specifying the specialized parameters of the bit rate (b/sec) and the number of bits taken together as a data symbol (b/symbol).

The Pulse Train token can provide a pulse of variable duration and frequency for logic clock and sample control signals. The Sawtooth token can be used to provide a linear ramp voltage input to sweep the output frequency of a frequency modulator. The Sinusoid token can provide the quintessential input voltage signal for a system.

Figure 1.9 shows the Noise/PN Token feature of the Source Library. The available noise source tokens include the Gaussian Noise token, the PN Sequence (*pseudonoise* sequence) token, the Thermal Noise token, and the Uniform Noise token. The Gaussian Noise token generates a random voltage output for which either the mean μ and the standard deviation σ in volts (V) or the power spectral density (PSD) N_o in W/Hz into either a normalized 1 Ω or 50 Ω load can be specified.

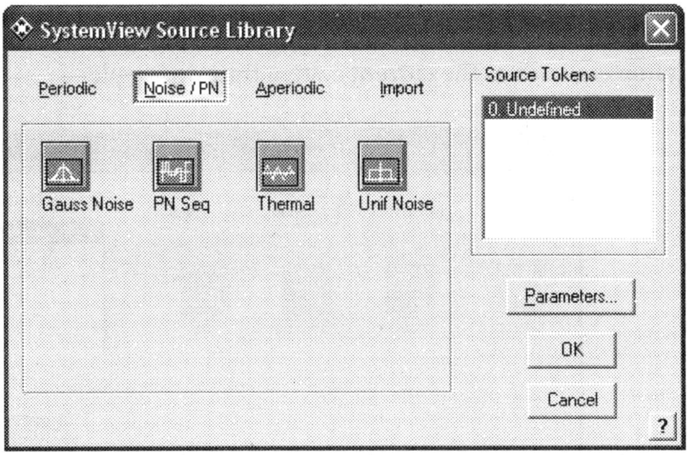

FIGURE 1.9 Noise/PN Token feature of the Source Library.

The PN Sequence token generates a pseudonoise (or pseudorandom) sequence of multilevel pulses at a specified rate Hz. The output voltage levels are nominally polar in the range $\pm A$ V, with levels separated by $2A/(N-1)$ V and where N is the number of levels and A is the amplitude. However, the voltage offset parameter can be set to $+A$ V and the PN Sequence token would output unipolar pseudonoise pulses of amplitude 0 V and $2A$ V. Specifying a rate = 1 kHz, $N = 2$, $A = 0.5$ V, the voltage offset = 0.5 V, and a pulse duration of 0.5 msec results in binary, unipolar, 50% duty cycle, pseudonoise pulse of 0 V and 1 V at 1 kb/sec.

The Thermal Noise token generates a random voltage output modeled as the random motion of electrons in a resistor of $R\ \Omega$, with a single-sided average power $P_{thermal}$ V^2 given by Equation 1.4.

$$P_{thermal} = 4\,k\,T\,R\,\Delta f \quad V^2 \tag{1.4}$$

In Equation 1.4, the Boltzmann constant $k = 1.38 \times 10^{-23}$ joule/ °K, T is the ambient temperature in °K, and Δf is the bandwidth of the system in Hz. If $T = 293$ °K or room temperature, $\Delta f = 10$ kHz and $R = 10$ KΩ, then $P_{thermal} = 1.6 \times 10^{-12}\ V^2$ or an RMS voltage across the resistor $V_{rms} \approx 1.3\ \mu V$.

The Uniform Noise token generates noise with voltage amplitude values that are uniformly distributed in the probability density function (pdf) between specified minimum and maximum values, or with a power spectral density (PSD) N_o in W/Hz, into either a normalized 1 Ω or 50 Ω resistive load that can be specified.

Figure 1.10 shows the Aperiodic Token feature of the Source Library. The available aperiodic source tokens include the Custom token, the Impulse token, the Step Function token, and the Time token. The Custom token generates an output that can be either a constant, such as a DC voltage, or a function of SystemVue parameters, such as the current system time or sample.

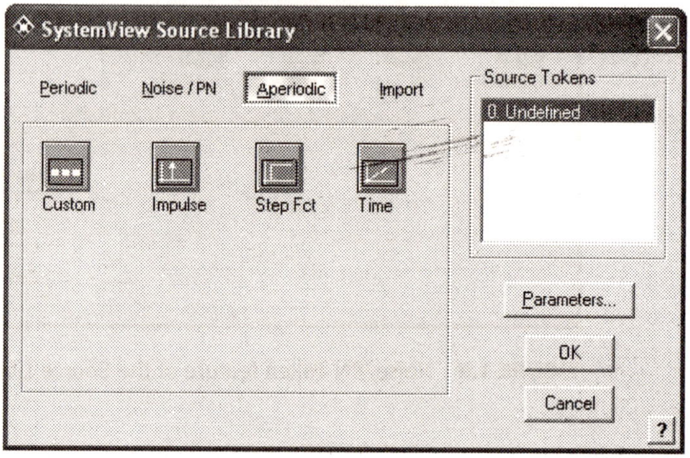

FIGURE 1.10 Aperiodic Token feature of the Source Library.

The Impulse token simulates the ideal unit impulse function, which can be used to excite a linear time-invariant (LTI) system. The Step Function token generates a voltage step of amplitude A V, which occurs at a parameter start time T_{start} sec. The Time token generates a voltage ramp with slope A V/sec at a fixed start time of 0 sec.

Figure 1.11 shows the Import Token feature of the Source Library. The available import source tokens include the External 1 Channel token, the External 2 Channel token, the Wave 1 Channel token, and the Wave 2 Channel token.

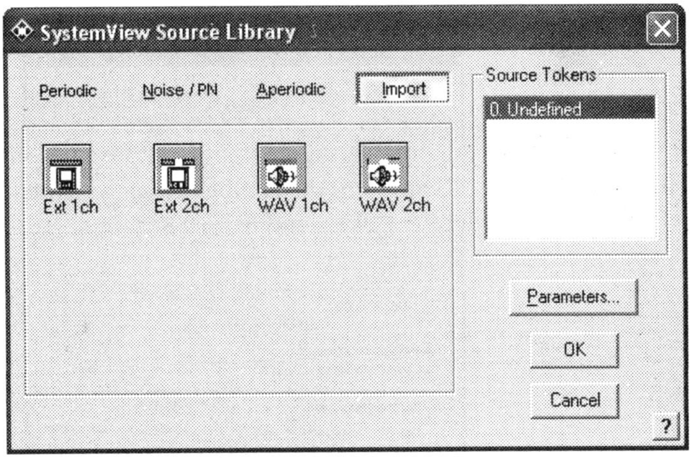

FIGURE 1.11 Import Token feature of the Source Library.

The External Channel tokens import either one or two streams of external data for input as a source token. Figure 1.12 shows the parameter window for the External Channel tokens, which opens an external file as text, 8- or 16-bit integers, or 32-bit single or 64-bit double floating point numbers.

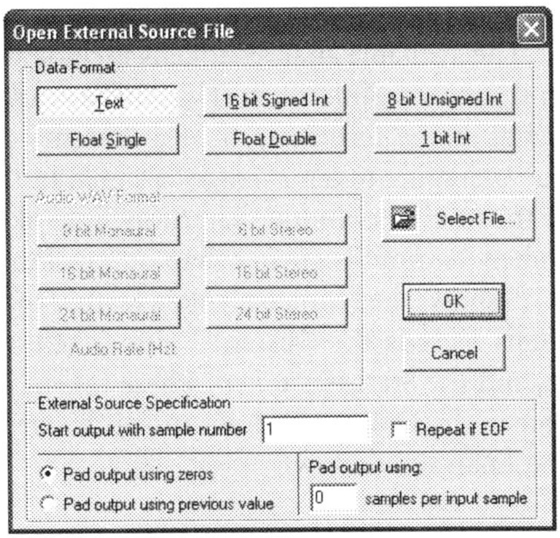

FIGURE 1.12 Open External Source File parameters for the External Channel tokens.

The Wave Channel tokens import either 8-, 16-, or 24-bit monaural or stereo streams of external .wav Windows sound files for input as a source token. Figure 1.13 shows the parameter window for the Wave Channel tokens, which can be selected to launch a Windows audio player.

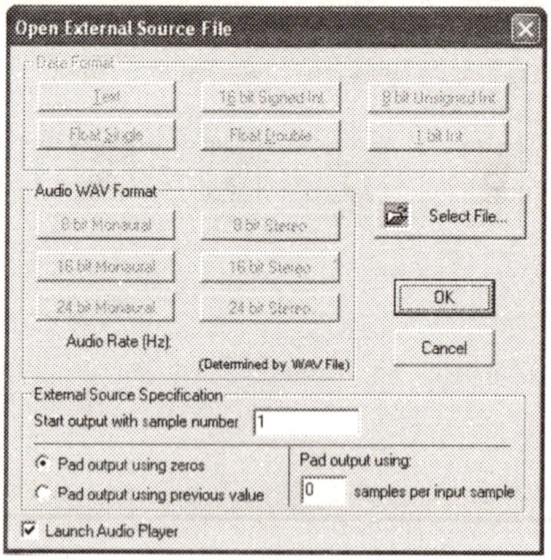

FIGURE 1.13 Open External Source File parameters for the Wave Channel tokens.

Sink Token Library

The Sink Token Library is the final generic token on the vertical Main Libraries toolbar (refer to Figure 1.7). Figure 1.14 shows the Analysis Token feature of the Sink Library. The available analysis tokens include the Analysis token, the Averaging token, and the Stop Sink token. The Analysis token provides the temporal display of a signal at the termination of a SystemVue simulation in the Analysis window. The Averaging token averages a signal in the SystemVue simulation and displays it in the Analysis window. The Stop Sink token can terminate a SystemVue simulation when its input is greater than or equal to the fixed threshold.

Figure 1.15 shows the Numeric Token feature of the Sink Library. The available numeric tokens include the Current Value token, the Data List token, the Final Value token, and the Statistics token. These tokens display their numeric value in both real-time in the Design window, and at the termination of a SystemVue simulation in the Analysis window.

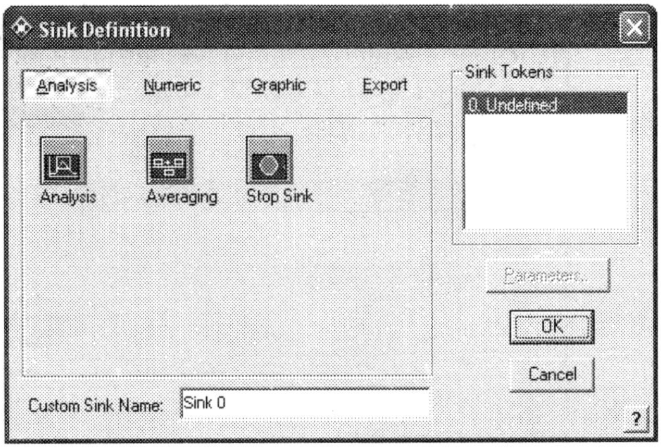

FIGURE 1.14 Analysis Token feature of the Sink Library.

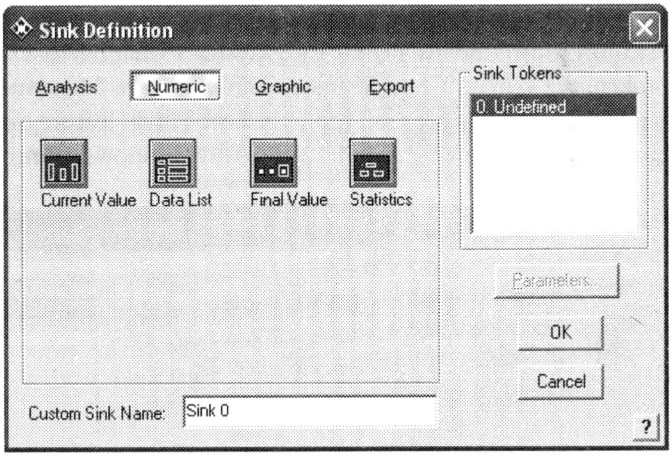

FIGURE 1.15 Numeric Token feature of the Sink Library.

The real-time display in the Design window is useful for probing the intermediate results of a simulation, but the graphical resolution is low and the cursor readout functions in the Analysis windows are not available. The Statistics token computes and displays the mean, the variance, the standard deviation, the adjacent sample correlation coefficient, and the minimum and maximum values of a signal.

Figure 1.16 shows the Graphic Token feature of the Sink Library. The available graphic tokens are the SystemVue token and the RealTime token. The SystemVue token is a simulation display that appears in the Analysis window. The RealTime token is the graphical version of the numeric tokens and displays both in real-time in the Design window, and at the termination of a SystemVue simulation in the Analysis window.

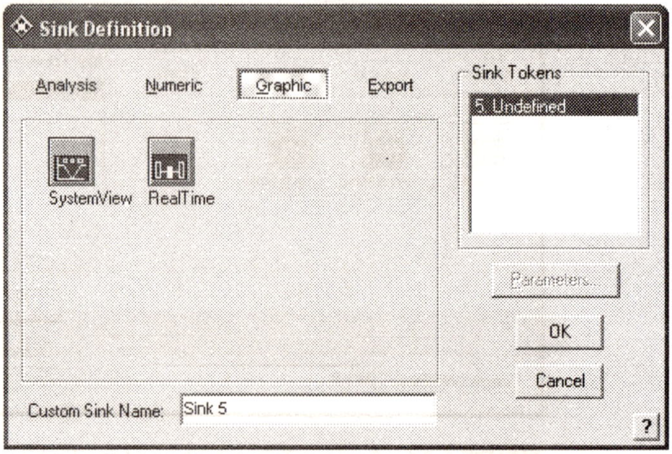

FIGURE 1.16 Graphic Token feature of the Sink Library.

Finally, Figure 1.17 shows the Export Token feature of the Sink Library. The available export source tokens include the External 1 Channel token, the External 2 Channel token, the Wave 1 Channel token, and the Wave 2 Channel token. These export tokens are similar in operation to the Import Token feature of the Source Library, as shown for source files in Figure 1.12 and .wav Windows sound files in Figure 1.13.

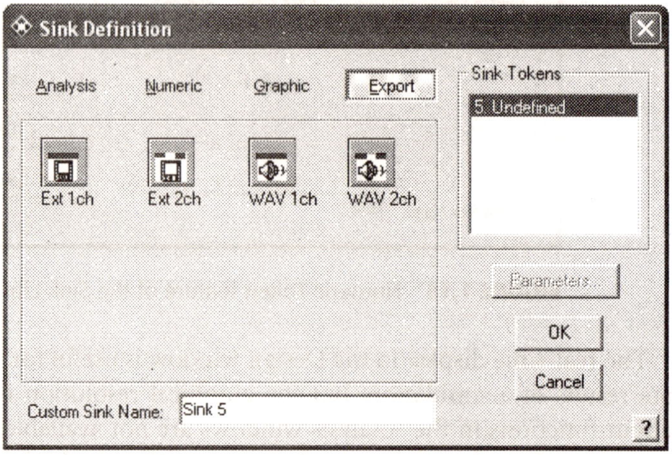

FIGURE 1.17 Export Token feature of the Sink Library.

Operator Token Library

The Operator Token Library is one of the generic tokens on the vertical Main Libraries toolbar (refer to Figure 1.7). Figure 1.18 shows the Filters/Systems Token feature of the Operator Library. The available filters and systems tokens include the Average token, the FFT (fast Fourier transform) token, the Linear Systems Filter token, and the OSF (order statistics filter) token. The Average token outputs the *moving average* $s_{MA}(t)$ of an input signal $s(t)$, as given by Equation 1.5.

$$s_{MA}(t) = \frac{1}{T}\int_{t-T}^{t} s(\alpha)\,d\alpha \tag{1.5}$$

In Equation 1.5, T is the *time window* of the moving average output.

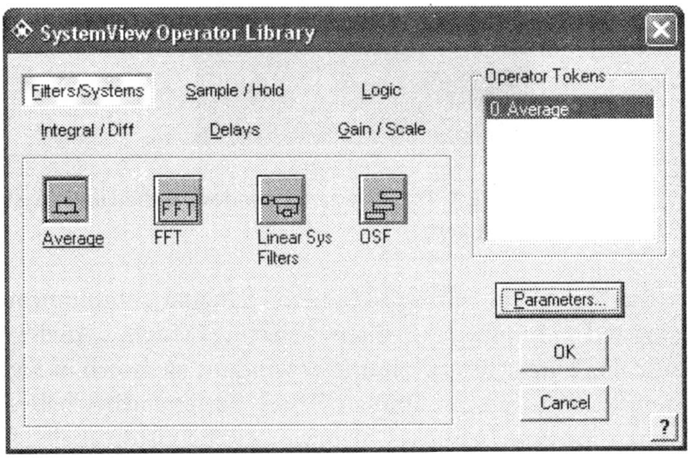

FIGURE 1.18 Filters/Systems Token feature of the Operator Library.

The FFT token performs the block-oriented FFT (spectrum) of an input signal. The FFT token output begins one SystemVue Simulation Time (sample) after the specified FFT sample size (FFT_{size}) and is continuous thereafter. The DC component of the output spectrum will occur at simulation sample number 1.5 FFT_{size}.

Each simulation sample represents a frequency resolution $f_{res} = f_{system}/FFT_{size}$, where f_{system} is the SystemVue Sample Rate Hz. The observed peak magnitude of the output of the FFT token is scaled by $FFT_{size}/2$ and is either the real or imaginary component of the complex spectrum, the magnitude, the phase, or the unwrapped (no rollover at 2) phase.

The Linear Systems Filter token implements the finite impulse response (FIR) and infinite impulse response (IIR) digital filters in the *z-domain*, the transfer function or Laplace filter in the *s-domain*, and the analog and communication filter system design. Figure 1.19 shows the parameter window for the Linear Systems Filter token. Chapter 6, "Filters and Linear Systems," in the *SystemVue User's Guide* (available from the Help menu) describes the complete operation of the Linear Systems Filter token.

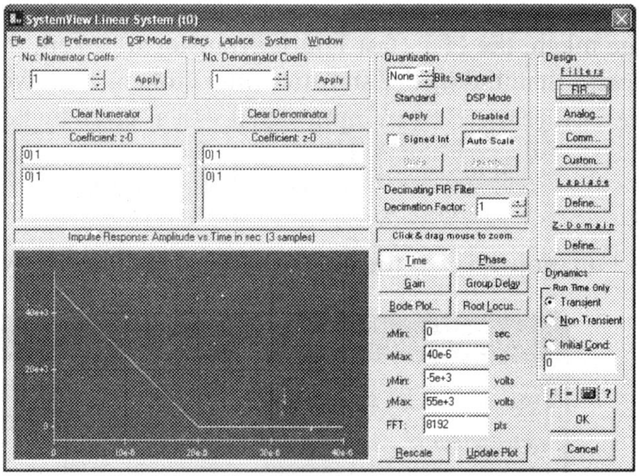

FIGURE 1.19 Parameter window of the Linear Systems Filter token.

The SystemVue simulations here in digital communication systems employ analog and communication filters. Selecting Design... Analog in Figure 1.19 opens the Analog Filter Library parameter window, as shown in Figure 1.20. The Linear Systems Filter token can simulate Bessel, Butterworth, Chebyshev, elliptic, and linear phase lowpass, bandpass, highpass, and bandstop analog filters.

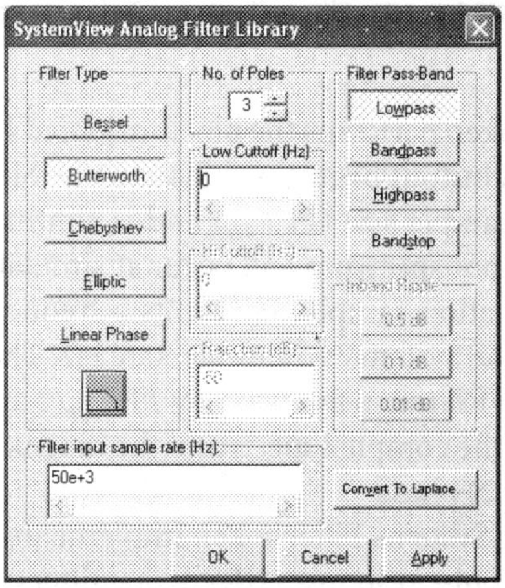

FIGURE 1.20 Analog Filter Library parameter window.

Selecting Design... Comm in Figure 1.19 opens the Communications Filter Library parameter window, as shown in Figure 1.21. The Linear Systems Filter token can simulate the root raised cosine, raised cosine, sinc (sin x/x) and Gaussian filters commonly used in digital communication systems [Lathi98].

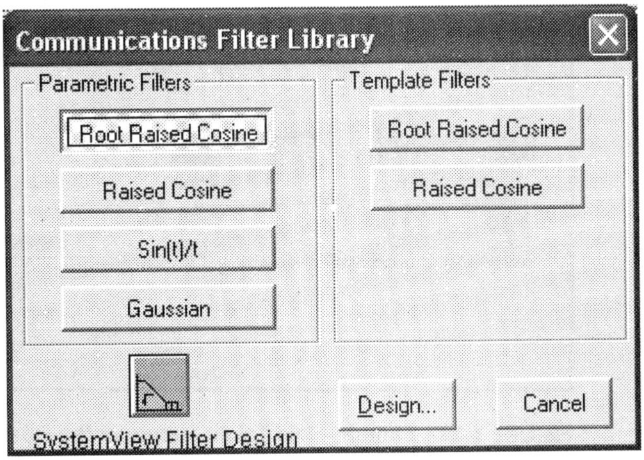

FIGURE 1.21 Communications Filter Library parameter window.

The OSF token performs order statistics filtering (OSF), where the output is the specified rank of the input signal within the specified time window. A parameter of rank = 50% produces the median filter and a rank = 100% produces the maximum value of the signal within the specified time window.

Figure 1.22 shows the Sample/Hold Token feature of the Operator Library. The available sample and hold tokens include the Decimate token, the Hold token, the Resample token, the Sampler token, the Peak Hold token, and the Sample/Hold token. The Sample/Hold Tokens manipulate the sampled signals of the SystemVue simulation. The Decimate token *downsamples* the simulation signal from the SystemVue Sample Rate f_{system} by an integer decimation factor. The Hold token is used after the Decimate token to hold either the last sample or insert zeros and restore the SystemVue Sample Rate.

The Resample token samples the input signal at a rate other than the SystemVue Sample Rate by holding a sample and producing an output at a new rate without interpolation. The Sampler token performs the same operation as the Resampler token, but either an average of the samples of the input signal within a specified *aperture* or a linear interpolation between sample points can be selected.

The Peak Hold token outputs the maximum value, minimum value, maximum location, or minimum value of the input signal. The Sample/Hold token *tracks* or outputs the input signal if the control signal is above or holds the input signal value if it is below the specified threshold voltage.

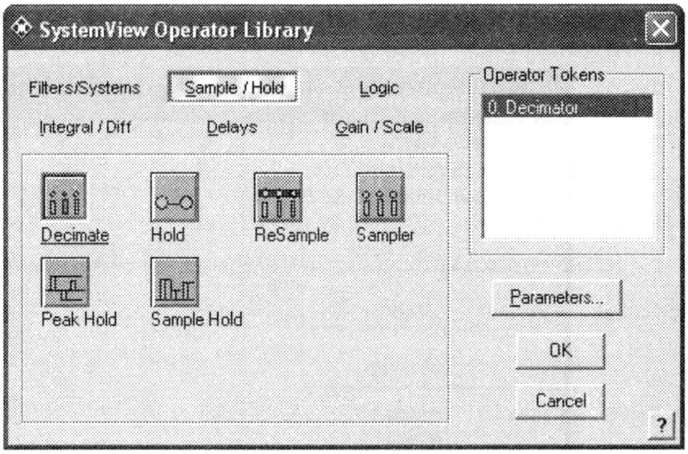

FIGURE 1.22 Sample/Hold Token feature of the Operator Library.

Figure 1.23 shows the Logic Token feature of the Operator Library. The available logic tokens include not only the standard AND token, the OR token, the NOT token, the NAND token, and the XOR token, but also the Compare token, the Pulse token, the Switch token, the Select token, and the Max Min token. These standard logic tokens are also available in the Logic Token Library of the Optional Libraries toolbar (refer to Figure 1.7).

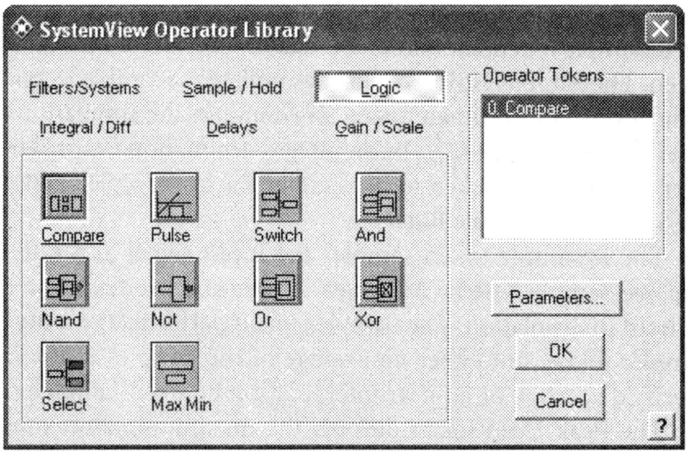

FIGURE 1.23 Logic Token feature of the Operator Library.

The input threshold voltage and the logic *true* output voltage and the logic *false* output voltage are not required to be the same as any standard logic families (such as TTL or CMOS) and can be specified to be any value. This feature is useful when interfacing logic tokens to the other types of available tokens.

The Compare token provides six selectable outputs: $A = B$, A != B, $A >= B$, $A > B$, $A <= B$, and $A < B$, where A and B are the two Boolean input signals. The Pulse token produces an output pulse equal to the false output voltage as the default, or a specified duration pulse at the logic true output voltage if the input crosses the specified threshold voltage parameter (from a low to high level).

The Switch token can select and output one of up to 20 input signals $x_n(t)$ as a function of the control voltage input and the specified control voltage, as given by Equation 1.6.

$$o(t) = x_n(t)$$
$$n = \text{int}\left[\frac{N(c(t) - C_{min})}{C_{max} - C_{min}}\right] \qquad 0 \le n \le N-1 \qquad (1.6)$$

In Equation 1.6, $o(t)$ is the output signal selected, $c(t)$ is the input control voltage, C_{max} and C_{min} are the specified maximum and minimum control voltages, N is the number of input signals $x_n(t)$, and int is the integer (whole number) function.

The Select token has a control input signal and two possible output signals. One output signal (y_0) tracks the input signal if the control input voltage is greater than or equal to the specified threshold voltage and 0 otherwise. The other output signal (y_1) tracks the input signal if the control input voltage is less than the specified threshold voltage and 0 otherwise, and is essentially the signal complement of the other output signal. The Max Min token can select up to 19 input signals and output the maximum value on 1 output signal and the minimum on another output signal, with specified gain and voltage offset.

Figure 1.24 shows the Integral/Differential Token feature of the Operator Library. The available logic tokens include the Integral token, the Derivative token, and the PID (proportional-integral-differential) token. The Integral token provides the zeroth order (rectangular) or first order (trapezoidal) integration of the input signal from the samples beginning at the start of the SystemVue Simulation Time, with a specified initial condition for the output signal. The Derivative token provides the derivative of the input signal from the samples beginning at the start of the SystemVue Simulation Time.

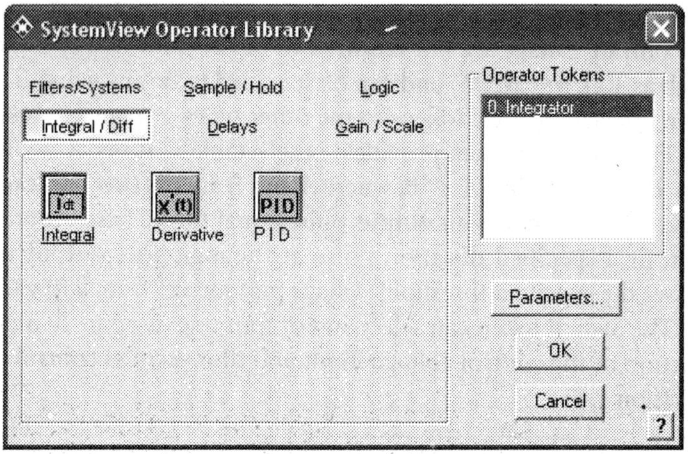

FIGURE 1.24 Integral/Differential Token feature of the Operator Library.

The PID token provides an output signal $o(t)$, which is the sum of proportional, derivative, and integral components of the input signal, with specified gains (G_p, G_d and G_I) for each, as given by Equation 1.7.

$$o(t) = G_p\, x(t) + G_I \int_{T_{start}}^{T_{stop}} x(\alpha)\, d\alpha + G_d \frac{d\,x(t)}{dt} \tag{1.7}$$

The proportional-integral-differential operation is the basis of a feedback control system.

Figure 1.25 shows the Delays Token feature of the Operator Library. The available logic tokens include the Delay token, the Sample Delay token, and the Variable Delay token. The Delay token delays the input signal by a fixed time in seconds and will linearly interpolate the output signal to the time delay specified, if selected. The Sample Delay token delays the input signal by a fixed number of samples. The Variable Delay token delays the input signal by a variable amount of time τ specified by a control voltage input signal, as given by Equation 1.8.

$$\tau = \frac{\left(\tau_{max} - \tau_{min}\right)}{\left(C_{max} - C_{min}\right)} \left(c(t) - C_{min}\right) + \tau_{min} \tag{1.8}$$

In Equation 1.8, τ is the delay in seconds, $c(t)$ is the input control voltage, C_{max} and C_{min} are the specified maximum and minimum control voltages, and τ_{max} and τ_{min} are the specified maximum and minimum delay times in seconds. If $C_{min} \geq C_{max}$ then $\tau = \tau_{max}$ and if $C_{in} \leq C_{min}$ then $\tau = \tau_{min}$. The Variable Delay token is useful in a SystemVue simulation of the effect of *jitter* on the performance of a digital communication system.

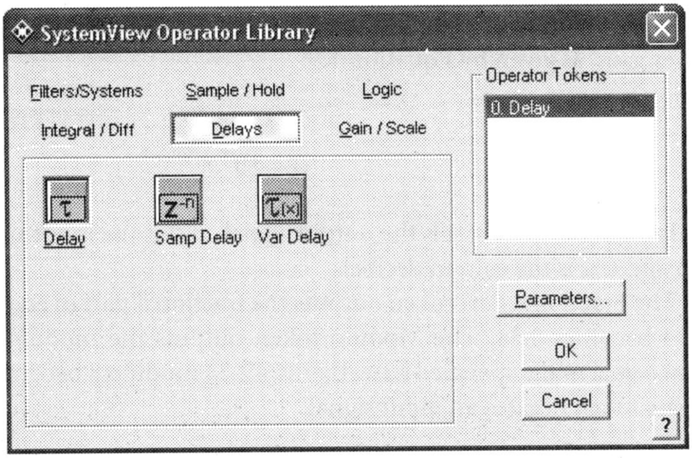

FIGURE 1.25 Delays Token feature of the Operator Library.

Figure 1.26 shows the Gain/Scale Token feature of the Operator Library. The available gain and scale tokens include the Digital Scale token, the Gain token, the Fractional Part token, the Modulo token, and the Negate token. The Digital Scale token extracts the specified number of significant bits from the input word as a signal with a representation of 1 V/bit. The input signal must be *positive integer valued* and, for example, with an input of 13 V representing binary 1101, 4 specified input bits (15 V maximum or binary 1111), and 2 as the specified number of extracted significant bits, the output signal is 3 V representing binary 11.

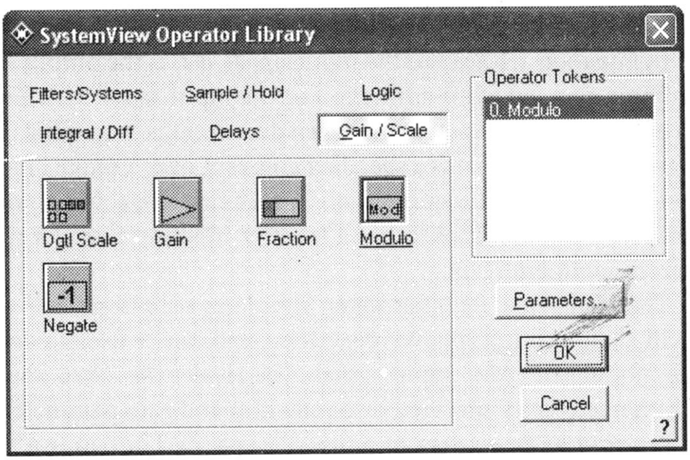

FIGURE 1.26 Gain/Scale Token feature of the Operator Library.

The Gain token multiplies the input signal by a fixed linear gain or *decibel* (dB) power gain, as given by Equation 1.9.

$$o(t) = G\,x(t)$$
$$G = 10^{G_{dB}/20} \tag{1.9}$$

In Equation 1.9, $o(t)$ is the output signal, $x(t)$ is the input signal, G is the linear gain and G_{dB} is the gain in decibels.

The Fractional Part token outputs the fractional part of an input signal, that is, 12.34 becomes 0.34. The Modulo token outputs the modulus operation for an input signal in the specified base, that is, 12.24 modulo 8 becomes 4.34. Finally, the Negate token negates an input signal.

Function Token Library

The Function Token Library is one of the generic tokens on the vertical Main Libraries toolbar (refer to Figure 1.7). Figure 1.27 shows the Nonlinear Token feature of the Function Library. The available nonlinear tokens are useful for digital communication system simulation and include the Block token, the Coulomb token, the Dead Band token, the Half Wave Rectifier token, the Hysteresis token, the Limit token, the Quantize token, the Rectifier token, and the Transfer Function token. The nonlinear tokens process an input signal by a voltage transfer function G_{NL}, as given by Equation 1.10.

$$o(t) = G_{NL}(x(t))\,x(t) \tag{1.10}$$

In Equation 1.10, $x(t)$ is the input signal, $o(t)$ is the output signal, and G_{NL} is the nonlinear gain, which is a function of the input signal $x(t)$. The transfer function can be inferred from each of the token descriptions or graphic. The Block token has a transfer function that is a specified constant over a range of input. The Coulomb token has a transfer function that is a specified slope, but a nonzero output intercept. The Dead Band token has a transfer function that is a specified slope, but a nonzero input intercept.

The Half Wave Rectifier token simulates a diode or half wave rectifier with a specified input intercept and ideal slope, and can be used to demodulate an amplitude modulated (AM) input signal. The Hysteresis token has a transfer function with characteristics specified by bandwidth and gain parameters. Hysteresis represents the *history* dependent response of a system. The bandwidth is indexed relative to the SystemVue Sample Rate, where a small relative bandwidth (~3%) produces a sluggish hysteresis response (lowpass filter), while a larger relative bandwidth (~30%) produces a stiff hysteresis response.

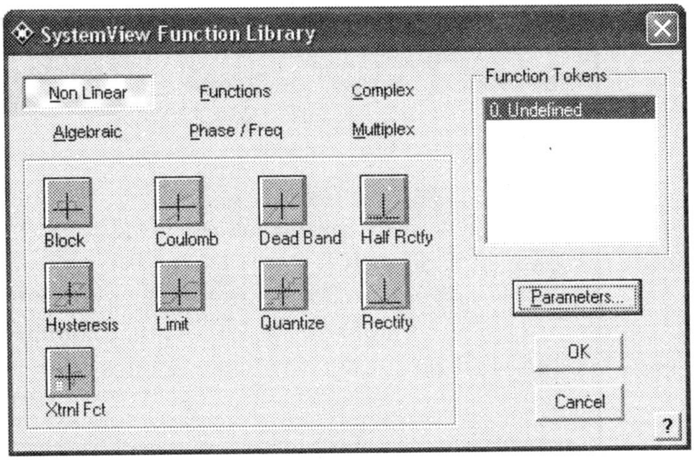

FIGURE 1.27 Nonlinear Token feature of the Function Library.

The Limit token simulates a linear response up to a maximum input, then a *saturated* output beyond that input. The limit function can be used to remove incidental amplitude variations, as in a frequency modulated (FM) input signal. The Quantize token simulates an analog-to-digital converter (ADC) with an output that is connected directly to a digital-to-analog converter (DAC), with a specified number of bits of resolution.

The Rectifier token simulates two diodes as a full wave rectifier with a specified input intercept and slope. Finally the Transfer Function token produces a custom transfer function that is specified in an external text file. The file data can be in two possible text formats: input and output pairs with no header, separated by a space, tab, or comma with the input in ascending order; or equally spaced data with a header, as described in the *SystemVue User's Guide.*

Figure 1.28 shows the Functions Token feature of the Function Library. A useful function token is the Custom token, which can be used to provide a fixed or variable voltage threshold in the detection of a signal. The remainder of the function tokens is described in the *SystemVue User's Guide.*

Figure 1.29 shows the Algebraic Token feature of the Function Library. Two useful algebraic tokens are the Power token, which can be used to square an input signal, and the Polynomial token, which can provide voltage offsets and gain for the processing of a signal. The remainder of the algebraic tokens is described in the *SystemVue User's Guide.*

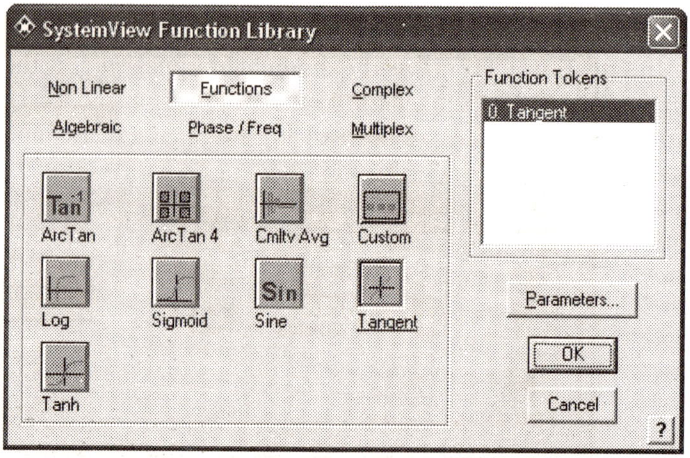

FIGURE 1.28 Functions Token feature of the Function Library.

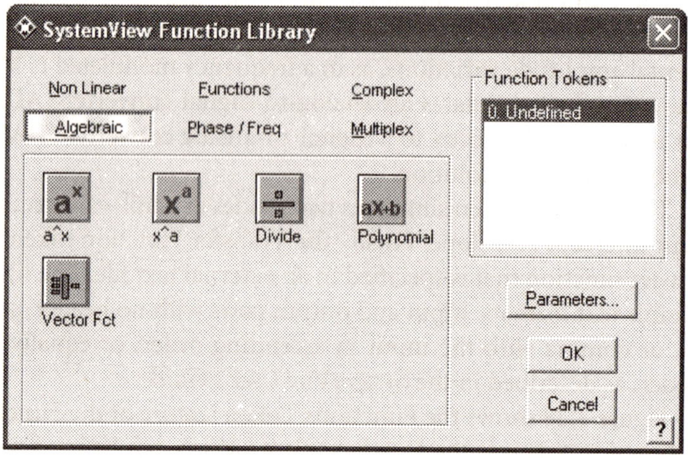

FIGURE 1.29 Algebraic Token feature of the Function Library.

Figure 1.30 shows the Phase/Frequency Token feature of the Function Library. The Phase Modulator token modulates the phase of a sinusoidal carrier, with a specified carrier amplitude A_c V, carrier frequency f_c Hz, modulation gain or *phase deviation factor* k_p 2π (radians)/V, and both sine and cosine outputs, as given by Equation 1.11.

$$\begin{aligned} s_0(t) &= A \sin(2\pi f_c t + k_p m(t) + \theta) \\ s_1(t) &= A \cos(2\pi f_c t + k_p m(t) + \theta) \end{aligned} \tag{1.11}$$

In Equation 1.11, $s_i(t)$ is the modulated signal, $m(t)$ is the baseband information signal, and θ is the fixed initial phase of the sinusoidal carrier. The Phase Modulator token is the basis for both analog and digital phase modulation (PM).

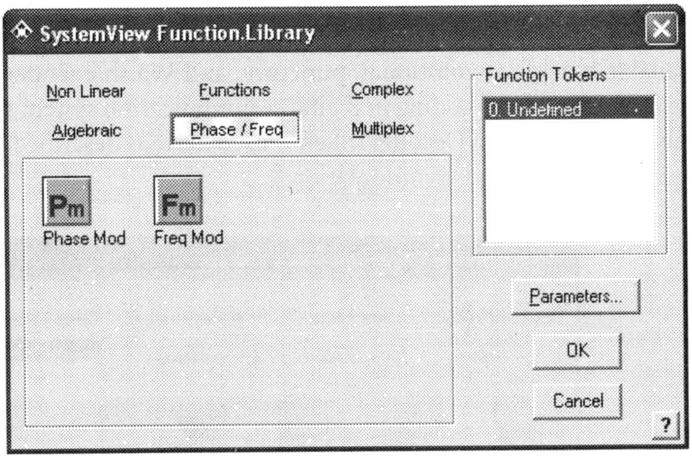

FIGURE 1.30 Phase/Frequency Token feature of the Function Library.

The Frequency Modulator token modulates the frequency of a sinusoidal carrier, with a specified carrier amplitude A_c V, carrier frequency f_c Hz, modulation gain or *frequency deviation factor* k_f Hz/V, and both sine and cosine outputs, as given by Equation 1.12.

$$\begin{aligned} s_0(t) &= A\ \sin(2\pi f_c t + k_f \int m(\alpha)\mathrm{d}\alpha + \theta) \\ s_1(t) &= A\ \cos(2\pi f_c t + k_f \int m(\alpha)\mathrm{d}\alpha + \theta) \end{aligned} \qquad (1.12)$$

In Equation 1.12, $s_i(t)$ is the modulated signal, $m(t)$ is the baseband information signal, and θ is the fixed initial phase of the sinusoidal carrier. The Frequency Modulator token is the basis for both analog and digital frequency modulation (FM). A comparison of Equation 1.11 and Equation 1.12 demonstrates that both PM and FM are forms of *angle modulation*, since the information signal $m(t)$ is contained within the angle of the sinusoidal carrier. PM and FM differ in that in FM, the information signal $m(t)$ is integrated before modulating the angle of the sinusoidal carrier [Lathi98].

Finally, the Multiplex Token and the Complex Token features of the Function Library are not used in the SystemVue simulation of the digital communication systems described here. These functions are useful for the simulations of advanced digital communication systems and are described in the *SystemVue User's Guide.*

Communications Token Library

The Communications Token Library is the one of the generic tokens on the vertical Optional Libraries toolbar. The Main Libraries and the Optional Libraries are toggled with the Change Library Reservoir button (refer to Figure 1.7). Figure 1.31 shows the Encode/Decode Token feature of the Communications Library. The available block, convolutional, puncture, and Walsh encoder and decoder tokens are useful for source coding in digital communication system simulation. However, only the Gray Coder/Decoder token is used here, as described in Chapter 3, "Bandpass Modulation and Demodulation."

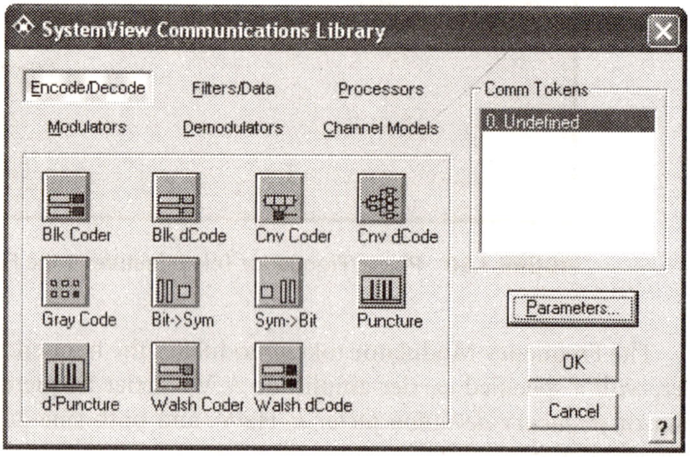

FIGURE 1.31 Encode/Decode Token feature of the Communications Library.

The Bit to Symbol token performs the aggregation of N binary bits into a 2^N-level baseband signal for multiple level (M-ary) signaling. The Symbol to Bit token performs the deaggregation of a 2^N-level baseband signal into N binary bits.

Figure 1.32 shows the Filters/Data Token feature of the Communications Library. The useful filters and data tokens include the Integrate and Dump token, the Gold Code Generator token, and the PN Sequence Generator token. The Integrate and Dump token is the basis for the optimum correlation receiver for baseband and bandpass signals, as described in Chapter 2, "Baseband Modulation and Demodulation," and Chapter 3, and given by Equation 1.13.

$$o(t) = \frac{1}{T}\int_{t-T}^{t} x(\alpha)\,d\alpha \tag{1.13}$$

In Equation 1.13, $o(t)$ is the output signal of the integrate-and-dump process, $x(\alpha)$ is the input signal, and T is the integration window.

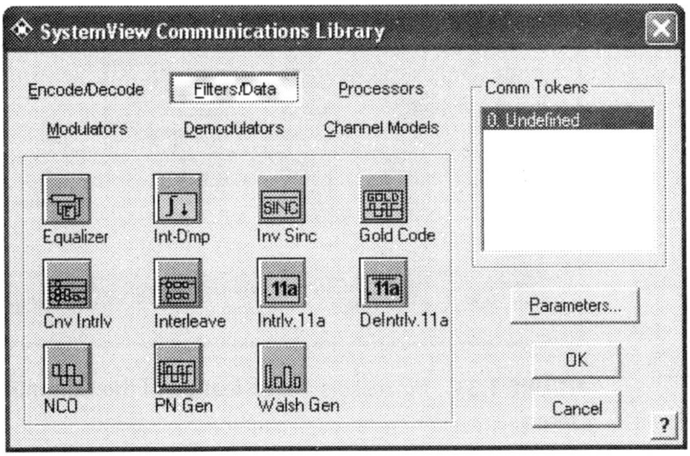

FIGURE 1.32 Filters/Data Token feature of the Communications Library.

The Gold Code Generator token and the PN Sequence Generator token are used to provide a pseudonoise sequence to spread a signal as the basis of both direct sequence spread spectrum (DSSS), and frequency hopping spread spectrum (FHSS), as described in Chapter 5, "Multiplexing."

Figure 1.33 shows the Processors Token feature of the Communications Library. The useful processor tokens include the BER (bit error rate) token, the Bit Sync (synchronization) token, the Compandor token, the Decompander token, the QAM (quadrature amplitude modulation) Mapper token, and the QAM Demapper token.

The BER token is used to observe the number of information bits received in error in a digital communication system by comparing the transmitted bits before modulation with the received demodulated bits, with error induced by additive noise or non-linearity in the communication channel. The BER token is used extensively in the SystemVue simulation and analysis of baseband and bandpass digital communication systems in Chapters 2 and 3; carrier, phase and symbol synchronization and equalization of bandlimited channels in Chapter 4, "Synchronization and Equalization"; and multiplexing in Chapter 5.

The Bit Sync token implements the recovery of the bit time T_b from a baseband signal, which is used in the detection of a transmitted information bits, as described in Chapter 4. The Compandor token and the Decompander token implement the μ-Law or the A-Law compression of a signal at the transmitter and expansion of a signal at the receiver (*companding*), as described in Chapter 6, "Sampling and Quantization."

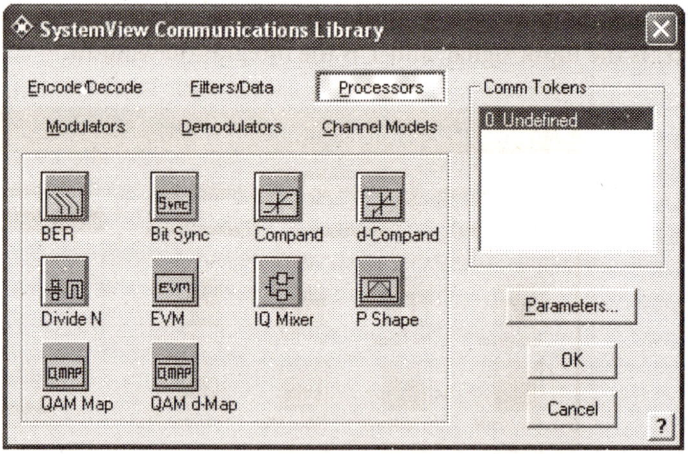

FIGURE 1.33 Processors Token feature of the Communications Library.

Finally, the QAM Mapper token assigns an *in-phase* (I) and *quadrature* (Q) symbol value to a multiple-level input signal by using an external *constellation* text file, producing a QAM signal, as described in Chapter 3. The QAM Demapper token performs the nearest neighbor detection of a QAM signal, producing a multiple-level output signal by again using an external constellation text file, as described in Chapter 3.

Figure 1.34 shows the Modulators Token feature of the Communications Library. The useful modulator tokens include the DSB-AM (double sideband, amplitude modulated) Modulator token, the MFSK (minimum frequency shift keying) Modulator token, the OFDM (orthogonal frequency division multiplexing) Modulator token, the QAM/Quadrature Modulator token, and the Time Division Multiplexer token. The DSB-AM Modulator token is an analog modulator that simulates an AM signal.

The MFSK Modulator token implements a minimum transmission bandwidth FSK signal, as described in Chapter 3. The OFDM Modulator token is an advanced modulation method with spectrum efficiency and several desirable properties for broadband digital communication, as described in Chapter 5. The QAM/Quadrature Modulator token inputs an in-phase (I) and quadrature (Q) baseband signal, and produces a modulated IQ passband carrier signal. Finally, the Time Division Multiplexer token combines multiple signals into an output that is time shared between the input signals, as described in Chapter 5.

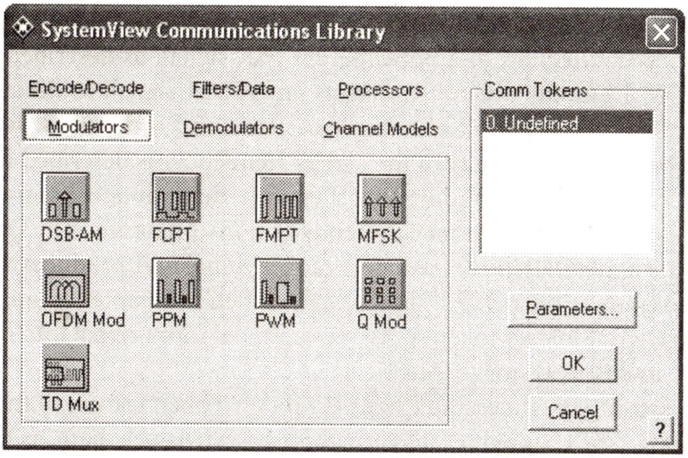

FIGURE 1.34 Modulators Token feature of the Communications Library.

Figure 1.35 shows the Demodulators Token feature of the Communications Library. The useful demodulator tokens include the Costas Loop token, the Noncoherent FSK (frequency shift keying) Demodulator token, the OFDM Demodulator token, the Phase Lock Loop token, the MPSK (multiple phase shift keying) Demodulator token, the QAM Detector token, and the Time Division Demultiplexer token. The Costas Loop token implements a third-order Costas loop that can be used to coherently demodulate PSK signals.

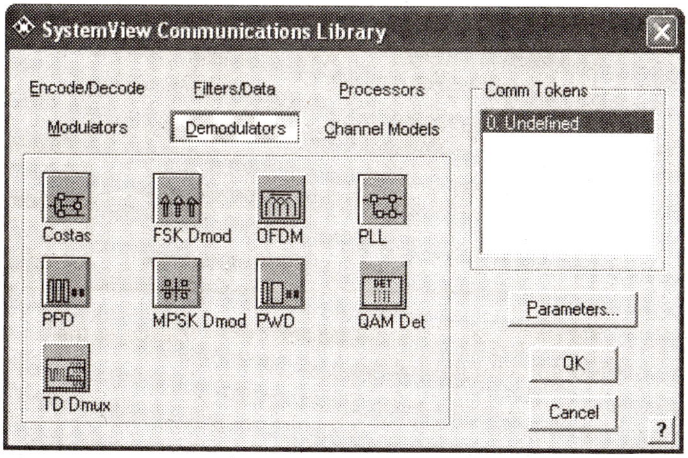

FIGURE 1.35 Demodulators Token feature of the Communications Library.

The Noncoherent FSK Demodulator token performs an optimum *maximum likelihood* detection of a noncoherent FSK signal, as described in Chapter 3. The OFDM Demodulator token detects an OFDM signal, as described in Chapter 5. The Phase Lock Loop token implements a third order phase lock loop (PLL) and can be used for recovery of the carrier frequency, as described in Chapter 4.

The MPSK Demodulator token can demodulate binary or multiple-level (M-ary) PSK signals either coherently or non-coherently. The QAM Detector token inputs a modulated IQ passband carrier signal and produces an in-phase (I) and quadrature (Q) baseband signal, as described in Chapter 3. Finally, the Time Division Demultiplexer token splits a time shared input signal into multiple output signals, as described in Chapter 5.

Figure 1.36 shows the Channel Models Token feature of the Communications Library. The channel models are not used extensively here, as the communication channel in the SystemVue simulation of a digital communication system here is the additive white Gaussian noise (AWGN) channel. The AWGN communication channel consists of an Adder token and a Gaussian Noise token from the Source Library. However, Chapter 4 considers the effect of the Multipath Channel token and the Fading Channel token as the communication channel on the performance of a digital communication system.

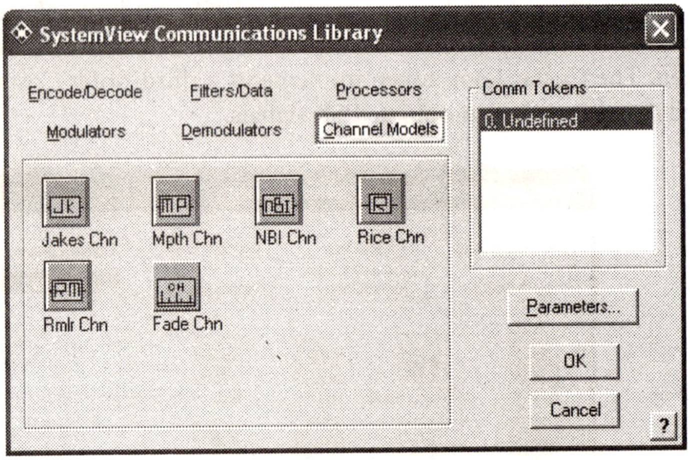

FIGURE 1.36 Channel Models Token feature of the Communications Library.

Logic Token Library

The Logic Token Library is one of the generic tokens on the vertical Optional Libraries toolbar. The Main Libraries and the Optional Libraries are toggled with the

Change Library Reservoir button (refer to Figure 1.7). The Logic Token Library is a more extensive set of tokens to those available in the Logic feature of the Operator Library (refer to Figure 1.23).

Figure 1.37 shows the Gates/Buffers Token feature of the Logic Library. The available gate and buffer tokens are standard combinational logic elements. As for the Logic tokens of the Operator Library, the input threshold voltage, the logic true output voltage, and the logic false output voltage here are also not required to be the same as any standard logic families (such as TTL or CMOS), and can be specified to be any value. The gate and buffer tokens are described in the *SystemVue User's Guide.*

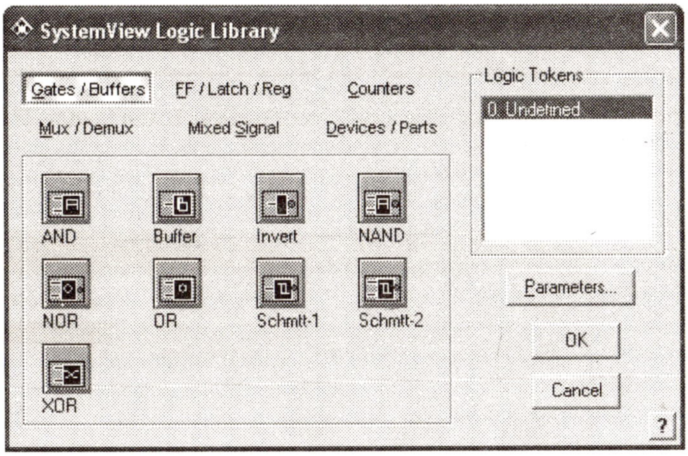

FIGURE 1.37 Gates/Buffers Token feature of the Logic Library.

Figure 1.38 shows the Flip Flop/Latch/Register Token feature of the Logic Library. The available flip flop, latch, and register tokens are standard sequential logic elements. The PROM (*programmable read-only memory*) token is organized as 8 addresses of 8 bits each (8×8), with the contents of the PROM entered as 4 parameters, where each parameter specifies the data at two address locations. The flip flop, latch, and register tokens are described in the *SystemVue User's Guide.*

Figure 1.39 shows the Counters Token feature of the Logic Library. The available counter tokens are three standard medium scale integration (MSI) devices. Although 4-bit and 12-bit counters are available, counters with register size less than that are available by using only the least significant bits (LSB) of the counter outputs.

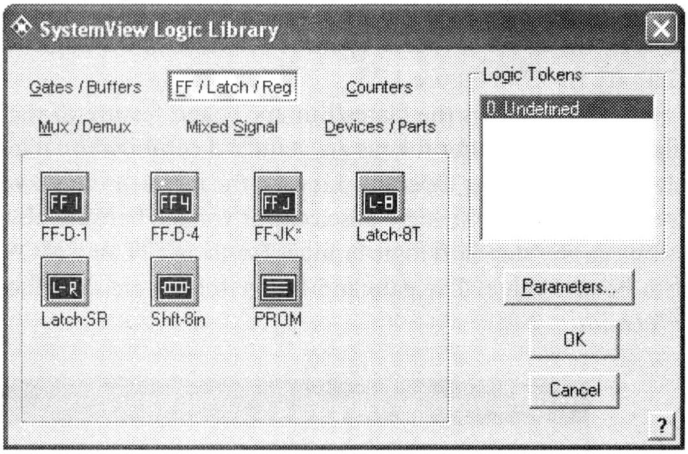

FIGURE 1.38 Flip Flop/Latch/Register Token feature of the Logic Library.

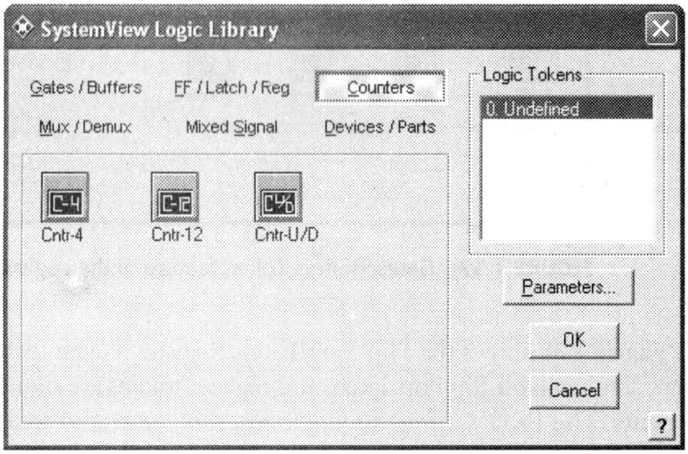

FIGURE 1.39 Counters Token feature of the Logic Library.

The Counter 4 token implements the MSI 74161 device, which is a synchronous, pre-settable 4-bit binary counter. The Counter 12 token implements the MSI MC4040 device, which is a synchronous 12-bit binary counter. The Counter Up/Down token implements the MSI 74191 device, which is a synchronous, pre-settable 4-bit counter with a control signal for incrementing or decrementing the counter. The counter tokens are described in the *SystemVue User's Guide.*

Figure 1.40 shows the Multiplexer/Demultiplexer Token feature of the Logic Library. The available multiplexer and demultiplexer tokens are two standard MSI devices. Although these devices include only an 8-input multiplexer and an 8-output demultiplexer, devices with 4 and 2 outputs are available by using only the LSBs of the 3-bit select inputs.

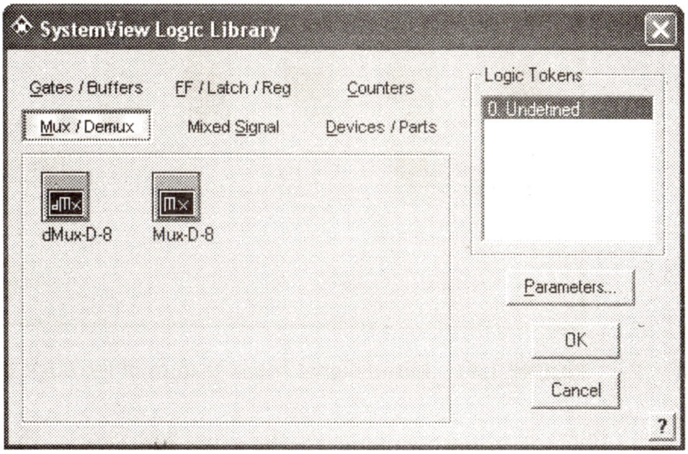

FIGURE 1.40 Multiplexer/Demultiplexer Token feature of the Logic Library.

The Multiplexer Digital 8 token implements the MSI 74151 device, which uses a 3-bit control signal to select and output one of eight digital input signals. The Demultiplexer Digital 8 token implements the MSI 74138 device, which uses a 3-bit control signal to select as logic false one of eight digital output signals. The multiplexer and demultiplexer tokens are described in the *SystemVue User's Guide.*

Figure 1.41 shows the Mixed Signal Token feature of the Logic Library. The available mixed signal tokens include the Analog to Digital Converter (ADC) token, the Analog Comparator token, the Digital to Analog Converter (DAC) token, the Double Pole Double Throw (DPDT) Analog Switch token, the Integer to Digital Converter (IDC) token, and the Single Pole Double Throw (SPDT) Analog Switch token. The ADC token converts an analog signal input to a parallel digital word with either a two's complement or an unsigned integer representation and up to 16 bits of resolution. The analog signal input voltage range, the logic true output voltage, and the logic false output voltage can be specified.

The Analog Comparator token provides a binary output with logic true indicating that the Input+ is greater than the Input– signal input. The logic true output voltage and the logic false output voltage can be specified. The DAC token provides an analog output voltage equivalent to a parallel digital word with either a two's complement or an unsigned integer representation and up to 16 bits of resolution. The logic true and logic false input voltage signals and the voltage range of the analog signal output can be specified.

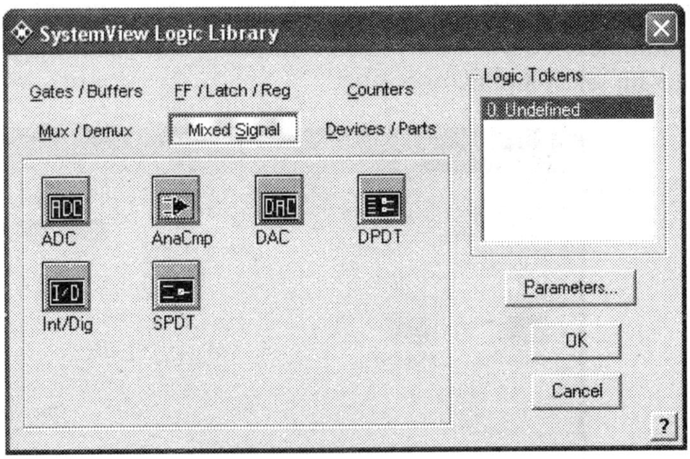

FIGURE 1.41 Mixed Signal Token feature of the Logic Library.

The IDC token performs nearly the same function as the ADC token, except that the input signal is not a specified voltage range but integers. The IDC token output is a parallel digital word with a two's complement representation and up to 16 bits of resolution. The logic true and logic false output voltage signals can be specified. Finally, the SPDT Analog Switch token and the DPDT Analog Switch token redirect either one or two analog signal inputs to one of two analog signal outputs by a logic control signal. The mixed signal tokens are described in the *SystemVue User's Guide.*

Figure 1.42 shows the Devices/Parts Token feature of the Logic Library. The available devices and parts tokens include the 8-Bit Identity Comparator token, the One Shot (retriggerable monostable multivibrator) token, and the Phase/Frequency Detector token. The 8-Bit Identity Comparator token implements the MSI 74521 device and compares two binary words of up to 8 bits each, providing a logic low signal when the two words match bit for bit.

The One Shot token implements the MSI 74123 device and provides a variety of triggering conditions and an output pulse width that can be specified. Finally, the Phase/Frequency Detector token implements the MSI MC4044 device for the recovery of a phase or frequency of a periodic signal. The devices and parts tokens are described in the *SystemVue User's Guide.*

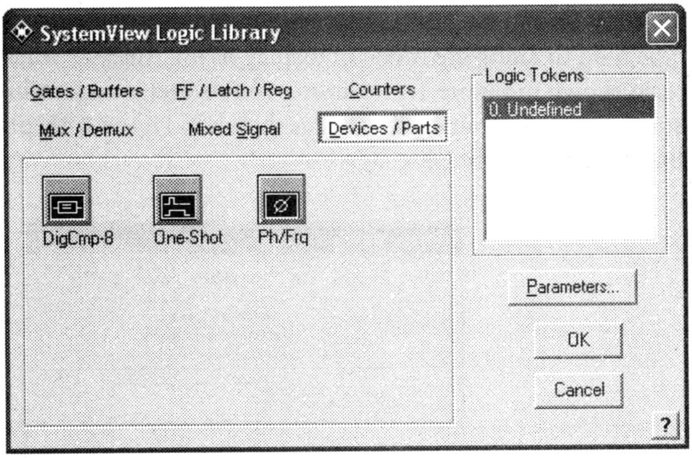

FIGURE 1.42 Devices/Parts Token feature of the Logic Library.

SYSTEMVUE SIMULATION DISPLAYS

The Analysis Token feature of the Sink Library provides the initial display of a signal at the termination of a SystemVue simulation in the Analysis window, as shown in Figure 1.43. The Analysis window consists of a standard Windows menu and a toolbar. The Windows menu provides the usual File operations of opening, saving, and printing a SystemVue plot, and should be familiar from experience in other Windows applications. The Edit operations enable you to copy the plot to a word processing application, such as Microsoft Word, to document the work.

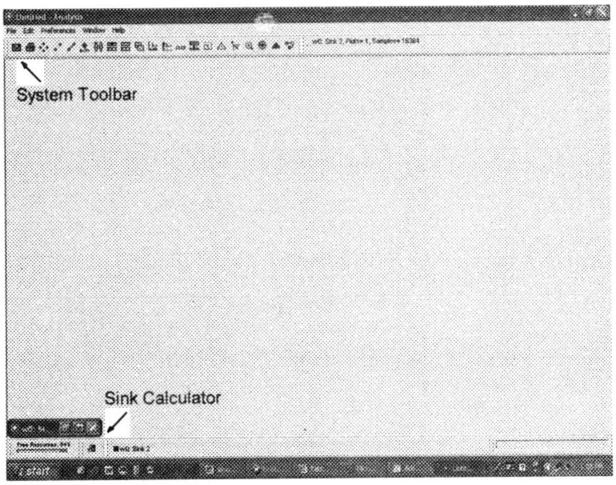

FIGURE 1.43 Analysis window of the SystemVue simulation environment.

The Preferences… Options menu provides several customization features that can be used to tailor the mode of display of the Analysis window. The Colors feature, as shown in Figure 1.44, allows you to select a background, border, grid, and plot color. The Animation feature, as shown in Figure 1.45, provides the useful capability to *animate* the plot.

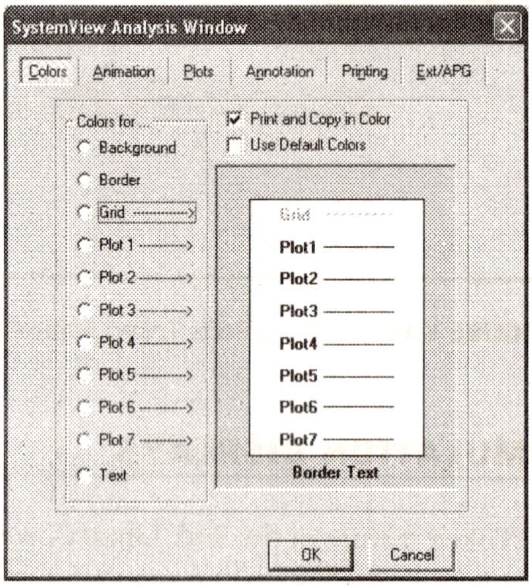

FIGURE 1.44 Analysis Colors feature of the Preferences… Options menu.

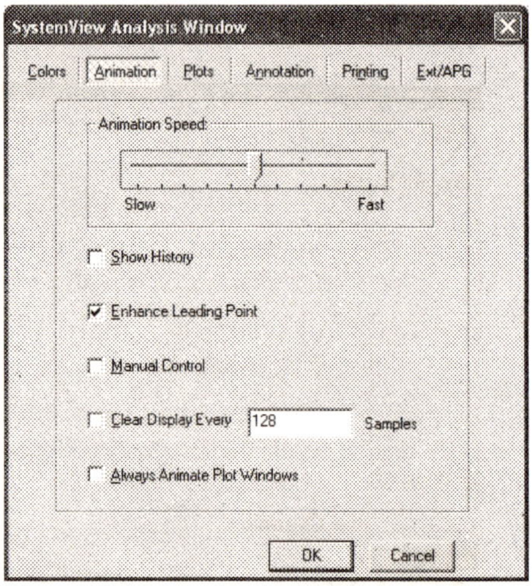

FIGURE 1.45 Analysis Animation feature of the Preferences… Options menu.

The Plots feature, as shown in Figure 1.46, enables a default setup for the display. Finally, the Annotation feature, as shown in Figure 1.47, enables a default setup for the vertical and horizontal axis labels, plot title, and a reminder if an interactive zoom plot display operation is in use. The Printing feature sets the default print intensity and External/APG is an advanced feature of the SystemVue simulation environment. The Automatic Program Generator (APG) is a tool for creating executable programs and compatible dynamic link libraries (DLL) from SystemVue MetaSystems.

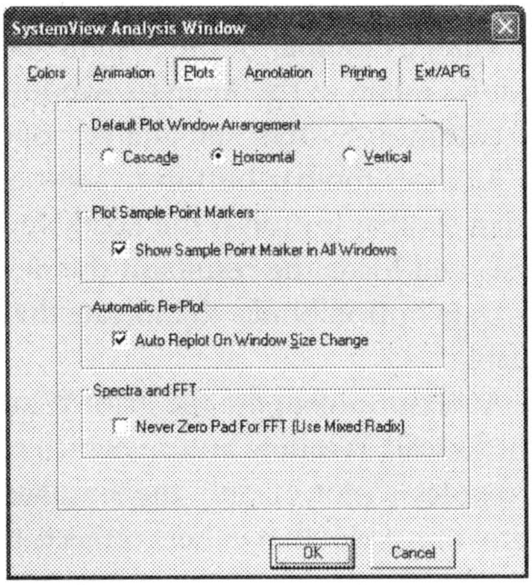

FIGURE 1.46 Analysis Plots feature of the Preferences... Options menu.

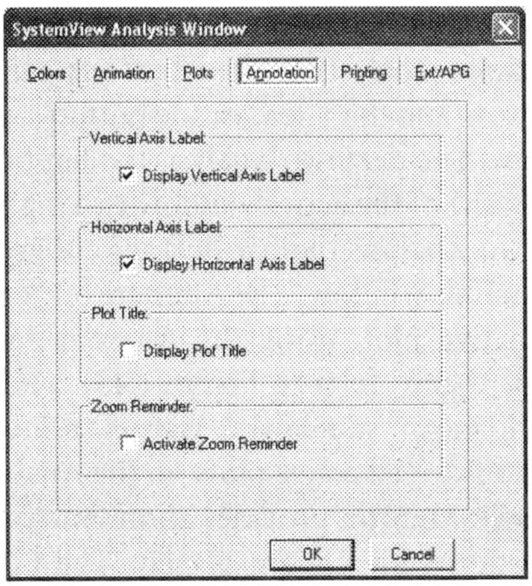

FIGURE 1.47 Analysis Annotation feature of the Preferences... Options menu.

The Window menu (refer to Figure 1.43) provides the capability to arrange, rescale, and plot the Analysis windows. The Help menu includes the Help Navigator, which launches the Adobe Acrobat viewer of the *SystemVue User's Guide*, and the Toolbar Help and Sink Calculator Help, which launches the Adobe Acrobat viewer of the particular section of the *SystemVue User's Guide* for the toolbar and the sink calculator.

Analysis Toolbar

The horizontal Analysis toolbar of the SystemVue simulation environment is just below the Windows menu in the Analysis window (refer to Figure 1.43), and presents the most common SystemVue analysis operations. Position the Windows cursor over an Analysis toolbar button to view a pop-up label that describes the function. The first toolbar button is the Load New Sink Data button, which clears the prior analysis and loads the results of the new simulation into the plots. The Load New Sink Data button blinks to indicate that updated SystemVue simulation results are available.

Next in the Analysis toolbar are the Print Window button, which prints the active plot, and the Rescale Window button, which rescales the active plot. The Points Only button provides a plot of only the simulation data points. The Connected Points button provides a plot that either connects the simulation data points with an interpolated straight line, or deletes the points and only shows the interpolated line.

The Differential X-Y button permits the dragging and dropping of an initial cursor, which then calculates and displays the differential between the initial cursor and a movable cursor in absolute units, decibels (dB), and simulation sample points. The X-Axis Segment Marker button specifies an x-axis range used for any subsequent processing, such as the voltage or power spectrum within that segment of the plot.

The next three Analysis toolbar buttons—the Tile Vertical button, the Tile Horizontal button, and the Tile Cascade button—manipulate the active plots in the Analysis window. The next two Analysis toolbar buttons change either the x-axis or the y-axis (or both) of the active window to a log axis plot.

The next two Analysis toolbar buttons either open or close all of the available plots on the Analysis window. The Animate button animates the SystemVue simulation of the active plot, using the Analysis Animation feature of the Preferences...Options menu (refer to Figure 1.45). The Plot Statistics button calculates and displays the mean, standard deviation, minimum, maximum, and linear regression equations for all active plots on the Analysis window.

The Microview button provides an interactive display of a section of the SystemVue simulation with a zoom display. The F11 function key on the keyboard zooms out (less magnification), and the F12 function key zooms in (more magnification) to obtain greater detail in the vicinity of the cursor. The Zoom button complements the operation of the Microview button and also provides an interactive display of a section of the SystemVue simulation. Right-click the mouse on the Analysis window to view a menu that includes a rescale plot option.

The Polar Grid button and the Load APG Results button are not used here, but are described in the *SystemVue User's Guide*. Finally, the Design Window button of the Analysis toolbar, the rightmost button, initiates the Design window where the design of the SystemVue simulation occurs.

Figure 1.48 shows the Sink Calculator window, which enables you to analyze one or more plots. Figure 1.48 shows the Operators feature of the Sink Calculator window, which can overlay plots, integrate or differentiate a plot, obtain the magnitude of a plot, decimate (*downsample*) a plot, smooth a plot with a moving average, overlay the statistics of a plot, or apply a tapered window to a plot.

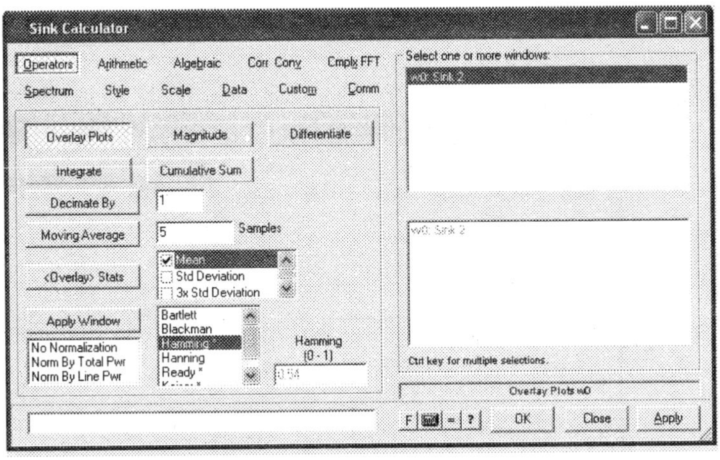

FIGURE 1.48 Operators feature of the Sink Calculator window.

Figure 1.49 shows the Arithmetic feature of the Sink Calculator window, which can add and multiply windows or plots; subtract, divide, negate, or reciprocate windows; or normalize windows or plots. The Algebraic feature of the Sink Calculator window can perform various algebraic transformations on a plot.

Figure 1.50 shows the Correlation Convolution feature of the Sink Calculator window, which can *autocorrelate* a plot, *cross correlate* two plots, or *convolve* two plots. The cross-correlation procedure is useful in determining the simulation delay between a transmitted data and received data in a bit error rate analysis. The Complex FFT (fast Fourier transform) feature of the Sink Calculator window can obtain the real component, the imaginary component, the magnitude, the magnitude squared, and the power and phase of the FFT. The *SystemVue User's Guide* provides a description of their implementation.

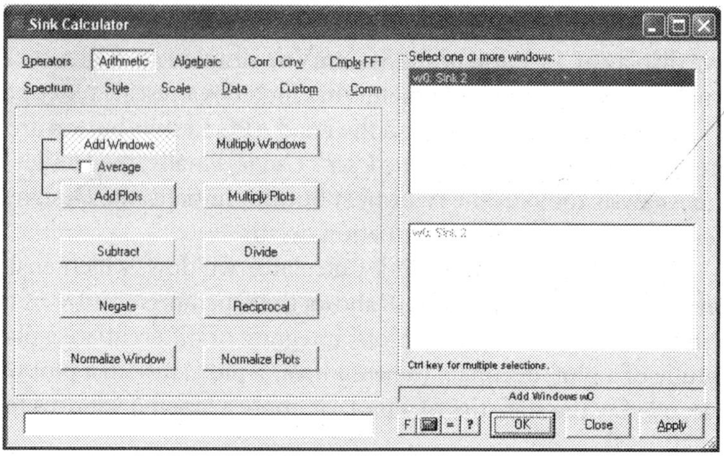

FIGURE 1.49 Arithmetic feature of the Sink Calculator window.

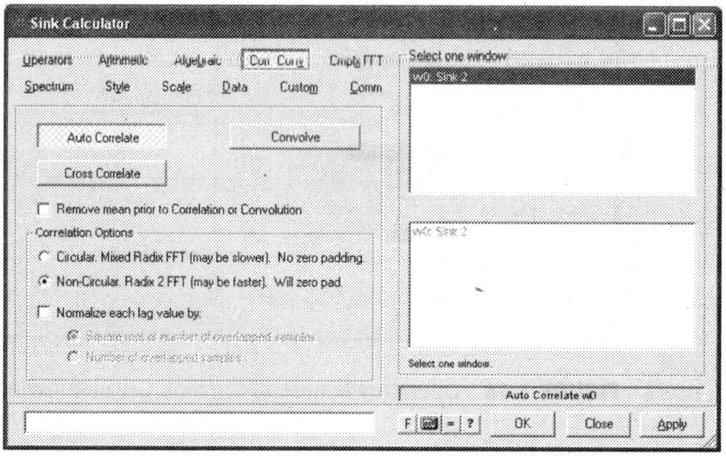

FIGURE 1.50 Correlation Convolution feature of the Sink Calculator window.

Figure 1.51 shows the Spectrum feature of the Sink Calculator window, which can obtain the power spectrum or the power spectral density (PSD) of a plot. The normalized (1 Ω load resistor) PSD in dB/Hz is used in the SystemVue analysis of the spectrum of a signal here.

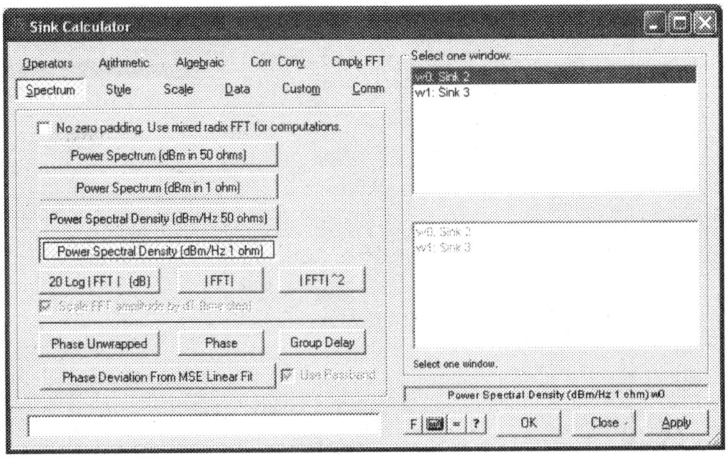

FIGURE 1.51 Spectrum feature of the Sink Calculator window.

Figure 1.52 shows the Style feature of the Sink Calculator window, which provides several types of displays, including the *scatter plot* that displays one plot against another. The scatter plot is the basis of the constellation plot, as described in Chapter 3.

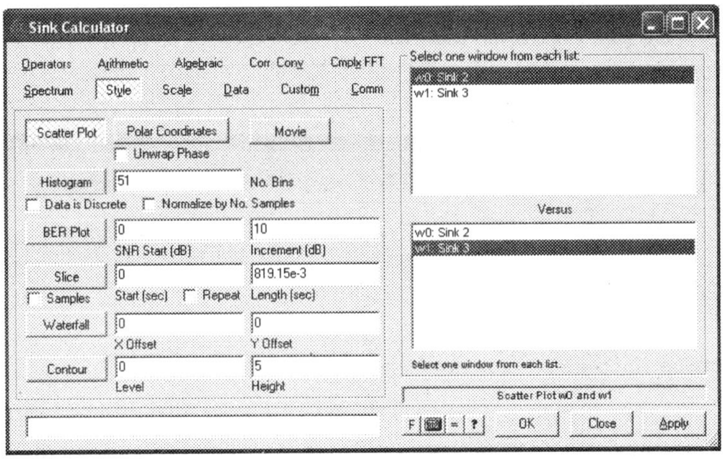

FIGURE 1.52 Style feature of the Sink Calculator window.

Figure 1.53 shows the Scale feature of the Sink Calculator window, which can set the minimum and maximum values of a plot or scale the axes. The Scale feature facilitates the SystemVue analysis of a signal and the preparation of documentation. Finally, the Data, Custom, and Communication features of the Sink Calculator window are not used here. The *SystemVue User's Guide* provides a description of their implementation.

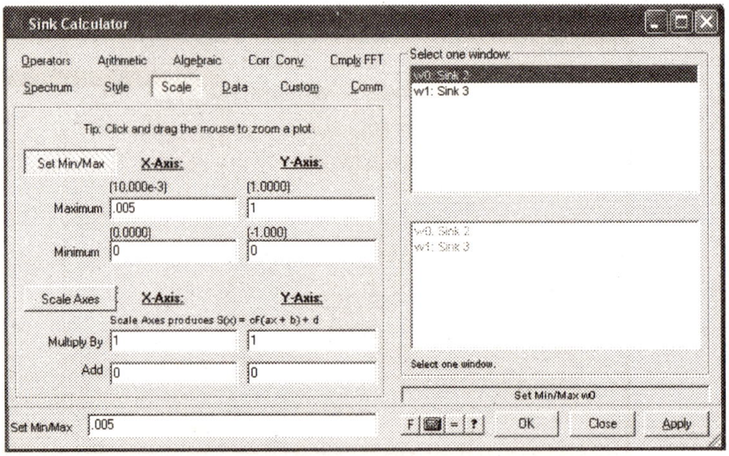

FIGURE 1.53 Scale feature of the Sink Calculator window.

Temporal Displays

The Sink tokens in a Design window provide a temporal display in the Analysis window. A half wave rectified sinusoid in the Design window, as a quick-start example of temporal displays in a SystemVue simulation, is shown in Figure 1.54. Figure 1.55 (see Fig1-55.svu on the CD-ROM) shows the SystemVue model of the half wave rectified sinusoid.

ON THE CD

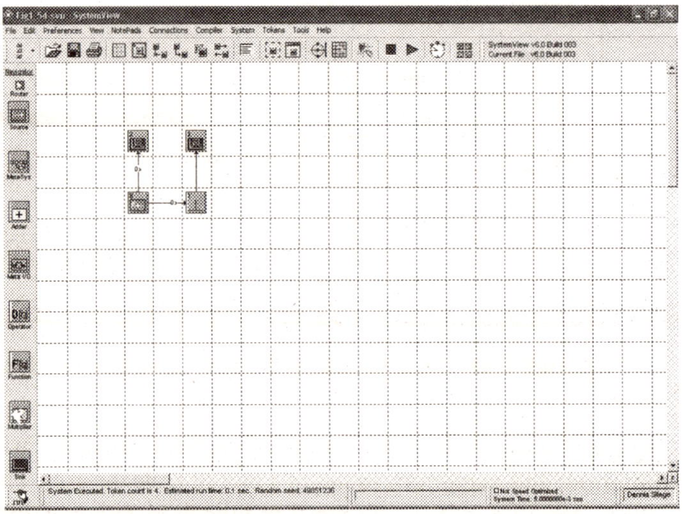

FIGURE 1.54 Design window of the half wave rectified sinusoid.

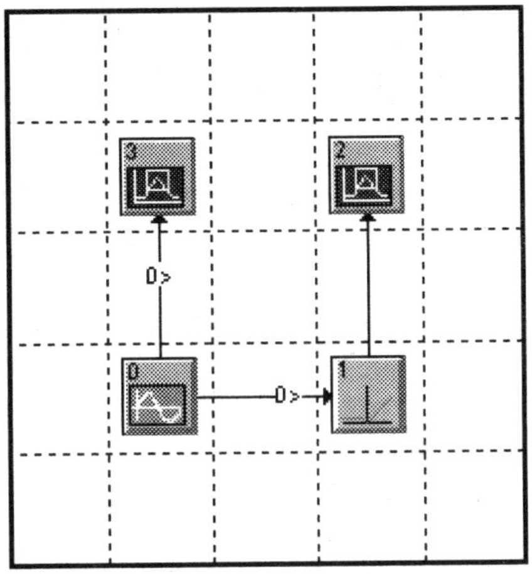

FIGURE 1.55 Half wave rectified sinusoid.

The sinusoidal source is the Sinusoid token from the Source Library, with parameters of an amplitude of 1 V, a frequency of 1 kHz, and 0° phase offset. The output of the Sinusoid token is inputted to the Half Wave Rectifier token from the Function Library, with a zero point of 0 V (ideal diode). The outputs of the Sinusoid token and the Half Wave Rectifier token are inputted to two Analysis tokens from the Sink Library.

The System Time: Sample Rate is set to 50 kHz, which is well above the 1 kHz sinusoidal frequency. The System Stop Time is initially set to 5 msec, and the resulting temporal displays of the two plots in the Analysis window are shown in Figure 1.56.

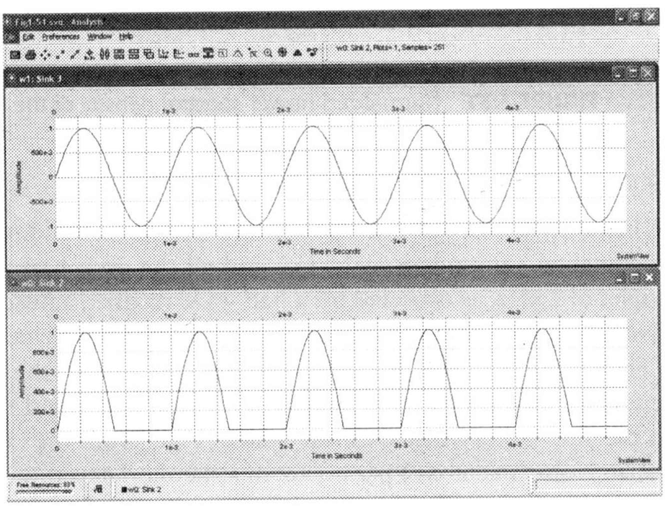

FIGURE 1.56 Analysis window of the half wave rectified sinusoid.

Spectral Displays

The spectral display facilitates analyzing the performance of a communication system. The Sink Calculator in the Analysis window can obtain the normalized power spectral density (PSD) of temporal display. The half wave rectified sinusoid SystemVue simulation (refer to Figure 1.55) is used as a quick-start example to demonstrate the calculation of the PSD. The System Number of Samples is set to 65 536 (2^{16}) and the resulting spectral resolution $\Delta f = 0.762$ Hz, as given by Equation 1.1.

The temporal display of the simulation of the half wave rectified sinusoid is indiscernible because a large number of sample points are displayed. The Spectrum feature of the Sink Calculation (refer to Figure 1.51) is used to calculate the normalized PSD of the half wave rectified sinusoid in Analysis window W1 Sink 2, as shown in Figure 1.57.

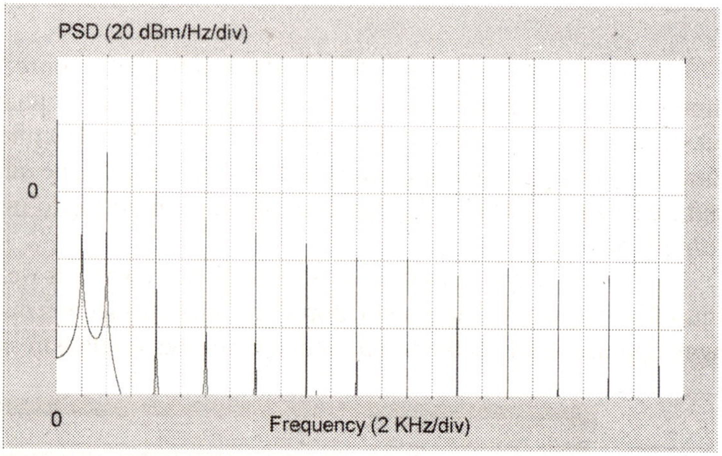

FIGURE 1.57 Normalized power spectral density of the half wave rectified sinusoid.

The normalized PSD of the half wave rectified sinusoid has a DC component, a fundamental component at 1 kHz, and all the even harmonic components at multiples of 2 kHz, as expected for a half wave rectified sinusoid [Carlson02].

Noise sources provide an interesting alternative test signal for linear system filters. Figure 1.58 (see Fig1-58.svu on the CD-ROM) shows the SystemVue model of a Gaussian Noise token from the Source Library inputted to an Analysis token from the Sink Library, and a Linear System Filter from the Operator Library. The Gaussian Noise token has a parameter of a standard deviation $\sigma = 1$ V and a mean $\mu = 0$ V. The Linear System Filter token is a 9-pole Butterworth analog lowpass

ON THE CD

filter (LPF) with a *cutoff* frequency of 1 kHz. The output of the Linear System Filter token is inputted to another Analysis token from the Sink Library.

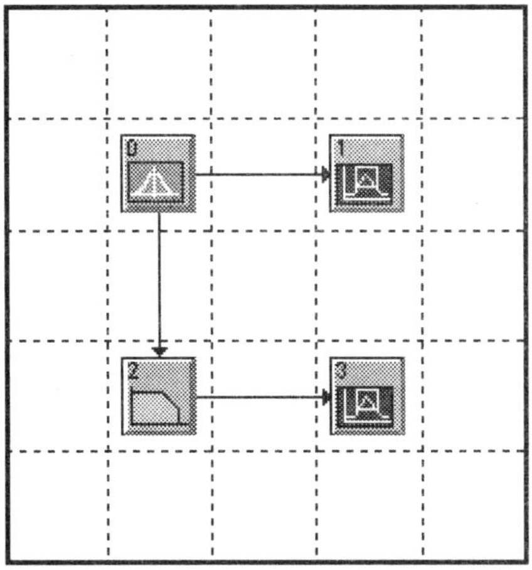

FIGURE 1.58 Gaussian noise source and lowpass filtered Gaussian noise.

The System Time: Sample Rate is set to 50 kHz, the System Number of Samples is set to 65 536 (2^{16}) and the resulting spectral resolution $\Delta f = 0.762$ Hz, as given by Equation 1.1. The normalized PSD of the Gaussian noise source is shown in Figure 1.59. The resulting PSD is *flat*, which is an expected characteristic of a *white noise* source [Haykin01].

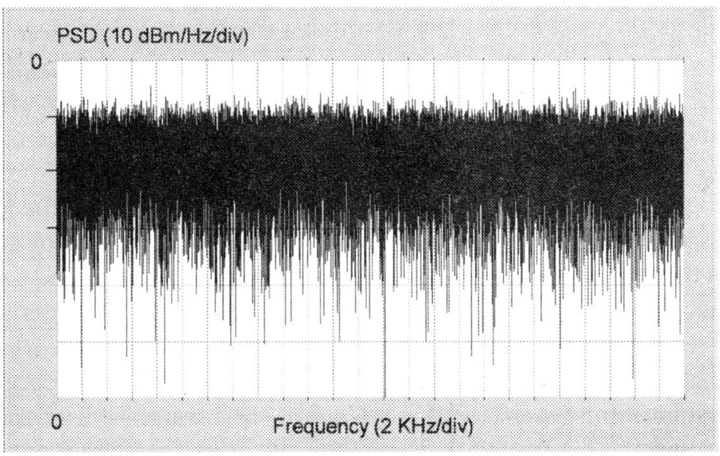

FIGURE 1.59 Normalized power spectral density of the Gaussian noise source.

The normalized PSD of the LPF Gaussian noise source is shown in Figure 1.60. The performance of the lowpass filter with a cutoff frequency of 1 kHz is readily seen.

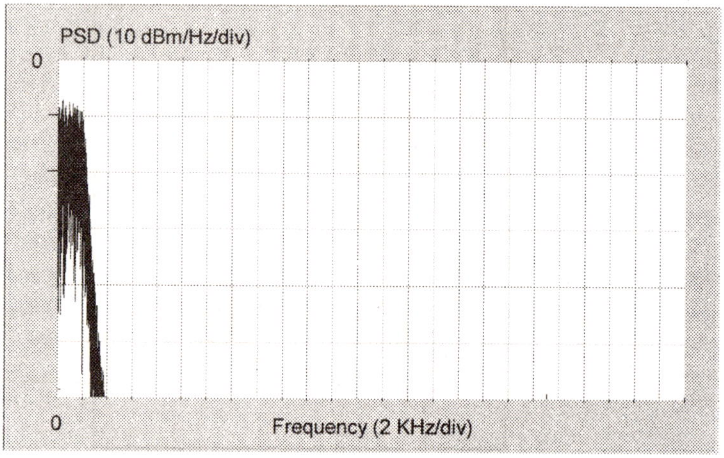

FIGURE 1.60 Normalized power spectral density of the lowpass filtered Gaussian noise source.

Correlation Display

The correlation display also facilitates analyzing the performance of a communication system. The Gaussian noise system (refer to Figure 1.58) is used as a quick-start example of the autocorrelation of a signal. The Gaussian Noise token parameters, the System Time: Sample Rate, and the System Number of Samples are the same.

The Correlation Convolution feature of the Sink Calculator button (refer to Figure 1.50) is used to calculate the autocorrelation of the Gaussian noise source in the Analysis window. The resulting scaled autocorrelation is shown in Figure 1.61. There is only a single autocorrelate point evident, which occurs when the shifting parameter $\tau = 0$. This autocorrelation is another characteristic of a white noise source [Haykin01].

Figure 1.62 shows the resulting scaled autocorrelation for the lowpass filtered Gaussian noise source in the Analysis window. The peak amplitude still occurs when the shifting parameter $\tau = 0$, but the amplitude is lower and a significant correlation is evident when $|\tau| > 0$. This autocorrelation is characteristic of a *colored noise* source [Haykin01].

The autocorrelation function is a basic premise of signal processing in digital communication systems, and is related to the inverse Fourier transform of the power spectral density (PSD) [Lathi98]. The cross-correlation of two signals emphasizes the similarity in their representation and can assess the apparent delay of one signal with respect to another. The cross-correlation is utilized in the bit error rate (BER) determination, where one signal to be correlated is the transmitted symbols before modulation and the other signal is the demodulated received symbols.

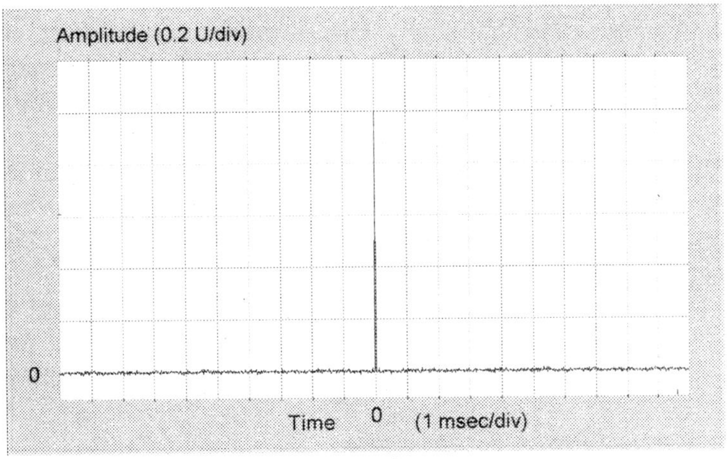

FIGURE 1.61 Autocorrelation of a Gaussian noise source.

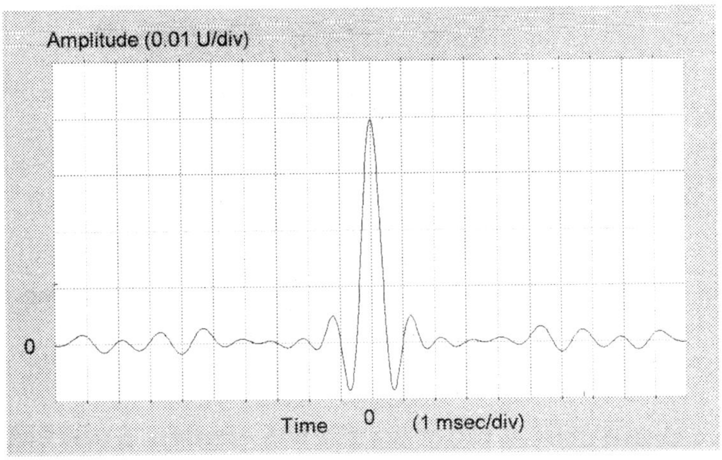

FIGURE 1.62 Autocorrelation of a lowpass filtered Gaussian noise source.

Dynamic System Probe

The Dynamic System Probe can be used to observe signals in the Design window during a SystemVue simulation. The Dynamic System Probe is both a virtual oscilloscope and spectrum analyzer and can utilize both an internal trigger source and external trigger token with specified threshold levels. The parameters of any SystemVue token can be changed dynamically to observe the effect on the simulation when the Dynamic System Probe is active. The Dynamic System Probe is first paused, and then a right click of the mouse on any token allows you to select the option to edit the token parameters.

Figure 1.63 shows the half wave rectified sinusoid of Figure 1.55, with a Dynamic System Probe monitoring the sinusoidal source, and the half wave rectifier output in real-time during the simulation. The Dynamic System Probe is described in the *SystemVue User's Guide.*

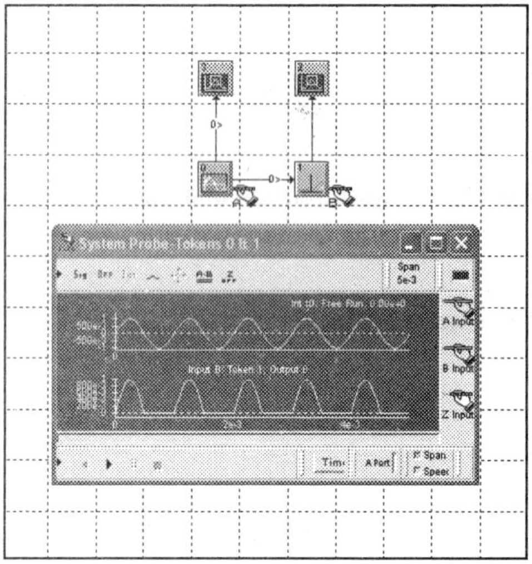

FIGURE 1.63 Dynamic System Probe monitoring the half wave rectified sinusoid simulation.

ANALOG COMMUNICATION SYSTEMS USING SYSTEMVUE

Analog communication systems modulate the continuous amplitude or frequency of a sinusoidal carrier with a baseband information signal $m(t)$. Although a digital communication system was the first mode of wireless transmission—in the amplitude shift keying (ASK) of a carrier using *on-off keying* and the Morse telegraphy code—analog communication held a prominent position for most of the last century.

Amplitude modulation (AM) was the first broadcast technology, beginning in the 1920s, followed by frequency modulation (FM) in the 1930s. Single sideband (SSB), a bandwidth-efficient version of double sideband AM (DSB-AM), was used for long-distance telephone communications beginning in the 1940s. Analog modulation techniques are simple to implement and relatively robust to noise and interference.

SystemVue can simulate analog communication systems and are used here as a quick-start example of the simulation environment. Although not a digital communication system, analog communication systems are familiar and provide a simple way to demonstrate the SystemVue simulation environment. SystemVue can import a *pseudo-analog* audio file (.wav) as the baseband information signal $m(t)$ to qualitatively assess the overall performance of the analog communication system.

Amplitude Modulation

Amplitude modulation (AM) modulates a baseband information signal $m(t)$ as the continuous amplitude of a sinusoidal carrier, as given by Equation 1.14.

$$s_{AM}(t) = A_c\,(1 + k_a\,m(t))\,\sin(2\pi f_c t) \qquad (1.14)$$

In Equation 1.14, A_c is the amplitude of the carrier, k_a is the amplitude modulation index in V^{-1}, and f_c is the carrier frequency. To avoid over modulation, $|\, k_a m(t)\, | \leq 1$ for all t. The AM signal in Equation 1.14 is also called a double sideband AM (DSB-AM) signal, because the carrier component is present even if $m(t) = 0$ and the spectrum has symmetrical upper and lower sidebands [Carlson02].

ON THE CD

Figure 1.64 (see Fig1-64.svu on the CD-ROM) shows the SystemVue simulation of an analog DSB-AM system. The Wave external file token from the Source Library inputs the audio file SVAudioIn.wav, which pronounces "SystemVue" at 8-bits of resolution with an initial 8 kHz sampling rate or a data rate of 64 kb/sec.

The System Time: Sample Rate is set to 800 kHz, well above the 8 kHz rate of the .wav file input. The System Stop Time is set to 620 msec, the length of the pronouncement, and the resulting spectral resolution $\Delta f = 1.61$ Hz, as given by Equation 1.1. The output of the Wave external file token is padded with 99 samples per input sample in the parameter specification to raise the 8 kHz source rate by a factor of 100 to the System Time: Sample Rate of 800 kHz

The output of the Wave external file token is inputted to the Polynomial token from the Function Library, which scales the SVAudioIn.wav file so that no distortion occurs in the DSB-AM modulator. The parameters of the Polynomial token are that the x^0 (offset) coefficient is 0.08, the x^1 (linear) coefficient is 0.02, and all other coefficients are zero.

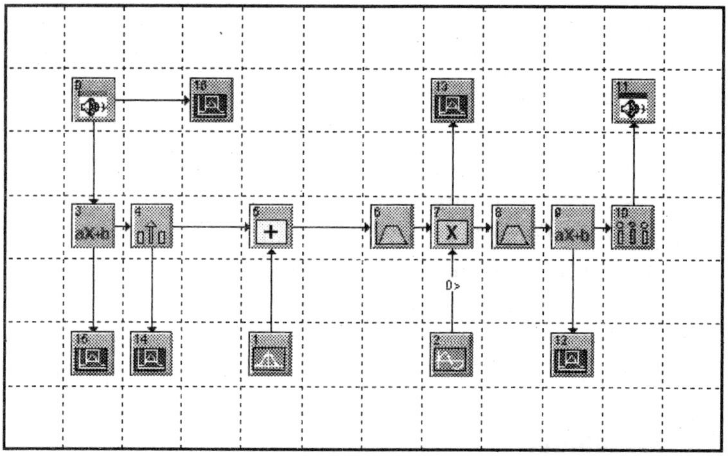

FIGURE 1.64 DSB-AM analog communication system with coherent demodulation.

The output of the Polynomial token, the baseband information signal $m(t)$, is inputted to the DSB-AM token from the Communications Library, with parameters of a carrier amplitude $A_c = 10$ V, a carrier frequency $f_c = 25$ kHz, an amplitude modulation index $k_a = 1$, and 0° phase offset. Although the amplitude modulation index could scale the input signal, the Polynominal token was used here to remove a DC bias in the SVAudioIn.wav file.

The communication channel is represented by the Adder token and the Gaussian Noise token from the Source Library, with initial parameters of a standard deviation $\sigma = 0$ V and mean $\mu = 0$ V. The output of the Adder token is inputted to the DSB-AM receiver beginning with the Linear System Filter token, which is a 9-pole Butterworth analog bandpass filter (BPF) with cutoff frequencies of 17 kHz and 32 kHz. The cutoff frequencies are centered ± 8 kHz about the carrier frequency of 25 kHz, and are set to the bandwidth of the DSB-AM signal [Carlson02].

The output of the BPF Linear System Filter token is inputted to the Multiplier token, the other input of which is the output of the Sinusoid token from the Source Library. The parameters of the Sinusoid token are an amplitude $A_o = 10$ V, a frequency $f_o = 25$ kHz, and 0° phase offset. The Multiplier token and the Sinusoid token form a *synchronous detector* for the coherent demodulation of the DSB-AM signal [Lathi98].

From Equation 1.14, if $f_c = f_o$ and there is no phase difference, then the output of the Multiplier token is given by Equation 1.15.

$$
\begin{aligned}
s_o(t) &= \gamma A_c\,(1 + k_a m(t))\,\sin(2\pi f_c t)\,A_o \sin(2\pi f_o t) \\
s_o(t) &= \frac{\gamma A_c A_o}{2}\,(1 + k_a m(t))\,(1 - \cos\,2\pi\,2 f_c t)
\end{aligned}
\tag{1.15}
$$

In Equation 1.15, γ is the attenuation of the communication channel and taken to be zero ($\gamma = 1$). The normalized power spectral density (PSD) of the output of the Multiplier token $s_o(t)$ is shown in Figure 1.65. From Equation 1.15, the spectral components of $s_o(t)$ are centered about a frequency of $2 f_c = 50$ kHz and at DC.

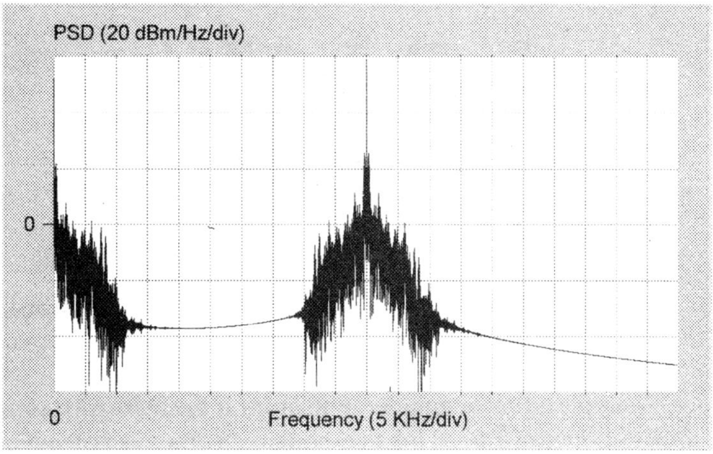

FIGURE 1.65 Power spectral density of the synchronous demodulated DSB-AM signal.

The unwanted DC component $\gamma A_c A_o/2$ and the spectral components centered at 50 kHz in the output of the Multiplier token are suppressed by being inputted to another Linear System Filter token, which is a 9-pole Butterworth analog BPF with cutoff frequencies of 80 Hz and 8 kHz.

The output of the Linear System Filter token is inputted to another Polynomial token, with parameters of the x^1 (linear) coefficient as 2.5 and all other coefficients as zero, to rescale the recovered .wav file. The output of the Polynomial token is inputted to the Decimator token from the Operator Library, with a decimation factor of 100, to restore the 8 kHz source rate from the System Time: Sample Rate of 800 kHz.

Figure 1.66 shows the Design window for the DSB-AM communication system of Figure 1.64 after the SystemVue simulation has been executed. The default Windows audio player is automatically evoked and the input SVAudioIn.wav file and the specified output SVAudioOut.wav can be heard. Initially no noise is added to the communication channel, but setting the standard deviation σ of the Gaussian Noise token to even 1 V produces a discernable deterioration of the received audio signal.

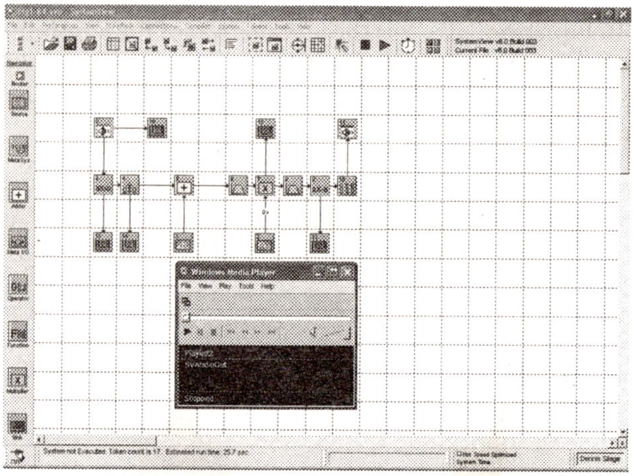

FIGURE 1.66 Design window for the DSB-AM analog communication system with Windows audio player.

A simple noncoherent demodulator for the DSB-AM communication system is shown in Figure 1.67. This demodulator is the *crystal radio* once constructed by many as part of an interest in electronics. The synchronous detector is replaced by the Half Wave Rectifier token from the Function Library.

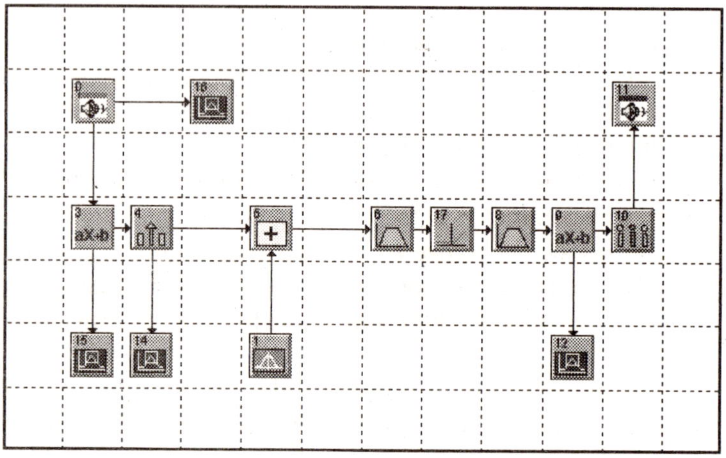

FIGURE 1.67 DSB-AM analog communication system with noncoherent demodulation.

Although it would be authentic for a crystal radio to drastically lower the number of poles used in the two BPFs, they remain the same here. If the number of poles in the second BPF is lowered, the DC component in the DSB-AM signal in Equation 1.15 produces an annoying whistle in the recovered audio output.

Frequency Modulation

Frequency modulation (FM) modulates a baseband information signal $m(t)$ as the continuous frequency of a sinusoidal carrier, as given by Equation 1.16.

$$s_{FM}(t) = A_c \sin(2\pi f_c t + k_f \int m(\alpha)\mathrm{d}\alpha) \tag{1.16}$$

In Equation 1.16, A_c is the amplitude of the carrier, k_f is the modulation gain or frequency deviation factor in Hz/V, and f_c is the carrier frequency. The baseband information signal $m(t)$ is integrated, since the *instantaneous frequency* f_i Hz of a signal is defined as the time derivative of the time-varying angle of the sinusoid [Stern04], as given by Equation 1.17.

$$f_i = \frac{\mathrm{d}\left[2\pi f_c t + k_f \int m(\alpha)\mathrm{d}\alpha\right]}{dt} = 2\pi f_c + k_f m(t) \tag{1.17}$$

ON THE CD

Figure 1.68 (see Fig1-68.svu on the CD-ROM) is the SystemVue simulation of an analog FM system. The Wave external file token from the Source Library inputs the audio file SVAudioIn.wav, which pronounces "SystemVue" at 8 bits of resolution with an initial 8 kHz sampling rate, or a data rate of 64 kb/sec.

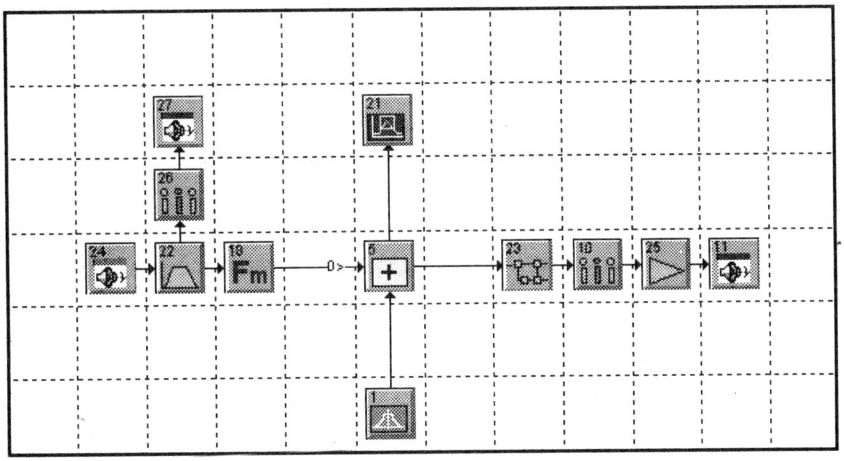

FIGURE 1.68 FM analog communication system.

The System Time: Sample Rate is set to 240 kHz, well above the 8 kHz rate of the .wav file input. The System Stop Time is set to 620 msec, the length of the pronouncement, and the resulting spectral resolution $\Delta f = 1.61$ Hz, as given by Equation 1.1. The output of the Wave external file source token is padded with 29 samples per input sample in the parameter specification, to raise the 8 kHz source rate by a factor of 30 to the System Time: Sample Rate of 240 kHz.

The output of the Wave external file token is inputted to the Linear System Filter token, which is a 3-pole Butterworth analog bandpass filter (BPF) with cutoff frequencies of 50 Hz and 5 kHz. The BPF is an alternative method to the Polynomial token used in the DSB-AM simulation to remove the unwanted DC bias in the SVAudioIn.wav file. The output of the BPF is inputted to the Decimator token from the Operator Library and the Frequency Modulator token from the Function Library.

The Decimator token has the parameter set to 10 to downsample the 240 kHz System Time: Sample Rate to 24 kHz for output to the Wave external file token from the Sink Library, which records the .wav source as SVAudioOut2.wav. The Frequency Modulator token has the parameters of a carrier amplitude $A_c = 5$ V, a carrier frequency $f_c = 25$ kHz, a modulation gain $k_f = 25$ Hz/V, and 0° phase offset.

The communication channel is represented by the Adder token and the Gaussian Noise token from the Source Library, with initial parameters of a standard deviation $\sigma = 0$ V and mean $\mu = 0$ V. The output of the Adder token is inputted to the PLL token from the Communications Library. The parameters of the PLL token are a VCO (voltage controlled oscillator) center frequency $f_{VCO} = 25$ kHz, a VCO phase of 0°, a modulation gain of 25 Hz/V, a lowpass filter bandwidth of 10 kHz, and both loop filter coefficients are set to zero. The complex operation of a PLL is described in more detail in Chapter 4.

The PLL token frequency output tracks the FM bandpass signal and outputs the demodulated analog signal. The PLL token frequency output is inputted to the Decimator token from the Operator Library. The Decimator token parameter is set to 10 to downsample the 240 kHz System Time: Sample Rate to 24 kHz.

The output of the Decimator token is inputted to the Gain token from the Operator Library, with a parameter of a linear gain of 10. The Gain token increases the amplitude of the demodulated analog signal for output to the Wave external file token from the Sink Library, which records the demodulated analog signal as SVAudioOut.wav.

Figure 1.69 shows the Design window for the analog FM communication system of Figure 1.68 after the SystemVue simulation has been executed. The default Windows audio player is automatically evoked, and the specified output SV AudioOut.wav file can be heard.

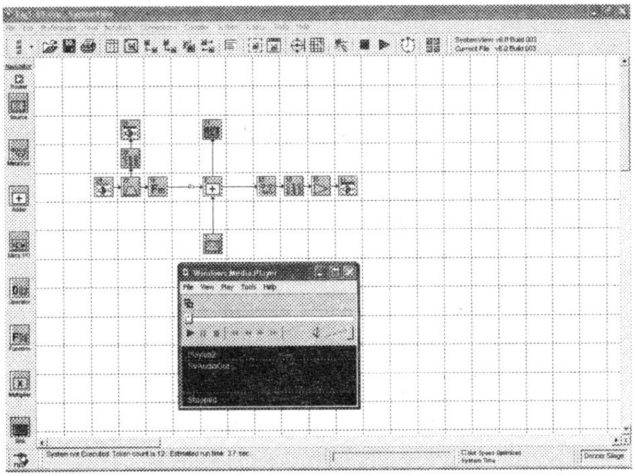

FIGURE 1.69 Design window for the analog FM communication system with Windows audio player.

Initially no noise is added to the communication channel, but setting the standard deviation σ of the Gaussian Noise token to 1 V produces a slight deterioration of the received audio signal. The PLL token parameter specifications include the capability of limiting the amplitude variation at the input, which improves the demodulated audio signal in the presence of additive white Gaussian noise [Stern04].

SUMMARY

In this chapter, the tenets of the simulation of a digital communication system were described. The SystemVue simulation environment was presented in a quick-start manner with a description of the Design window, the toolbars, the Analysis window, and a brief description of the available tokens. The implementation of temporal and spectral displays and the dynamic system probe in the SystemVue simulation environment facilitates the design and analysis of communication systems. Before describing digital communication systems, analog amplitude and frequency modulation was presented in the SystemVue simulation environment to illustrate SystemVue simulation procedures. Audio input files provided a qualitative assessment of the performance of these analog modulation and demodulation systems in the presence of additive white Gaussian noise.

Analyzing digital communication systems using SystemVue begins in Chapters 2 and 3, with the related foundations of baseband and bandpass modulation and demodulation. Chapter 4 considers the synchronization of the received signal to the carrier, phase, and symbol timing of the transmitter and equalization of bandlimited channels. Chapter 5 presents time division, frequency division, and code division multiplexing as another configuration of a digital communication system. Finally, Chapter 6 describes the process of sampling and quantization of analog signals, the basic premise for the transcription of real-world analog signals to a digital communication system.

REFERENCES

[Carlson02] Carlson, A. Bruce, et al., *Communication Systems.* McGraw-Hill, 2002.

[Gardner97] Gardner, Floyd, et al., *Simulation Techniques: Models of Communication Signals and Processes.* Wiley, 1997.

[Haykin01] Haykin, Simon, *Communication Systems.* Wiley, 2001.

[Lathi98] Lathi, B.P., *Modern Digital and Analog Communication Systems.* Oxford University Press, 1998.

[Stern04] Stern, Harold, et al., *Communication Systems Analysis and Design.* Prentice Hall, 2004.

2 Baseband Modulation and Demodulation

In This Chapter

- Rectangular Pulse Amplitude Modulation
- Sinc Pulse Amplitude Modulation
- Raised Cosine Pulse Amplitude Modulation
- Optimum Binary Baseband Receiver: The Correlation Receiver
- Multilevel (M-ary) Pulse Amplitude Modulation
- Partial Response Signaling
- Delta Modulation
- Eye Diagrams

Baseband modulation techniques encode information directly as the amplitude, width, or position of a pulse. They do not utilize a sinusoidal carrier. Some of these modulation schemes are known by the acronyms PWM (*pulse width modulation*), PPM (*pulse position modulation*), and PAM (*pulse amplitude modulation*). PWM is often used as the process control voltage for DC servomotors, and PPM is employed in optical data communication systems.

PAM was originally implemented as analog modulation, where the pulse amplitude is linearly related to the sampled analog signal. PAM is easily adapted to the binary transmission of information. PAM utilizes several types of signaling pulse waveforms with different transmission bandwidths, and simple and optimum receivers that can be conveniently designed and analyzed in SystemVue.

Binary digital data can be represented as either a symmetrical polar (± 3.3 V or ± 5 V) or asymmetrical unipolar PAM signal (for example, 0 and +3.3 V or 5 V).

PAM signals can be a portion τ of the bit time T_b (a duty cycle of less than 100%), the entire bit time, or even extend beyond a single bit time without causing *intersymbol interference* (ISI). PAM is used to introduce the concept of the baseband bandwidth needed to transmit pseudonoise (PN) random data, and the reception of data with bit error in the presence of additive white Gaussian noise (AWGN).

Shaping of the PAM signal from the simple rectangular pulse improves the spectral efficiency of baseband modulation to the theoretical minimum transmission bandwidth of half the binary data rate r_b. The optimum pulse shape is the sinc (sin x/x) pulse, but the raised cosine (or damped sinc-shaped) pulse is a common compromise solution. Binary PAM can be extended to multilevel (M-ary) PAM to increase spectral efficiency to an even greater extent. A controlled amount of ISI in partial response signaling or duobinary PAM can also produce a theoretical minimum, but with an increased complexity of the digital communication system.

The *quantitative measurement* of the bit error rate (BER) for a limited trial in a SystemVue simulation for several PAM baseband digital communication systems with simple and optimum PAM receivers is compared to the theoretical probability of bit error (P_b). Digital communication systems are subject to performance degradations with AWGN. SystemVue simulations of PAM are used to investigate the effect upon BER of lowpass filtering in the simple receiver, the performance of the optimum correlation receiver, the reduction in BER with Gray-coding of M-ary data, and partial response binary signaling.

SystemVue simulations of PAM digital communication systems and investigations of their characteristics and performance are provided and are the starting point for the what-ifs of baseband digital communication system design. These simulations confirm the theoretical expectation for P_b.

Delta modulation (DM) is an early digital communication system that can be implemented with simple equipment. A SystemVue simulation of DM presents its characteristics and performance tradeoffs. Finally, the *eye diagram* is a technique for the *qualitative assessment* of the performance of a baseband digital communication system. The effect of additive noise, communication channel bandwidth, and distortion and jitter on the binary or M-ary baseband signal can be readily seen.

The source files for these SystemVue simulations in baseband modulation and demodulation are located in the Chapter 2 folder on the CD-ROM, and are identified by the figure number (such as Fig2-1.svu). Appendix A includes a complete description of the contents of the CD-ROM.

ON THE CD

RECTANGULAR PULSE AMPLITUDE MODULATION

Rectangular pulse amplitude modulation (PAM) is convenient to generate in practice and represents binary logic signals, but displays the maximum bandwidth for the transmission of information. The rectangular PAM signal occupies one bit time T_b and has a constant amplitude.

SystemVue Simulation of Rectangular PAM

ON THE CD

A simple binary rectangular PAM digital communication system with AWGN is shown in Figure 2.1 (see Fig2-1.svu on the CD-ROM). The data source and PAM transmitter is a polar pseudonoise PN Sequence token from the Source Library, with parameters of a binary amplitude $A = \pm 5$ V, a data rate $r_b = 1$ kb/sec, and 0 V voltage and 0° phase offset. This PN Sequence token will transmit 100% duty cycle, polar rectangular pulses, and is the initial prototype pulse for PAM.

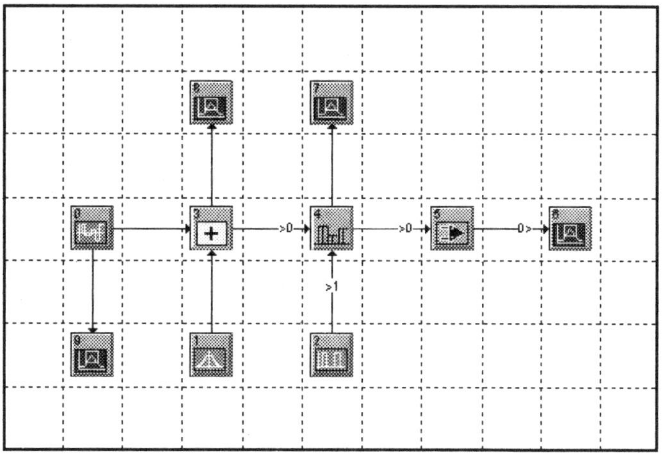

FIGURE 2.1 A simple binary rectangular PAM digital communication system.

The communication channel is represented by an Adder token and a Gaussian Noise token from the Source Library, with initial parameters of a standard deviation $\sigma = 0.1$ V and $\mu = 0$ V mean.

The simple PAM receiver is a Sample and Hold token from the Operator Library, with parameters of a control voltage of 2.5 V. The control voltage is provided by a Pulse Train token from the Source Library, with parameters of a unipolar amplitude $A = 0$ V or 5 V, a frequency $f_o = 1$ kHz (bit time $T_b = 1$ msec), a pulse width $\tau = 20$ μsec, a voltage offset of 0 V, but a phase offset of 180°.

The System Time: Sample Rate is set to 50 kHz, well above the nominal 1 kb/sec data rate, with 2500 samples and a resulting simulation time of 50 msec. This number of samples is used to display the signal data at a scale that can conveniently verify the performance of the SystemVue simulation. The spectral resolution is then 20

Hz (the sampling rate divided by the number of samples, or 50 kHz/2500), but only temporal displays will initially be used in this analysis.

The Pulse Train token provides the Sample and Hold token control voltage with a single point sample (T_{system} = 20 μsec) at a frequency f_o = 1 kHz and a 180° phase offset, which places the data sampling exactly at the midpoint of the bit time T_b. The output of the Sample and Hold token is inputted to the Analog Comparator token from the Logic Library, with parameters of a binary 1 output as 5 V, a binary 0 output as –5V, and a delay of 0 seconds. The negative comparator input is left unconnected, which is equivalent to a 0 V input.

The four Analysis tokens from the Sink Library display the data source as shown in Figure 2.2, the data source with AWGN in Figure 2.3, the sample and hold output in Figure 2.4, and the comparator output in Figure 2.5. This simple PAM receiver does not display any phase jitter, and is perfectly synchronized with the data source because of the exact parameter specification of the Pulse Train token control voltage source. Synchronization to the bit time T_b of the demodulation of a signal is considered in Chapter 4, "Synchronization and Equalization."

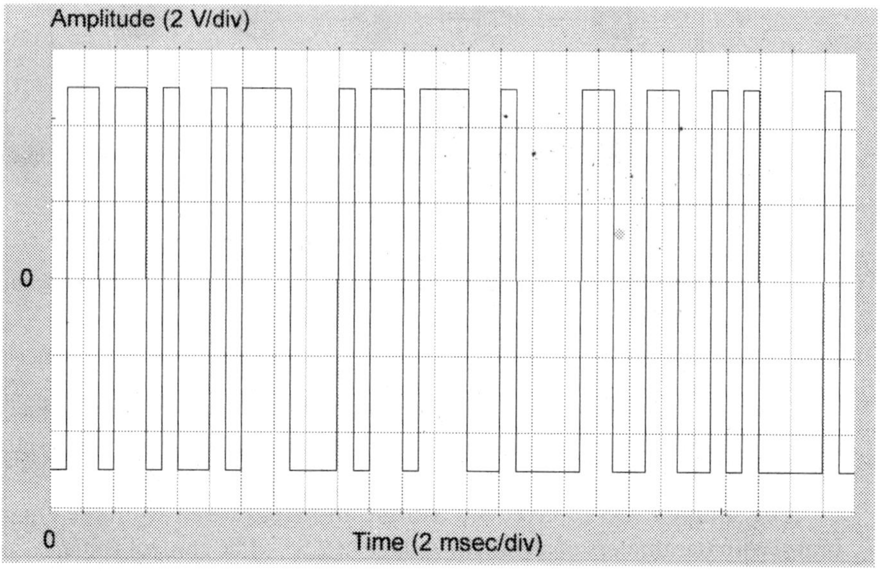

FIGURE 2.2 Binary rectangular PAM pseudonoise data source at 1 kb/sec.

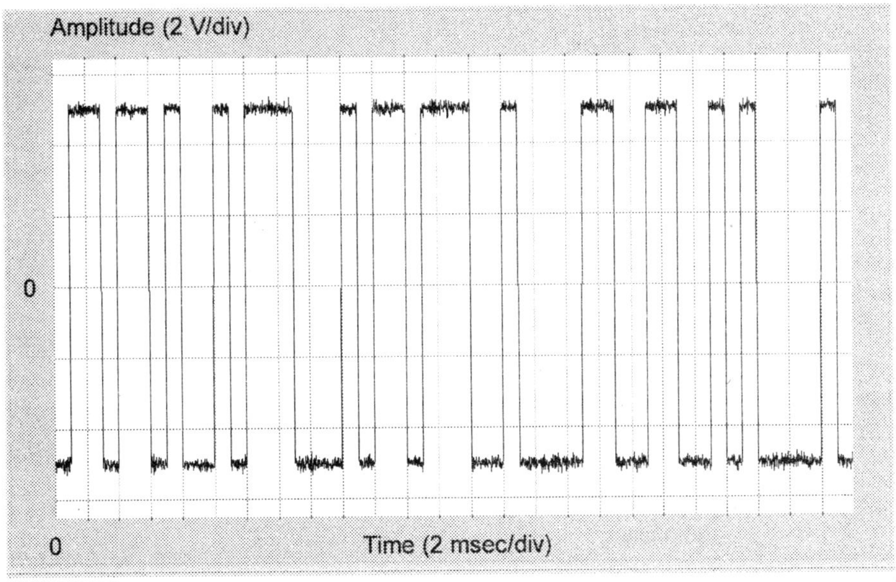

FIGURE 2.3 Binary rectangular PAM data source at 1 kb/sec with AWGN.

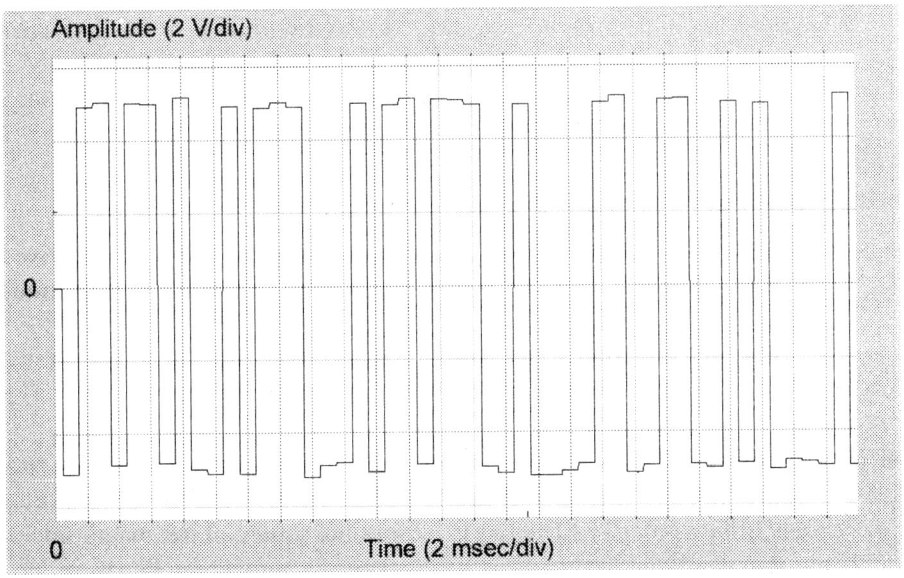

FIGURE 2.4 Sample and hold recovered binary rectangular PAM data.

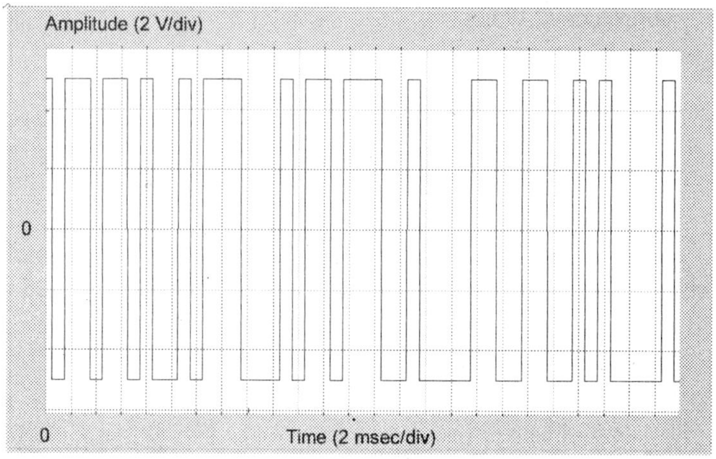

FIGURE 2.5 Analog comparator output of the recovered binary PAM data.

The rectangular PAM pseudonoise data source in Figure 2.2 is a random signal with no discernable period. Adding white Gaussian noise with a zero mean and a variance σ of only 0.1 volt in Figure 2.3 does not change the ± 5 V PAM signal appreciably, and certainly not enough to cause data errors by a crossing of the 0 V threshold in reception. The sample and hold output in Figure 2.4 shows a $T_b/2$ (0.5 msec) delay, and amplitude variation due to the additive noise on the received rectangular PAM signal. The analog comparator restores the original ± 5 V amplitude. The recovered signal is an exact duplicate, except for the $T_b/2$ (0.5 msec) sampling delay of the original pseudonoise PAM data source.

Rectangular PAM Power Spectral Density

The magnitude of the frequency domain representation $P_{PAM}(f)$ (derived from the Fourier transform) of a single rectangular pulse of width $\pm\ T_b/2$ seconds and amplitude A V is given by Equation 2.1.

$$P_{PAM}(f) = AT_b \operatorname{sinc}\left(2\pi\left(T_b/2\right)f\right) \tag{2.1}$$

This is the bi-sided voltage spectrum, extending from all negative to all positive frequencies ($-\infty < f < +\infty$). The bi-sided energy spectral density $E_{PAM}(f)$ of n pseudonoise polar PAM pulses is n times the square of the magnitude of the voltage spectrum divided by the load resistance R_L, as given by Equation 2.2.

$$E_{PAM}(f) = \frac{nA^2T_b^{\ 2}}{R_L} \operatorname{sinc}^2\left(2\pi\left(T_b/2\right)f\right) \tag{2.2}$$

The bi-sided power spectral density (PSD) $PSD_{PAM}(f)$ is the energy spectral density divided by the time to transmit the n PAM pulses or nT_b, as given by Equation 2.3.

$$PSD_{PAM}(f) = \frac{A^2 T_b}{R_L} \operatorname{sinc}^2\left(2\pi\left(T_b/2\right)f\right) \tag{2.3}$$

Setting the noise standard deviation $\sigma = 0$ V and increasing the System Time: Number of Samples specification to 262 144 (2^{18}) points results in a spectral resolution of approximately 0.19 Hz. This spectral resolution is an improvement over that specified previously (20 Hz), and is required for the detailed power spectral density analysis.

The normalized ($R_L = 1\ \Omega$) PSD of the analog comparator output without AWGN is shown unscaled and as a single-sided spectrum in Figure 2.6, in dBm/Hz (decibels referenced to 1 milliwatt per Hertz). The maximum frequency displayed is 25 kHz, the Nyquist frequency or half of the system simulation sampling rate (50 kHz/2). Because the rectangular pulse width τ is the bit time T_b, there are approximate nulls in the power spectral density that occur at the data rate $r_b = 1/T_b$ (1 kb/sec or 1 kHz) and multiples of the data rate due to the sinc^2 term in Equation 2.3. The PSD is not smooth because, although there are a large number of data, the ensemble is finite, but the spectral envelope clearly demonstrates the sinc^2 behavior.

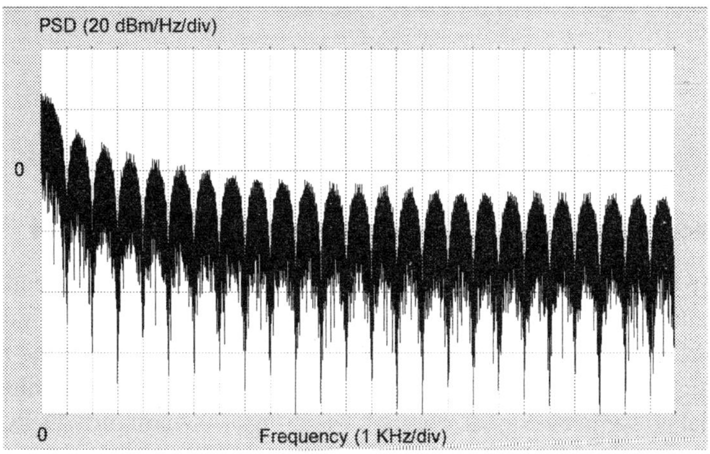

FIGURE 2.6 Power spectral density of a 1 kb/sec pseudonoise polar rectangular PAM signal.

The bandwidth of the binary rectangular PAM signal is somewhat problematical. Defining the bandwidth as a percentage of the total power of the rectangular

PAM signal is the usual course, because all of the available power would require an infinite bandwidth. The total power is easily calculated in the temporal domain. The power P in the rectangular PAM signal is the square of the amplitude A divided by R_L, as given by Equation 2.4.

$$P = \frac{A^2}{R_L} \tag{2.4}$$

The rectangular PAM signal has an amplitude $A = \pm 5$ V and the normalized ($R_L = 1\ \Omega$) power is 25 W = 43.98 dBm ($10 \log_{10}[25000]$). The data in Table 2.1 results from numerically integrating the power spectral density, shown in Equation 2.3, from 0 to an arbitrary frequency and comparing the resulting power to the expected total power.

TABLE 2.1 Bandwidth of a Binary Rectangular PAM Signal as a Percentage of the Total Power

Bandwidth (Hz)	Percentage of Total Power
$1/T_b$	90%
$1.5/T_b$	93%
$2/T_b$	95%
$3/T_b$	96.5%
$4/T_b$	97.5%
$5/T_b$	98%

Performance of Rectangular PAM in a Simple Receiver in AWGN

The binary rectangular PAM transmitter and simple receiver in Figure 2.1 can be evaluated for its performance in the presence of AWGN by including a bit error rate (BER) analysis and increasing σ, the standard deviation of the zero mean ($\mu = 0$) Gaussian noise source. The simple PAM receiver is naively unfiltered to capture the total power in the signal. Figure 2.7 (see Fig2-7.svu on the CD-ROM) shows the addition of the BER token from the Communications Library. To facilitate the simulation in SystemVue, all of the Analysis tokens, except that for the BER token output as the Final Value token, are deleted in the simulation model Fig2-7DT. As described in Chapter 1, "Communication Simulation Techniques," the SystemVue Textbook Edition does not permit tokens to be deleted.

ON THE CD

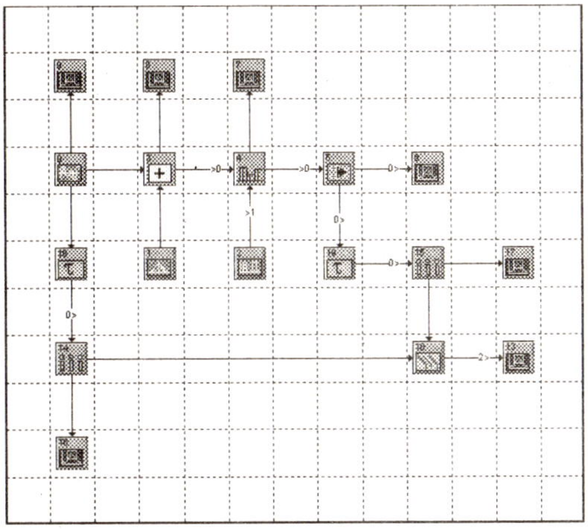

FIGURE 2.7 Rectangular PAM digital communication system with BER analysis.

The two comparison inputs to the BER token are single simulation samples per bit time T_b, derived from the PAM transmitter polar PN Sequence and PAM receiver Analog Comparator tokens. Because these signals are sampled at a 50 kHz System Time rate with $T_b = 1$ msec, each signal has 50 simulation-samples per bit time T_b.

Two Sampler tokens are used to *downsample* the 50 kHz System Time rate to the 1 kb/sec binary data rate or 1 kHz. The Sampler token parameters are set to a sample rate of 1 kHz, an aperture time of 20 μsec (T_{system}, a single point sample). Each Sampler input is delayed by 0.5 msec to acquire the data at $T_b/2$, the center of the rectangular PAM signal.

The BER token parameters are set to output on every trial (the parameter Number of Trials = 1), with a threshold of 0 V for polar rectangular PAM, and a 0 sec offset. The BER Analysis token displays the total error for the BER averaging performed over 1 bit, and the output then is the total observed errors. The two Analysis tokens, which display the two comparison inputs to the BER token, are to ensure that the error rate determination is proceeding correctly. Setting the Gaussian noise standard deviation at $\sigma = 0$ V should produce 0 observed errors.

The BER analysis of rectangular PAM in AWGN for the simple receiver requires a measure of signal and noise to assess relative performance. Although the total available signal is not used in detection in the simple PAM receiver, the signal power in Equation 2.4 provides one of the measures in the BER analysis for the eventual comparison to the performance of the optimum PAM receiver in AWGN.

The other measure is the noise power of the zero mean Gaussian noise source. The noise power P_n is the square of the standard deviation σ or the variance divided by R_L, as given by Equation 2.5.

$$P_n = \frac{\sigma^2}{R_L} \tag{2.5}$$

The signal-to-noise power ratio (SNR) or the ratio of the signal power, shown in Equation 2.4, to the noise power, shown in Equation 2.5, is given by Equation 2.6. SNR is usually reported in dB (decibels), as shown in Equation 2.7, and is not a function of R_L, because the signal and noise powers are measured at the same impedance.

$$SNR = \frac{A^2 / R_L}{\sigma^2 / R_L} = \frac{A^2}{\sigma^2} \tag{2.6}$$

$$SNR = 10 \log_{10}\left[\frac{A^2}{\sigma^2}\right] \tag{2.7}$$

The System Stop Time is set to 10 sec for convenience, which would process 10,000 information bits in possible error. The Gaussian noise standard deviation is increased to $\sigma = 2$ V, and the normalized noise power is 4 W with an SNR of 7.96 dB. A portion of the resulting rectangular PAM signal with AWGN is shown in Figure 2.8.

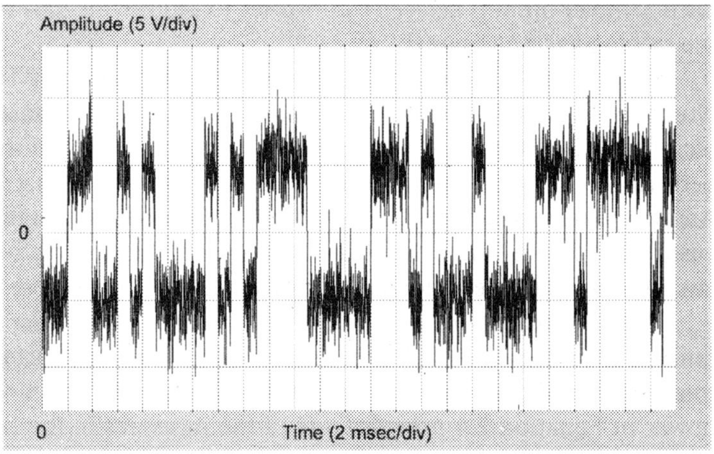

FIGURE 2.8 Binary rectangular PAM data source with AWGN, $\sigma = 2$ V, and SNR = 7.96 dB.

Figure 2.9 displays the total observed errors = 67, or an error rate of 6.7×10^{-3} with an SNR = 7.96 dB. However, the observed statistical process for BER is subjected to fluctuations each time the simulation is executed. Table 2.2 is a tabular list of the observed BER in a single trial of 10 000 information bits as a function of the SNR for the unfiltered rectangular PAM communication system of Figure 2.7. The minimal discernable 1 bit in error produces a BER of 10^{-4}.

TABLE 2.2 Observed BER as a Function of SNR in an Unfiltered Rectangular PAM Digital Communication System, Normalized Signal Power = 25 W

SNR dB	AWGN σ V	BER
∞	0	0
13.98	1	10^{-4}
10.46	1.5	8×10^{-4}
7.96	2	6.7×10^{-3}
6.02	2.5	2.33×10^{-2}
4.44	3	4.89×10^{-2}
1.94	4	1.061×10^{-1}

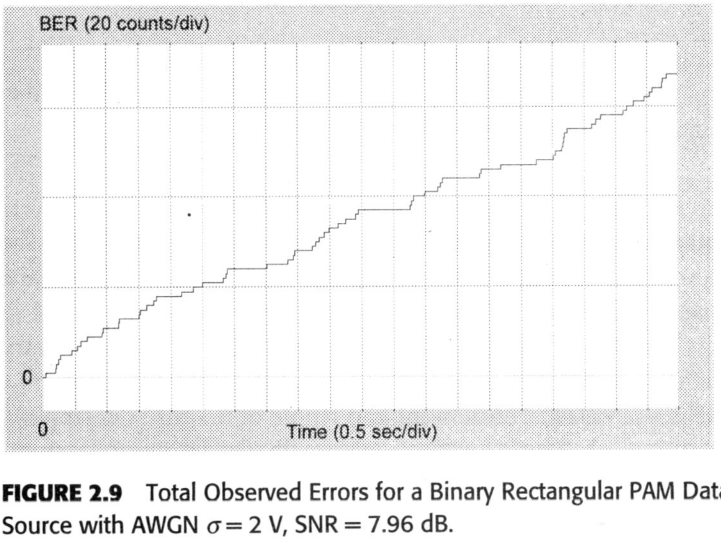

FIGURE 2.9 Total Observed Errors for a Binary Rectangular PAM Data Source with AWGN $\sigma = 2$ V, SNR = 7.96 dB.

A more accurate analysis would increase the number of information bits, perform multiple simulations, and average the BER at each SNR value. However, this brief BER analysis provides a comparison of the performance of unfiltered and lowpass filtered, rectangular PAM for the simple receiver in AWGN.

Performance of Filtered Rectangular PAM in a Simple Receiver in AWGN

The BER performance of the binary rectangular PAM transmitter and unfiltered simple receiver (refer to Figure 2.1) can be improved by including a lowpass filter (LPF) after the AWGN channel model and at the receiver input, as shown in Figure 2.10 (see Fig2-10.svu on the CD-ROM). To facilitate the simulation in SystemVue, all of the Analysis tokens, except that for the BER token output as the Final Value token, are deleted in the simulation model Fig2-10DT. As described in Chapter 1, the SystemVue Textbook Edition does not permit tokens to be deleted.

ON THE CD

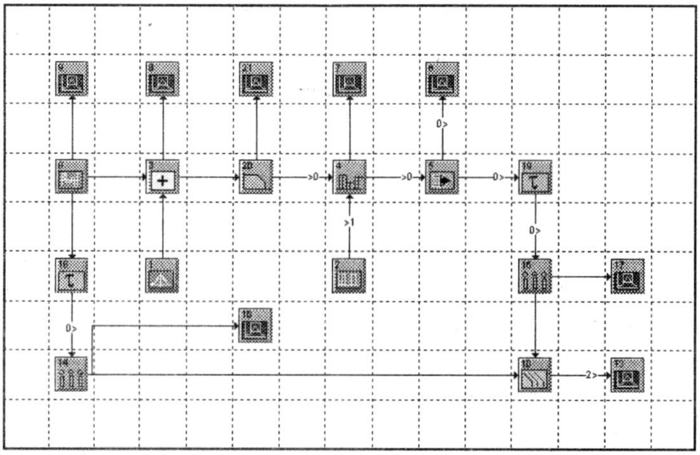

FIGURE 2.10 Lowpass filtered binary rectangular PAM digital communication system with BER analysis.

The LPF passes a selected amount of the total signal power, but suppresses the undesired white noise spectral components above the cutoff frequency. The initial compromise in performance is a slight loss of signal power and an increase in the rise and fall times of the received originally rectangular PAM signal.

The Linear System Filter token from the Operator Library provides the LPF, which is selected to be a 9-pole Chebyshev filter, with a cutoff frequency f_{cutoff} of 1.2 kHz ($1.2 \times 1/T_b$), and an inband ripple of 0.1 dB (1% ripple). Although a 9-pole Butterworth filter provides a maximally flat frequency response in the baseband, the LPF here is to have a steep response slope in the vicinity of the cutoff frequency to eliminate noise. The Chebyshev filter performs better than the Butterworth filter in that regard, with the tradeoff of an inband ripple.

A LPF with a cutoff frequency f_{cutoff} of 1.2 kHz would pass at least 90% of the total signal power for a binary data rate of 1 kb/sec, as given by Table 2.1. Figure 2.11 shows the original pseudonoise rectangular PAM data source overlaid with the

LPF rectangular PAM signal, with no AWGN and a System Stop Time of 25 msec for a convenient data display.

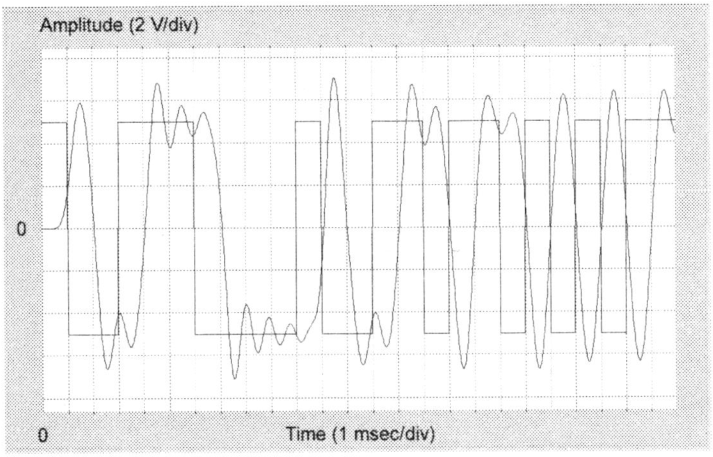

FIGURE 2.11 A binary rectangular PAM pseudonoise data source overlaid with LPF binary rectangular PAM with no AWGN.

The data signal reference BER Sampler input is delayed here by 1.5 msec to acquire the data at $T_b/2$ (0.5 msec), the center of the rectangular PAM signal, and to compensate for the delay (1 msec). The value of this delay can be observed by the cross-correlation of the two sampler signals at the input to the BER token.

The noise power of the AWGN signal at the output of the LPF is not proportional to the square of the standard deviation σ or the variance, as given by Equation 2.5. If an AWGN signal is passed through an LPF, or any linear time-invariant (LTI) filter, the output still displays a zero mean Gaussian probability density function, but the variance will be different. The average normalized ($R_L = 1\ \Omega$) noise power at the output of the LPF σ_o^2 is given by Equation 2.8.

$$\sigma_o^2 = N_o f_{cutoff} \tag{2.8}$$

$N_o/2$ is the nearly uniform amplitude of the AWGN power spectral density [Stern04]. However, the Gaussian noise source power at the input to the simple filtered PAM receiver is still given by the square of the standard deviation σ in Equation 2.5, and the SNR is given by Equation 2.7.

The System Stop Time is set to 10 sec for convenience, which would process 10,000 information bits in possible error. Setting the Gaussian noise source standard deviation $\sigma = 0$ V should produce 0 observed errors. Increasing to $\sigma = 2$ V results in the portion of the rectangular PAM pseudonoise data source overlaid with LPF rectangular PAM signal with AWGN, as shown in Figure 2.12.

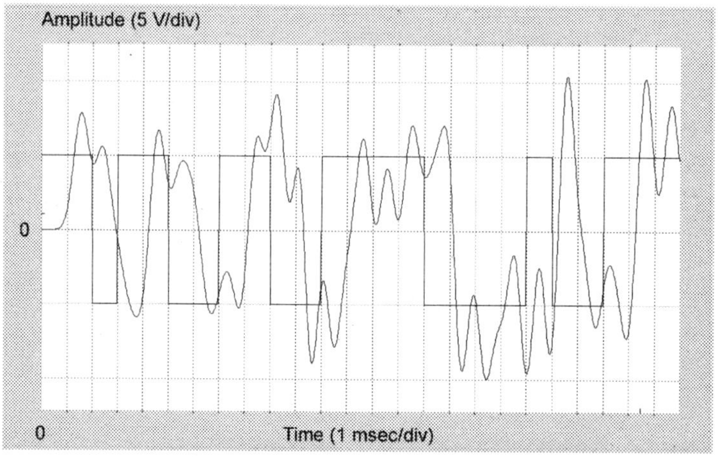

FIGURE 2.12 A rectangular PAM pseudonoise data source overlaid with LPF rectangular PAM signal with AWGN $\sigma = 2$ V, SNR = 7.96 dB.

A comparison of Figures 2.11 and 2.12 shows the effect on the signal of AWGN at this SNR with a LPF PAM receiver. Data is sampled at an initial delay of 1.5 msec to acquire the data at $T_b/2$ (0.5 msec), and to compensate for the group delay (1 msec) at intervals of 1 msec afterwards. Noise has increased the possibility of a bit error occurring at the sampling interval because of the obvious distortion of the received filtered rectangular PAM signal. The rectangular PAM pseudonoise data sources in Figures 2.11 and 2.12 are different because they are the result of two different random simulations.

A BER analysis is conducted with the same parameters for the simple unfiltered PAM receiver, as given in Table 2.2. Table 2.3 is a tabular list of the observed BER in a single trial of 10 000 information bits, as a function of the SNR for the LPF rectangular PAM communication system (refer to Figure 2.10). A comparison of Tables 2.2 and 2.3 shows an approximate +14 dB SNR increase in BER performance for LPF rectangular PAM over unfiltered rectangular PAM in the simple receiver.

Baseband PAM digital communication systems use an LPF at the input of the simple receiver to eliminate the error contribution of any noise beyond the bandwidth required to process the data source. A rectangular PAM data source has a wide bandwidth and even 90% of the total transmitted power requires an LPF with a cutoff frequency f_{cutoff} of at least $r_b = 1/T_b$ Hz. Shaping of the PAM data source as a sinc or a raised cosine pulse can lower the required cutoff frequency of the LPF.

TABLE 2.3 Observed BER as a Function of SNR in an LPF (9-pole Chebyshev, 0.1 dB ripple, f_{cutoff} = 1.2 kHz) Binary Rectangular PAM Digital Communication System, Normalized Signal Power = 25 W

SNR dB	AWGN σ V	BER
∞	0	0
6.02	2.5	0
0	5	0
−3.52	7.5	6×10^{-4}
−6.02	10	4.5×10^{-3}
−7.96	12.5	1.89×10^{-2}

SINC PULSE AMPLITUDE MODULATION

Sinc pulse amplitude modulation is somewhat difficult to generate in practice, but it displays the minimal bandwidth to transmit the data source. The sinc PAM signal extends beyond one bit time T_b, but avoids intersymbol interference (ISI) by having a zero value at $\pm nT_b$ (n is an integer and $n \neq 0$). This requisite is *Nyquist's First Criterion* to avoid ISI [Lathi98].

SystemVue Simulation of Sinc PAM

ON THE CD

A binary sinc PAM transmitter is shown in Figure 2.13 (see Fig2-13.svu on the CD-ROM). The data source and PAM transmitter consists of a periodic Pulse Train token and a PN Sequence token from the Source Library, a Multiplier token, and a Linear System Filter token from the Operator Library. The PN Sequence token has parameters of binary amplitude $A = \pm 5$ V, a data rate $r_b = 1$ kb/sec, and 0 V voltage and 0° phase offset.

The Pulse Train token is configured as a periodic impulse source with a frequency $f_o = 1$ kHz (bit time $T_b = 1$ msec), 0 V voltage and 0° phase offset, a pulse width of 20 μsec, and a unipolar amplitude of 0 to 47.6 V peak. This seemingly large and unusual peak amplitude is required to excite the Linear System Filter token sinc Communications Filter to the nominal 5 V peak output voltage. The Pulse Train impulse source is multiplied by the PN Sequence token output and results in a peak impulse amplitude of 47.6 V × (± 5 V) = 238 V.

The sinc Communications Filter has a bit rate r_b =1 kHz and its impulse response is designed to have a width of $\pm 4T_b = \pm 4$ msec, so that there are eight complete zero-crossing cycles. The sinc filter is implemented as an FIR digital filter with 401 taps to provide an 8 msec response (400/50 kHz = 8 msec), with a odd number of total taps for symmetry.

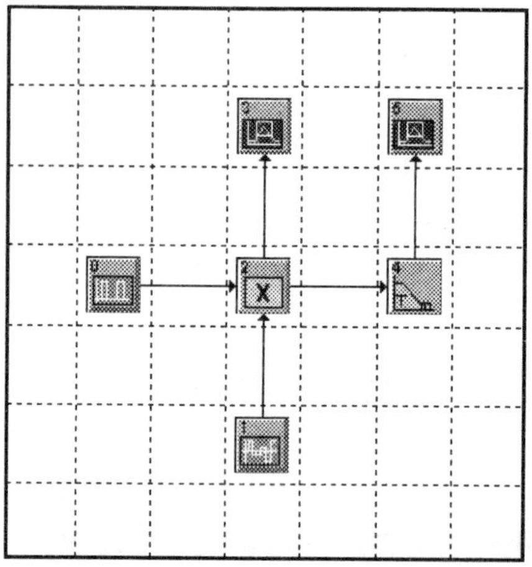

FIGURE 2.13 A binary sinc PAM transmitter.

The peak amplitude of the impulse response of the sinc Communications Filter is 1.05×10^3, as determined by the Linear System Filter design parameters. The excitation of the sinc filter is determined by the strength (area) of the periodic impulse train input, which is 4.76×10^{-3} V-sec (238 V for 20 μsec). The peak output voltage of the sinc filter is then nominally 5 V ($[1.05 \times 10^3/\text{sec}] \times [4.76 \times 10^{-3}$ V-sec$] \approx 5$ V). Figure 2.14 shows the two initial polar sinc PAM signals representing a binary 0 and binary 1, with a temporary binary data rate $r_b = 100$ b/sec to adequately separate and display the $\pm 4T_b$ sinc pulse. The bit time T_b remains at 1 msec and the sinc pulse here avoids ISI by having a zero value at $\pm T_b$, $\pm 2T_b$, $\pm 3T_b$, and $\pm 4T_b$.

The System Time: Sample Rate is set to 50 kHz, well above the nominal 1 kb/sec data rate. The Pulse Train token output simulates a periodic impulse train, because the pulse width parameter of 20 μsec is one simulation-sampling interval (T_{system} = 20 μsec). The output of the Multiplier token is a periodic, polar impulse train that excites the sinc Communications Filter token.

The communication channel is represented by an Adder token and a Gaussian Noise token from the Source Library, with the initial parameters of the standard deviation $\sigma = 0$ (no noise).

The data rate is then reset to 1 kb/sec, and with the System Time: Sample Rate at 50 kHz with 2500 samples, results in a simulation time of 50 msec. This number of samples is used to display the signal data at a scale that can conveniently verify the performance of the SystemVue simulation. The spectral resolution then is 20 Hz (the sampling rate divided by the number of samples or 50 kHz/2500), but only temporal displays will be used initially in the analysis. Figure 2.15 shows the summated sinc PAM signal, which is somewhat difficult to interpret and to ascertain that it encodes the original binary data.

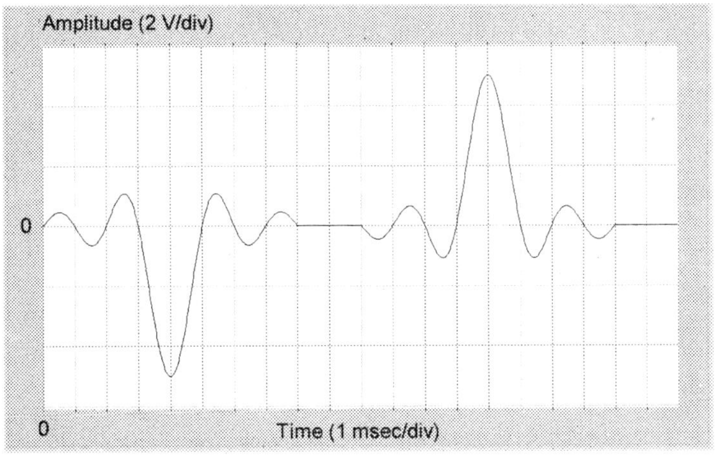

FIGURE 2.14 Two sinc PAM signals at a temporary data rate of 100 b/sec.

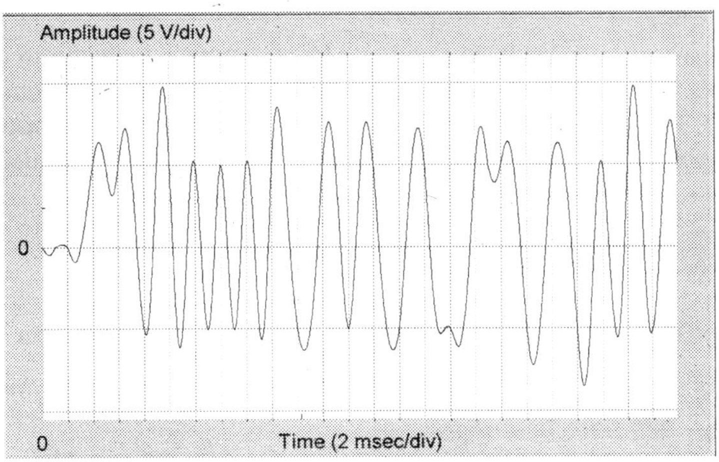

FIGURE 2.15 Summated sinc PAM signals at a data rate of 1 kb/sec.

Figure 2.14 indicates that the sinc PAM signal is delayed by 4 msec, as determined by the Linear System Filter design parameters. The transmitted sinc PAM signal is sampled at $t = 4$ msec and every 1 msec after that, with a binary 1 indicated by +5 V (or a positive voltage) and a binary 0 by – 5V (or a negative voltage). The transmitted binary data from the summated sinc PAM signal then is 111101101010... and this same data can be verified by inspecting the non-delayed, polar periodic impulse source that excites the sinc filter, as shown in Figure 2.16.

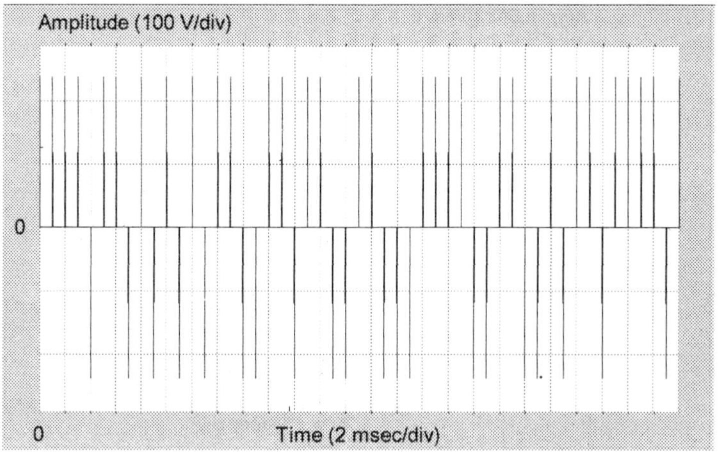

FIGURE 2.16 Polar periodic impulse source that excites the sinc filter.

Sinc PAM Power Spectral Density

Binary rectangular PAM requires a large channel bandwidth to transmit a reasonable portion of the original signal without significant distortion, as given by Table 2.1. The temporal domain description of a sinc pulse $p_{SINC}(t)$ centered at $t = 0$, with a peak amplitude A and zero crossings at $\pm nT_b$ is given by Equation 2.9.

$$p_{SINC}(t) = A\,\text{sinc}\left(\pi t/T_b\right) = A\,\frac{\sin\left(\pi t/T_b\right)}{\pi t/T_b} \tag{2.9}$$

The magnitude of the frequency domain representation of the sinc pulse $P_{SINC}(f)$ is given by Equation 2.10.

$$\begin{aligned} P_{SINC}(f) &= A\,T_b \quad -1/2T_b \le f \le 1/2T_b \\ P_{SINC}(f) &= 0 \qquad \text{otherwise} \end{aligned} \tag{2.10}$$

This rectangular voltage spectrum is a bi-sided spectrum and the bandwidth BW_{SINC} is given by Equation 2.11.

$$BW_{SINC} = \frac{1}{2T_b} = \frac{r_b}{2} \quad \text{Hz} \tag{2.11}$$

The bi-sided energy spectral density $E_{SINC}(f)$ of n pseudonoise polar sinc PAM pulses is n times the square of the magnitude of the frequency domain representation divided by the load resistance R_L, as given by Equation 2.12.

$$E_{SINC}(f) = \frac{n\,A^2 T_b^{\,2}}{R_L} \quad -1/2T_b \le f \le 1/2T_b \tag{2.12}$$

The bi-sided power spectral density (PSD) $PSD_{SINC}(f)$ is the energy spectral density divided by the time to transmit the n sinc PAM pulses or nT_b, as in Equation 2.13.

$$PSD_{SINC}(f) = \frac{A^2 T_b}{R_L} \quad -1/2T_b \le f \le 1/2T_b \tag{2.13}$$

Setting the noise standard deviation $\sigma = 0$ V, and increasing the System Time: Number of Samples specification to 262 144 (2^{18}) points, results in a spectral resolution of approximately 0.19 Hz. This spectral resolution is an improvement over that specified previously (20 Hz) and is required for the detailed power spectral density analysis.

The normalized PSD of the sinc PAM signal is shown unscaled in Figure 2.17, in dBm/Hz. The maximum frequency displayed is 25 kHz, the Nyquist frequency or half of the system simulation sampling rate (50 kHz/2). A comparison of the PSD of the sinc PAM signal in Figure 2.17 to that of the rectangular PAM signal in Figure 2.6 shows some significant differences.

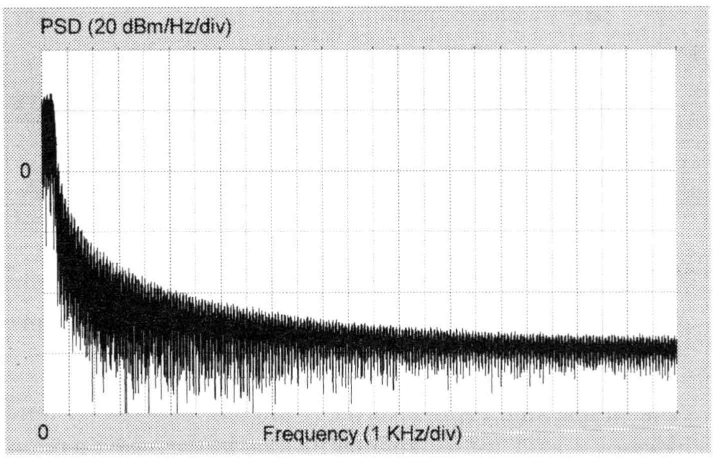

FIGURE 2.17 Power spectral density of a 1 kb/sec pseudonoise polar sinc PAM signal.

There are no approximate nulls in the PSD in Figure 2.17 that had occurred at the data rate $r_b = 1/T_b$ (1 kb/sec or 1 kHz), and multiples of the data rate in Figure

2.6, due to the sinc^2 term in Equation 2.3. The PSD here is approximately rectangular, with a cutoff frequency $f_{cutoff} = 1/2T_b$ or 500 Hz, and rolling off quite sharply thereafter.

The bandwidth of the sinc PAM signal is therefore taken to be $f_c = 1/2T_b$ or 500 Hz, and less problematical than the bandwidth of the rectangular PAM signal. This is the theoretical minimum for the binary transmission of data at a rate r_b b/sec in a bandwidth of $r_b/2$ Hz as postulated by Nyquist.

Performance of Sinc PAM in a Simple Receiver in AWGN

ON THE CD

The binary sinc PAM transmitter and filtered receiver, shown in Figure 2.18 (see Fig2-18.svu on the CD-ROM), can be evaluated for its performance in the presence of AWGN by including a bit error rate (BER) analysis and increasing the standard deviation σ of the Gaussian noise source. To facilitate the simulation in SystemVue, all of the Analysis tokens, except that for the BER token output as the Final Value token, are deleted in the simulation model Fig2-18DT. As described in Chapter 1, the SystemVue Textbook Edition does not permit tokens to be deleted.

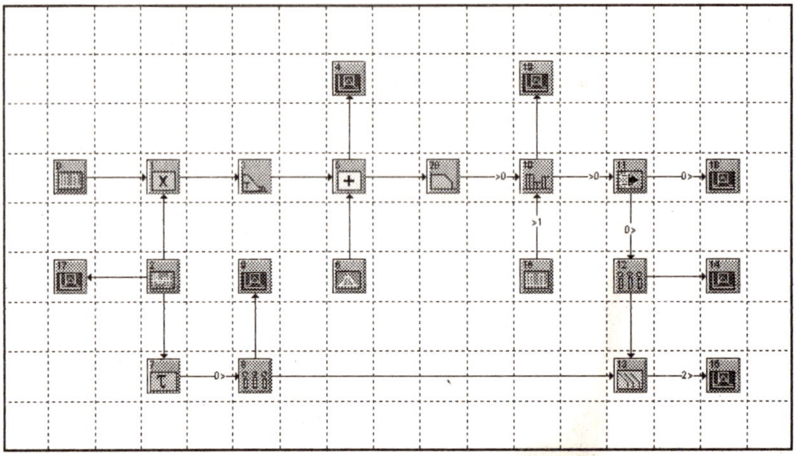

FIGURE 2.18 Binary sinc PAM digital communication system with BER analysis.

The simple sinc PAM receiver is filtered, because an unfiltered receiver has been demonstrated, for rectangular PAM at least, to have a poor BER performance. It is reasonable that a lowpass filter (LPF) would capture the total power in the sinc PAM signal here, because of the sharp cutoff frequency of its PSD. This was not the case for rectangular PAM.

The sinc PAM receiver Pulse Train token provides the Sample and Hold token control voltage with a single point sample ($T_{system} = 20$ µsec) at a 180° phase offset, which places the data sampling exactly at the midpoint of the bit time T_b. The output of the Sample and Hold token is inputted to an Analog Comparator token from the Logic Library, with parameters of a binary 1 output as 5 V, a binary 0 output as –5V, and a delay of 0 seconds. The negative comparator input is left unconnected, which is equivalent to a 0 V input. This simple PAM receiver does not display any phase jitter and is perfectly synchronized with the data source because of the exact parameter specification of the Pulse Train token control voltage source.

The two comparison inputs to the BER token are single simulation samples per bit time T_b derived from the PAM transmitter polar PN Sequence and PAM receiver Analog Comparator tokens. Because these signals are sampled at a 50 kHz System Time rate with $T_b = 1$ msec, each signal has 50 simulation-samples per bit time T_b.

The Linear System Filter token from the Operator Library provides the LPF, which is selected to be a 9-pole Chebyshev filter, with a cutoff frequency $f_{cutoff} = 600$ Hz ($1.2 \times 1/2T_b$), and an inband ripple of 0.1 dB. An LPF with a cutoff frequency of 600 Hz would pass nearly all of the total signal power for a binary data rate of 1 kb/sec, as in Equation 2.12.

Two Sampler tokens are used to downsample the 50 kHz System Time rate to the 1 kb/sec binary data rate or 1 kHz. The Sampler token parameters are set to a sample rate of 1 kHz, an aperture time of 20 μsec (T_{system}, a single point sample). The PN Sequence token data Sampler input is delayed by 6.5 msec to correlate the data with the center of the received sinc PAM signal. The value of this delay can be observed by the cross-correlation of the two sampler signals at the input to the BER token, as described in Chapter 1. The nominal 6.5 msec group delay consists of the 4 msec sinc PAM signal delay, 0.5 msec sinc PAM receiver sampling delay, and 2 data samples or 2 msec of additional processing delay.

The BER token parameters are set to output on every trial (Number of Trials = 1), with a threshold of 0 V for polar sinc PAM, and a 6.5 msec offset. The BER Analysis token displays the cumulative average error, but the BER averaging is performed over 1 bit and the output is the total observed errors. The two Analysis tokens, which display the two comparison inputs to the BER token, are to ensure that the error rate determination is proceeding correctly.

The System Stop Time is set to 10 sec for convenience, which would process 10,000 information bits in possible error. The Gaussian Noise token is specified by its standard deviation σ. Setting the Gaussian noise source standard deviation $\sigma = 0$ V should produce 0 observed errors.

As for LPF rectangular PAM, the noise power of a Gaussian noise source at the output of the LPF is not proportional to the square of the standard deviation or variance, as given by Equation 2.5. However, the Gaussian noise source power at the input to the simple filtered PAM receiver is still given by the square of the standard deviation σ in Equation 2.5, and the SNR is given by Equation 2.7.

The signal power, which contributes to the SNR, is problematical because the sinc PAM signal.extends beyond one bit time. Although there is no ISI at the data sampling interval $T_b/2$ in Figure 2.15, adjacent sinc PAM signals contribute or remove signal power everywhere else over the bit time. The sinc PAM signal has the same voltage (± 5 V) as the rectangular PAM signal at the midpoint of the bit time T_b, with synchronized data sampling and no jitter in the simple receiver.

The normalized power in the sinc PAM signal is the average of the square of the sinc Communications Filter output and can be computed by the system shown in Figure 2.19 (see Fig2-19.svu on the CD-ROM), which is an extension of Figure 2.13. The additional SystemVue tokens are the Algebraic x^a token from the Function Library, which squares the signal with $a = 2$, and the Average token from the Operator Library. The Average token time parameter is set to 10 sec, the System Stop Time. The computed average normalized power in the sinc PAM signal is approximately 24.4 W, and the power in the rectangular PAM signal is 25 W exactly.

ON THE CD

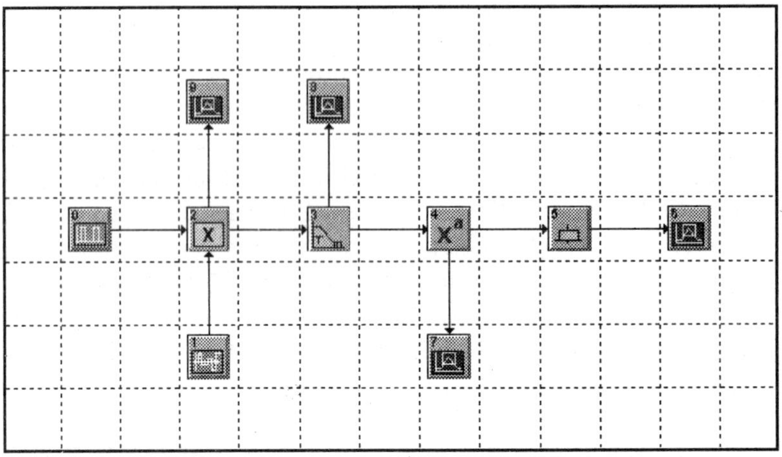

FIGURE 2.19 Computation of the normalized power in the sinc Communications Filter output.

A BER analysis is conducted with approximately the same parameters as the simple filtered rectangular PAM communication system, as given in Table 2.3. Table 2.4 is a tabular list of the observed BER in a single trial of 10 000 information bits as a function of the SNR for the LPF sinc PAM communication system shown in Figure 2.18. A comparison of Tables 2.3 and 2.4 shows an approximate –3 dB SNR decrease in BER performance for LPF sinc PAM over that of LPF rectangular PAM in the simple receiver.

TABLE 2.4 Observed BER as a Function of SNR in an LPF (9-pole Chebyshev, 0.1 dB ripple, f_C = 600 Hz) Binary Sinc PAM Digital Communication System, Normalized Signal Power ≈ 24.4 W

SNR dB	AWGN σ V	BER
∞	0	0
5.91	2.5	0
−0.01	5	1.3×10^{-3}
−3.63	7.5	7.3×10^{-3}
−6.13	10	1.66×10^{-2}
−8.06	12.5	3.09×10^{-2}

If the LPF filter is removed, the group delay is reduced to 4.5 msec, and there is a further approximate −12 dB decrease in BER performance for sinc PAM compared to LPF sinc PAM in the simple receiver. This confirms that the LPF is providing a BER performance increase for sinc PAM. The communication system design tradeoff here for LPF sinc PAM is a reduction in transmission bandwidth at a cost of a slight decrease in BER performance compared to LPF rectangular PAM.

RAISED COSINE PULSE AMPLITUDE MODULATION

Raised cosine pulse amplitude modulation, also known as the damped sinc PAM, is less difficult to generate in practice than a sinc pulse, and displays a compromise bandwidth between that of the rectangular and sinc pulses to transmit the information source. The raised cosine PAM signal extends beyond one bit time T_b, but again avoids intersymbol interference (ISI) by having a zero value at $\pm nT_b$ (n is an integer and $n \neq 0$).

SystemVue Simulation of Raised Cosine PAM

A binary raised cosine PAM transmitter and AWGN channel, shown in Figure 2.20 (see Fig2-20.svu on the CD-ROM), is conceptually the same as the sinc PAM system shown in Figure 2.13. The parameters of the data source and raised cosine PAM transmitter are identical to that described for the sinc PAM transmitter in Figure 2.13, except for the inclusion here of the Linear System Filter token raised cosine Communications Filter, and the slight adjustment of the strength of the periodic impulse source.

ON THE CD

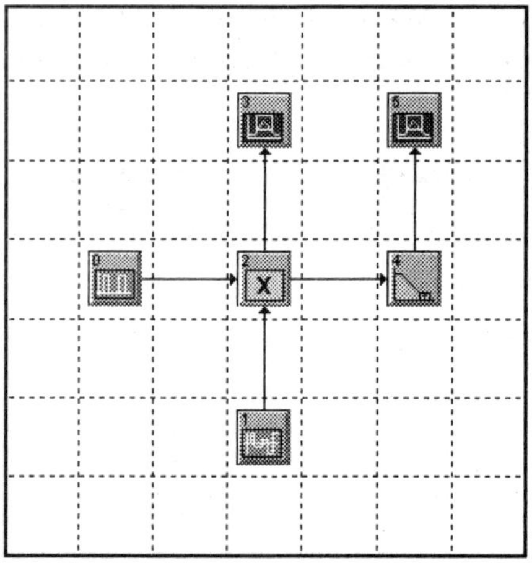

FIGURE 2.20 A binary raised cosine PAM transmitter and AWGN channel.

The data source and PAM transmitter consists of a periodic Pulse Train token and a PN Sequence token from the Source Library, a Multiplier token, and a Linear System Filter token from the Operator Library. The PN Sequence token has parameters of binary amplitude $\pm$ 5 V, a data rate r_b of 1 kb/sec, and 0 V voltage and 0° phase offset.

The communication channel is represented by an Adder token and a Gaussian Noise token from the Source Library, with initial parameters of a standard deviation $\sigma = 0.1$ V and $\mu = 0$ V mean.

The Pulse Train token is configured as a periodic impulse source with a frequency f_o = 1 kHz (bit time T_b = 1 msec), 0 V voltage and 0° phase offset, a pulse width of 20 μsec, and a unipolar amplitude of 0 to 50.2 V peak. The Pulse Train impulse source is multiplied by the PN Sequence token output and results in a peak impulse amplitude of 50.2 V × (± 5 V) = 251 V.

The raised cosine Communications Filter has a bit rate of 1 kHz, and its impulse response is designed to have a width of $\pm 4T_b = \pm 4$ msec, the same as the sinc PAM pulse. There are eight complete zero-crossing cycles, although the zero-crossing at $\pm$ $3T_b$ and $\pm$ $4T_b$ have negligible amplitude. The raised cosine filter is implemented as an FIR digital filter with 401 taps, to provide an 8 msec response (400/50 kHz = 8 msec), with an odd number of total taps for symmetry. The raised cosine Communications Filter is also specified by the *rolloff factor* α, $0 \leq \alpha \leq 1$, which determines the bandwidth of pulse, and here $\alpha = 0.5$, an arbitrary intermediate value.

The peak amplitude of the impulse response of this raised cosine Communications Filter is 9.96×10^2, as determined by the Linear System Filter design parameters. The peak amplitude for the raised cosine pulse is slightly less than that of the sinc pulse (1.05×10^3). The excitation of the raised cosine filter is determined by the strength (area) of the periodic impulse train input, which is 5.02×10^{-3} V-sec (251 V for 20 μsec). The peak output voltage of the raised cosine filter is then nominally 5 V ($[9.96 \times 10^2/\text{sec}] \times [5.02 \times 10^{-3}\ \text{V-sec}] \approx 5$ V). Figure 2.21 shows the two initial polar raised cosine PAM signals representing a binary 0 and binary 1, with a temporary data rate of 100 b/sec to adequately separate and display the $\pm 4T_b$ raised cosine pulse. The bit time T_b remains at 1 msec, and the raised cosine pulse here avoids ISI by having a zero value at $\pm T_b$, $\pm 2T_b$, $\pm 3T_b$, and $\pm 4T_b$.

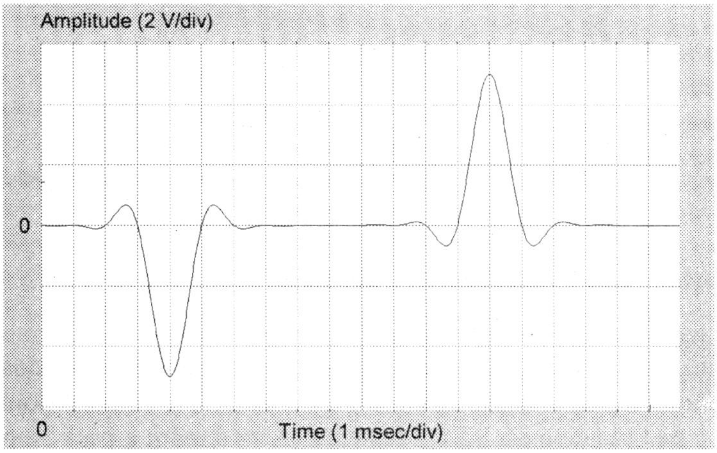

FIGURE 2.21 Two raised cosine $\alpha = 0.5$ PAM signals at a temporary data rate of 100 b/sec.

A comparison of Figure 2.14 for the sinc pulse and Figure 2.21 demonstrates the limited temporal response of the raised cosine pulse and the reduced susceptibility of the transmitted signal to ISI and bit errors due to jitter.

The System Time: Sample Rate is set to 50 kHz, well above the nominal 1 kb/sec data rate. The data rate is then reset to 1 kb/sec and with the System Time: Sample Rate at 50 kHz with 2500 samples, results in a simulation time of 50 msec. Figure 2.22 shows the summated raised cosine PAM signal, which can be interpreted, as was the sinc PAM signal, to ascertain that it does encode the original binary data correctly.

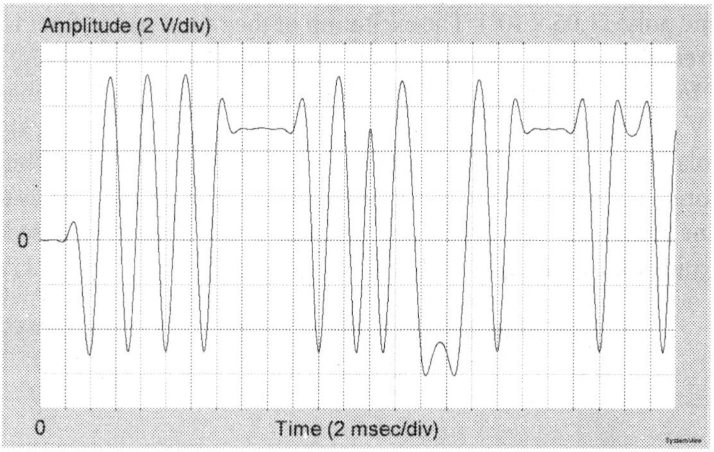

FIGURE 2.22 Summated binary raised cosine PAM signals at a data rate of 1 kb/sec.

Raised Cosine PAM Power Spectral Density

Binary raised cosine PAM requires a channel bandwidth intermediate to that of the binary rectangular and sinc PAM systems, to transmit a reasonable portion of the original signal without significant distortion. The temporal domain description of a raised cosine pulse $p_{RC}(t)$ centered at $t = 0$, with a peak amplitude A and zero crossings at $\pm nT_b$, is given by Equation 2.14.

$$p_{RC}(t) = A\,\text{sinc}\left(\pi t / T_b\right)\left[\frac{\cos\left(2\pi\beta t\right)}{1-\left(4\beta t\right)^2}\right] \tag{2.14}$$

The term β is the *damping factor*, $0 \le \beta \le 1/2T_b$. As β approaches the maximum value of $1/2T_b$, the *main lobe* of the raised cosine pulse narrows, the *tails* flatten, and the transmitted signal is less susceptible to ISI and timing jitter. The rolloff factor β and the damping factor α are related by Equation 2.15.

$$\alpha = 2T_b\beta = \frac{2\beta}{r_b} \tag{2.15}$$

The magnitude of the frequency domain representation of the raised cosine pulse $P_{RC}(f)$ is given by Equation 2.16.

$$
\begin{aligned}
P_{RC}(f) &= AT_b & |f| &\le \frac{r_b}{2} - \beta \\
P_{RC}(f) &= 2AT_b\left[1+\cos\left\{\frac{\pi}{2\beta}\left(|f| - \frac{r_b}{2} + \beta\right)\right\}\right] & \frac{r_b}{2} - \beta &\le |f| \le \frac{r_b}{2} + \beta \qquad (2.16) \\
P_{RC}(f) &= 0 & |f| &\le \frac{r_b}{2} + \beta
\end{aligned}
$$

The shape of this voltage spectrum is called a *raised cosine* because of the $1 + \cos(x)$ term for a portion of the frequency range. The data rate r_b, rather than the bit time T_b, often is used to describe the frequency domain limits. The raised cosine voltage spectrum is a bi-sided spectrum and the bandwidth BW_{RC} is given by Equation 2.17.

$$
BW_{RC} = \frac{1}{2T_b}(1+\alpha) = \frac{r_b}{2}(1+\alpha) \quad \text{Hz} \qquad (2.17)
$$

The maximum bandwidth for the raised cosine pulse is $1/T_b$ Hz, because the rolloff factor α is at most 1. For the rectangular pulse, a bandwidth of $1/T_b$ Hz would only contain 90% of the total transmitted power, as shown in Table 2.1. The sinc pulse has a bandwidth of $1/2T_b$ Hz, given by Equation 2.11, which is the minimum bandwidth of the raised cosine pulse when $\alpha = 0$. Of course, if the *rolloff factor* $\alpha = 0$, then the damping factor $\beta = 0$, and the raised cosine pulse is a sinc pulse.

The bi-sided energy spectral density of n pseudonoise polar raised cosine PAM pulses $E_{RC}(f)$ is n times the square of the magnitude of the voltage spectrum, divided by the load resistance R_L, as given by Equation 2.18.

$$
\begin{aligned}
E_{RC}(f) &= nA^2T_b^2 / R_L & |f| &\le \frac{r_b}{2} - \beta \\
E_{RC}(f) &= \frac{4n\,A^2T_b^2}{R_L}\left[1+\cos\left\{\frac{\pi}{2\beta}\left(|f| - \frac{r_b}{2} + \beta\right)\right\}\right]^2 & \frac{r_b}{2} - \beta &\le |f| \le \frac{r_b}{2} + \beta \qquad (2.18) \\
E_{RC}(f) &= 0 & |f| &\le \frac{r_b}{2} + \beta
\end{aligned}
$$

The bi-sided power spectral density (PSD) $PSD_{RC}(f)$ is the energy spectral density divided by the time to transmit the n raised cosine PAM pulses or nT_b, as given by Equation 2.19.

$$PSD_{RC}(f) = A^2T_b / R_L \qquad |f| \le \frac{r_b}{2} - \beta$$

$$PSD_{RC}(f) = \frac{4A^2T_b}{R_L}\left[1 + \cos\left\{\frac{\pi}{2\beta}\left(|f| - \frac{r_b}{2} + \beta\right)\right\}\right]^2 \qquad \frac{r_b}{2} - \beta \le |f| \le \frac{r_b}{2} + \beta \tag{2.19}$$

$$PSD_{RC}(f) = 0 \qquad |f| \le \frac{r_b}{2} + \beta$$

Setting the noise standard deviation $\sigma = 0$ V and increasing the System Time: Number of Samples specification to 262 144 (2^{18}) points results in a spectral resolution of approximately 0.19 Hz. The normalized PSD of the raised cosine $\alpha = 0.5$ PAM signal is shown unscaled in Figure 2.23, in dBm/Hz. The maximum frequency displayed is 25 kHz, the Nyquist frequency or half of the system simulation sampling rate (50 kHz/2). A comparison of the PSD of the raised cosine PAM signal in Figure 2.23 to that of the sinc PAM signal in Figure 2.17 and the rectangular PAM signal in Figure 2.6 shows some significant differences.

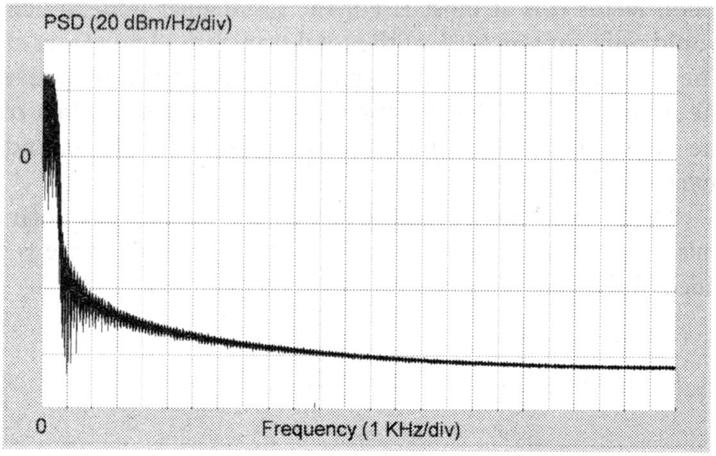

FIGURE 2.23 Power spectral density of a 1 kb/sec pseudonoise polar raised cosine $\alpha = 0.5$ PAM signal.

There are no approximate nulls in the PSD in Figure 2.23 that had occurred at the data rate and multiples of the data rate in Figure 2.6, due to the sinc^2 term in Equation 2.3. The PSD here is approximately squared raised cosine, with a cutoff frequency $f_{cutoff} = (1/2T_b)(1 + \alpha)$ or 750 Hz, and rolling off very sharply thereafter.

Although the cutoff frequency here (750 Hz) is higher than that of the sinc PAM signal (500 Hz), the raised cosine PAM PSD has significantly less power for frequencies beyond its f_c, compared to the sinc PAM PSD in Figure 2.17. As shown in Figure 2.14, the sinc PAM pulse is only an approximation, and at $t = \pm 4T_b$ has certainly not decayed to zero amplitude, as the raised cosine PAM pulse shown in Figure 2.21.

Performance of Raised Cosine PAM in a Simple Receiver in AWGN

ON THE CD

The raised cosine $\alpha = 0.5$ PAM transmitter and filtered receiver, shown in Figure 2.24 (see Fig2-24.svu on the CD-ROM), can be evaluated for its performance in the presence of AWGN by including a bit error rate (BER) analysis and increasing the standard deviation σ of the Gaussian noise source. To facilitate the simulation in SystemVue, all of the Analysis tokens, except that for the BER token output as the Final Value token, are deleted in the simulation model Fig2-24DT. As described in Chapter 1, the SystemVue Textbook Edition does not permit tokens to be deleted. The simple raised cosine PAM filtered receiver in Figure 2.24 is identical to the sinc PAM receiver in Figure 2.18, except for the cutoff frequency, inband ripple of the LPF, and the BER sampled signal correlation delay.

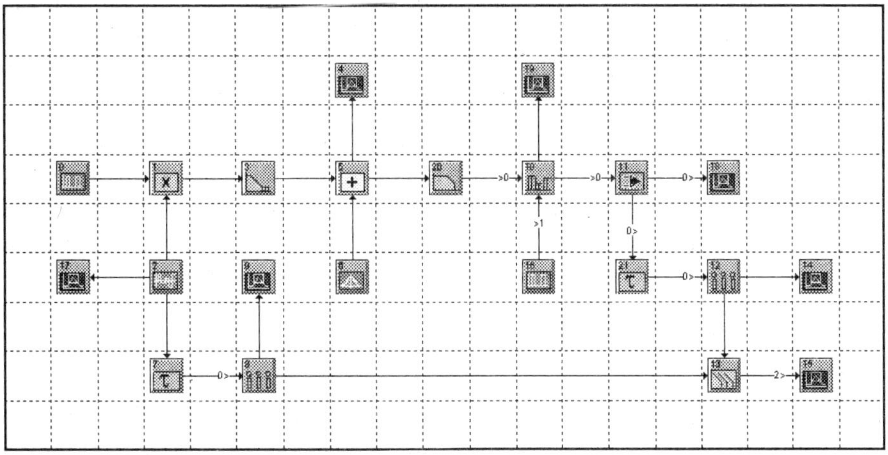

FIGURE 2.24 Raised cosine PAM digital communication system with BER analysis.

The Linear System Filter token from the Operator Library provides the LPF, which is selected to be a 9-pole Chebyshev filter, with a cutoff frequency $f_{cutoff} = 900$ Hz $(1.2 \times (1/2T_b)(1 + \alpha) = 1.2 \times 750$ Hz) and an inband ripple of 0.1 dB. An LPF with a cutoff frequency of 900 Hz would pass nearly all of the total signal power for a binary data rate of 1 kb/sec, as in Equation 2.19.

The PN Sequence token data Sampler input is delayed by 5.5 msec to correlate the data with the center of the received raised cosine PAM signal. The value of this delay can be observed by the cross-correlation of the two sampler signals at the input to the BER token. The nominal 5.5 msec delay consists of the 4 msec raised cosine PAM signal delay, 0.5 msec sinc PAM receiver sampling delay and 1 data sample or 1 msec of additional processing delay.

Although setting the Gaussian noise standard deviation $\sigma = 0$ V should produce zero observed bit errors, it does not. The initial 0.1 dB inband ripple of the 9-pole Chebyshev filter, with a cutoff frequency $f_{cutoff} = 900$ Hz induces bit errors even with no noise. The effect on BER is similar to that observed for bit synchronization error, as described in Chapter 4. Changing the inband ripple to 0.01 dB ($\approx 0.1\%$ ripple) produced 0 observed bit errors consistently.

This is due to the extent of the low frequency components in the PSD of the raised cosine PAM signal in Figure 2.23, and the obvious sensitivity of the BER analysis to inband ripple in the LPF. As a what-if of digital communication system design, a 9-pole Butterworth with a cutoff frequency $f_{cutoff} = 900$ Hz also produced zero observed bit errors consistently. However, the Chebyshev filter with reduced inband ripple was used here for comparison with the other BER analyses.

The tradeoff is that the reduced inband ripple of the Chebyshev LPF slightly decreases the frequency response beyond the cutoff frequency. At a frequency of 1.5 kHz, the response is approximately –65 dB for the 0.1 dB ripple, but –60 dB for the 0.01 dB ripple, 9-pole Chebyshev LPF with $f_c = 900$ Hz. The comparable 9-pole Butterworth LPF with $f_{cutoff} = 900$ Hz has a response of only –40 dB at a frequency of 1.5 kHz.

As for LPF rectangular and sinc PAM, the noise power of a Gaussian noise source at the output of the LPF is not proportional to the square of the standard deviation or variance, as given by Equation 2.5. However, the Gaussian noise source power at the input to the simple filtered PAM receiver is still given by the square of the standard deviation σ in Equation 2.5, and the SNR is given by Equation 2.7.

The signal power, which contributes to the SNR, is less problematical here because the raised cosine PAM signal extends beyond one bit time, with reduced (damped) amplitude compared to the sinc PAM signal. The raised cosine PAM signal has the same voltage (± 5 V) as the rectangular and sinc PAM signal at the midpoint of the bit time T_b, and there is no ISI with synchronized data sampling and no jitter in the simple receiver.

The normalized power in the raised cosine PAM signal is the average of the square of the Communications Filter output, and can be computed by a system nearly identical to that in Figure 2.19, except for the substitution of the raised cosine $\alpha = 0.5$ Communications Filter. The Average token time parameter is set to 10 sec, the System Stop Time. The computed average normalized power in the raised cosine $\alpha = 0.5$ PAM signal is approximately 20.7 W for a time average of 10 sec. The power in the sinc and rectangular PAM signals are 24.4 W and 25 W, respectively.

A BER analysis is conducted with approximately the same parameters as the simple filtered rectangular and sinc PAM communication systems, as given in Tables 2.3 and 2.4. Table 2.5 is a tabular list of the observed BER in a single trial of 10 000 information bits as a function of the SNR for the LPF raised cosine $\alpha = 0.5$ PAM communication system, shown in Figure 2.24. A comparison of Tables 2.3 and 2.5 shows an approximate –7 dB SNR decrease in BER performance for LPF raised cosine PAM over that of LPF rectangular PAM in the simple receiver.

TABLE 2.5 Observed BER as a Function of SNR in an LPF (9-pole Chebyshev, 0.01 dB ripple, $f_{cutoff} = 900$ Hz) Binary Raised Cosine $\alpha = 0.5$ PAM Digital Communication System, Normalized Signal Power ≈ 20.7 W

SNR dB	AWGN σ V	BER
∞	0	0
9.63	2.5	1.3×10^{-3}
–0.82	5	1.54×10^{-2}
–4.34	7.5	4.01×10^{-2}
–6.84	10	6.68×10^{-2}
–8.78	12.5	9.44×10^{-2}

If the LPF filter is removed and the delay is reduced to 4.5 msec, there is a further approximate –11 dB decrease in BER performance for raised cosine PAM, compared to LPF raised cosine PAM in the simple receiver. This confirms that the LPF is providing a BER performance increase for raised cosine PAM. The communication system design tradeoff here for LPF raised cosine PAM is a reduction in transmission bandwidth at a cost of a substantial decrease in BER performance compared to LPF rectangular PAM.

OPTIMUM BINARY BASEBAND RECEIVER: THE CORRELATION RECEIVER

The improved, but still simple, structure for the binary baseband PAM receiver has included a LPF and a sampler at the midpoint of each bit time T_b. Can a receiver be built with even better performance? Actually, not only better performance, but optimum performance can be achieved with the *correlation receiver*.

The Correlation Receiver for Baseband Symmetrical Signals

If the processing is to be linear and restricted to the entire bit time, then the optimum processing of a received signal in the presence of AWGN has been shown to be a filter with a temporal impulse response $h_o(t)$ that is matched to the expected signal, as given by Equation 2.20 [Stern04].

$$h_o(t) = k\ s_1(iT_b - t) \quad (i-1)\,T_b \le t \le iT_b \tag{2.20}$$

In Equation 2.20, k is an arbitrary gain constant, $s_1(t)$ is the signal that arbitrarily represents a binary 1, and the impulse response and the signal is defined over one bit time (i is an integer). The gain constant k is usually taken to be 1, because it processes both the received signal and noise. The signal $s_0(t)$ represents a binary 0 and these signals are initially assumed to be equally likely to occur (equal *apriori* probability) and symmetrical, as given by Equation 2.21.

$$s_1(t) = -s_0(t) \quad (i-1)T_b \le t \le iT_b \tag{2.21}$$

The output of the matched filter $z_j(t)$ is the *convolution* of the received signal $s_j(t)$ ($j = 0, 1$) and the impulse response, as given by Equation 2.22. Here it is assumed that the received signal is attenuated by the communication channel by an amount γ, and there is no additive noise.

$$\begin{aligned} z_j(t) &= \frac{1}{T_b}\int_{(i-1)\,T_b}^{t} \gamma s_j(\tau)\,h_o(t-\tau)\,d\tau \quad (i-1)T_b \le t \le iT_b \\ z_j(t) &= \frac{1}{T_b}\int_{(i-1)\,T_b}^{t} \gamma s_j(\tau)\,s_1[iT_b-(t-\tau)]\,d\tau \quad (i-1)T_b \le t \le iT_b \end{aligned} \tag{2.22}$$

The output of the matched filter $z_j(t)$ ($j = 0, 1$) is sampled at the end of each bit time, $t = iT_b$, as given by Equation 2.23.

$$\begin{aligned} z_j(iT_b) &= \frac{1}{T_b}\int_{(i-1)\,T_b}^{iT_b} \gamma s_j(\tau)\,s_1[iT_b-(iT_b-\tau)]\,d\tau \quad (i-1)T_b \le t \le iT_b \\ z_j(iT_b) &= \frac{1}{T_b}\int_{(i-1)\,T_b}^{iT_b} \gamma s_j(\tau)\,s_1(\tau)\,d\tau \quad (i-1)T_b \le t \le iT_b \quad j = 0,1 \end{aligned} \tag{2.23}$$

Equation 2.23 indicates that the optimum receiver can be implemented, not merely as a matched filter, but equivalently as a *correlator* or correlation receiver, as shown in Figure 2.25 (see Fig2-25.svu on the CD-ROM).

ON THE CD

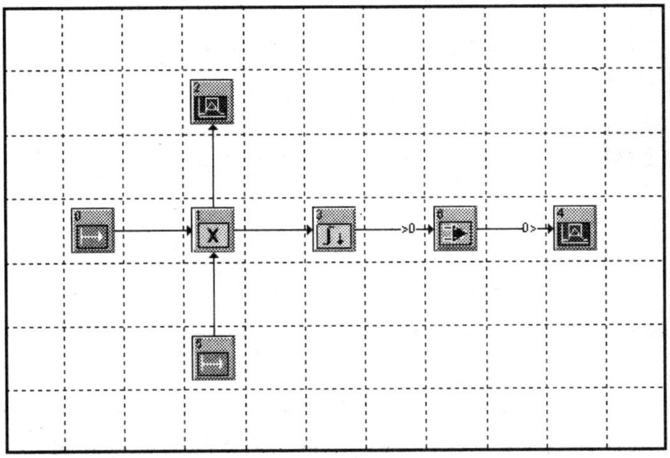

FIGURE 2.25 The matched filter implemented as a correlation receiver for baseband symmetrical signals.

The received signal $s_j(t)$ $(j = 0, 1)$ is first multiplied by the reference signal $s_1(t)$, synchronized to the start of the bit time and then integrated over that one bit time T_b. The output of the Integrate and Dump token from the Communications Library is held for each bit time, and compared to the optimum threshold τ_{opt}, which is 0 V (or unconnected) for symmetrical signals, to decide if the received signal is a binary 1 or binary 0.

The Integrate and Dump token is reset to zero internally at the start of the next bit time, and is assumed to be perfectly synchronized to the start and duration of the data source. The operation is called *integrate-and-dump* and the matched filter implemented as a correlation receiver in Figure 2.25 can replace the simple PAM receiver.

The Analog Comparator token from the Logic Library has an optimum threshold for equally probable binary baseband signals, given by Equation 2.24.

$$\tau_{opt} = \frac{z_0(iT_b) + z_1(iT_b)}{2} \tag{2.24}$$

Because the signals are symmetrical, the optimum threshold is $\tau_{opt} = 0$ V.

Probability of Bit Error for Baseband Symmetrical Signals

For AWGN, the noise processed by the optimum receiver also has a Gaussian probability distribution. The received signal $s_j(t)$ is attenuated by the communication channel by an amount γ, and has a normalized *energy per bit* E_b^j given by Equation 2.25.

$$E_b^{\,j} = \int_{(i-1)T_b}^{iT_b} \gamma^2 s_j^{\,2}(t)\, dt \quad (i-1)T_b \le t \le iT_b \quad j = 0,1 \tag{2.25}$$

Because the signals are symmetrical, from Equation 2.21, the E_b^j are equal ($E_b^0 = E_b^1 = E_b$). The probability of bit error P_b^j due to either a binary 1 being transmitted and a binary 0 being received (P_b^1), or a binary 0 being transmitted and a binary 1 being received (P_b^0) is given by Equation 2.26.

$$P_b^j = Q\left(\sqrt{\frac{2E_b}{N_o}}\right) \qquad (2.26)$$

$N_o/2$ is the nearly uniform amplitude of the AWGN power spectral density [Stern04]. The *Q function* is also called the *complementary error function* or *co-error function.* The Q function cannot be evaluated in closed form, and is usually presented in tabular form for zero mean μ and unit standard deviation σ Gaussian distributions.

The overall probability of bit error P_b includes the apriori probability of a binary 1 (P_1) or binary 0 (P_0) being transmitted, as given by Equation 2.27 for the binary transmission of data ($j = 0,1$).

$$P_b = \sum P_j P_b^j = P_0 P_b^0 + P_1 P_b^1 \qquad (2.27)$$

Because the normalized energy per bit is equal ($E_b^0 = E_b^1$), the probabilities of bit error are equal ($P_b^0 = P_b^1$) and, regardless of the value of the apriori probabilities, the overall probability of bit error for baseband symmetrical signals is given by Equation 2.28.

$$P_b = Q\left(\sqrt{\frac{2E_b}{N_o}}\right) \qquad (2.28)$$

Performance of Symmetrical PAM for the Optimum Receiver in AWGN

The binary rectangular PAM transmitter and simple filtered PAM receiver in Figure 2.10 can be evaluated for its performance in the presence of AWGN with the optimum receiver, as shown in Figure 2.26 (see Fig2-26.svu on the CD-ROM). To facilitate the simulation in SystemVue, all of the Analysis tokens, except that for the BER token output as the Final Value token, are deleted in the simulation model Fig2-26DT. As described in Chapter 1, the SystemVue Textbook Edition does not permit tokens to be deleted.

ON THE CD

The matched filter implemented as a correlation receiver or correlator replaces the LPF and Sample and Hold token in Figure 2.7. The reference signal $s_1(t)$ here for the correlation receiver is the Pulse Train token from the Source Library, a 5 V rectangular pulse with a duration that is $T_b - T_{system}$, where T_{system} is the System Time interval. Here, the bit time T_b is 1 msec and the System Time T_{system} is 20 μsec. A Pulse

Train token cannot have a 100% duty cycle. The attenuation of the communication channel is taken to be zero ($\gamma = 1$) in Figure 2.26.

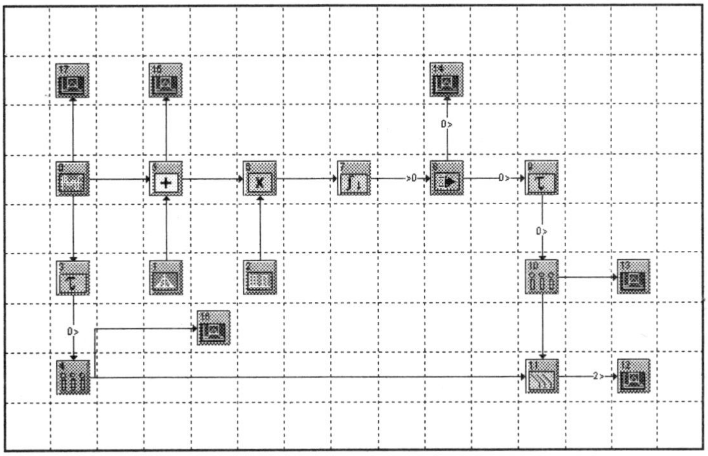

FIGURE 2.26 Binary rectangular symmetrical PAM digital communication system with BER analysis and the optimum receiver.

A BER analysis is conducted with the same parameters for the simple filtered PAM receiver as given in Table 2.3. Table 2.6 is a tabular list of the observed BER in a single trial of 10 000 information bits, as a function of the SNR for the rectangular PAM communication system with the optimum receiver shown in Figure 2.26. A comparison of Tables 2.3 and 2.6 shows an approximate +3 dB SNR increase in BER performance for rectangular PAM using the optimum receiver, over rectangular PAM using the simple filtered receiver.

TABLE 2.6 Observed BER as a Function of SNR in a Binary Rectangular PAM Digital Communication System with Optimum Receiver, Normalized Signal Power = 25 W

SNR dB	AWGN σ V	BER
∞	0	0
6.02	2.5	0
0	5	0
−3.52	7.5	0
−6.02	10	1×10^{-4}
−7.96	12.5	2.8×10^{-3}
−9.54	15	8.8×10^{-3}

The Gaussian noise source is specified by its standard deviation σ and the resulting noise power is given by Equation 2.5. The Gaussian noise source can also be specified by the power spectral density N_o W/Hz. The energy per bit E_b for the rectangular PAM signal from Equation 2.25 is 2.50×10^{-2} V^2-sec, because the magnitude of the rectangular PAM and reference signals are 5 V, and the bit time T_b is 1 msec. The attenuation of the communication channel is assumed to be zero ($\gamma = 1$).

A bit error analysis for the binary symmetrical rectangular PAM communication system with the optimum receiver can be obtained by varying N_o and using the ratio of E_b/N_o as the SNR parameter. Table 2.7 is the result, and compares the BER obtained for a small, single trial of 10 000 information bits to the theoretical P_b, given by Equation 2.28.

TABLE 2.7 Observed BER and Theoretical P_b as a Function of E_b/N_o, in a Binary Symmetrical Rectangular PAM Digital Communication System with Optimum Receiver

E_b/N_o dB	No V2-sec	BER	Pb
∞	0	0	0
10	2.50×10^{-3}	0	4.05×10^{-6}
8	3.96×10^{-3}	0	2.06×10^{-4}
6	6.28×10^{-3}	2.2×10^{-3}	2.43×10^{-3}
4	9.96×10^{-3}	1.21×10^{-2}	1.25×10^{-2}
2	1.58×10^{-2}	3.91×10^{-2}	3.75×10^{-2}
0	2.50×10^{-2}	8.13×10^{-2}	7.93×10^{-2}

The Correlation Receiver for Baseband Asymmetrical Signals

The signals $s_0(t)$ and $s_1(t)$ that represent a binary 0 and binary 1 need not be symmetrical. Asymmetrical PAM can transmit a binary 1 as a rectangular pulse, and a binary 0 as no pulse at all. If the binary baseband signals remain equally probable, then Equation 2.24 still determines the optimum threshold value. If the baseband signals are not equally probable ($P_1 \neq P_0$), then the optimum threshold is given by Equation 2.29.

$$\tau_{opt} = \frac{2\sigma_o^2 \ln \frac{P_1}{P_0} + z_0(iT_b)^2 - z_1(iT_b)^2}{2\left[z_0(iT_b) - z_1(iT_b)\right]} \tag{2.29}$$

If $P_1 = P_0$ then the optimum threshold τ_{opt} in Equation 2.29 reduces to Equation 2.24. The output of the correlation receiver $z_j(iT_b)$ is assumed to have no additive noise. The additive Gaussian noise is zero mean with a variance of σ^2, but σ_o^2 is the variance of the noise processed by the integrate-and-dump implementation of the matched filter as a correlation receiver. The Gaussian noise variance σ_o^2 can be determined by the test system shown in Figure 2.27 (see Fig2-27.svu on the CD-ROM).

ON THE CD

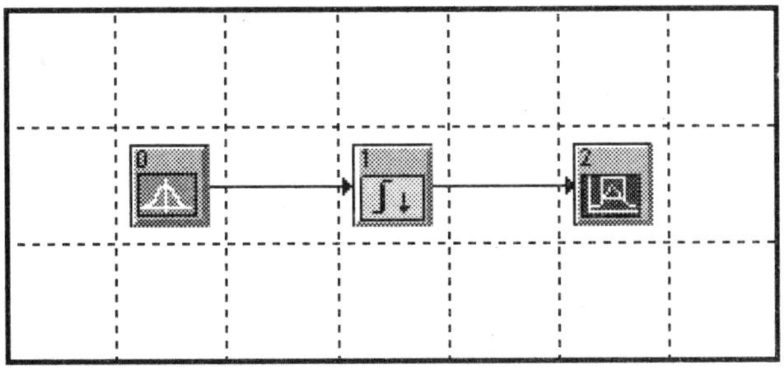

FIGURE 2.27 Test system for the determination of the Gaussian noise variance σ_o^2.

The Gaussian noise source can also be specified by the power spectral density N_o W/Hz. The parameters of the Integrate and Dump token are specified as an integration time equal to the bit time T_b of 1 msec and a continuous output. The resulting simulation for an arbitrary $N_o = 6.28 \times 10^{-3}$ W/Hz or $\sigma = 12.53$ V is shown in Figure 2.28.

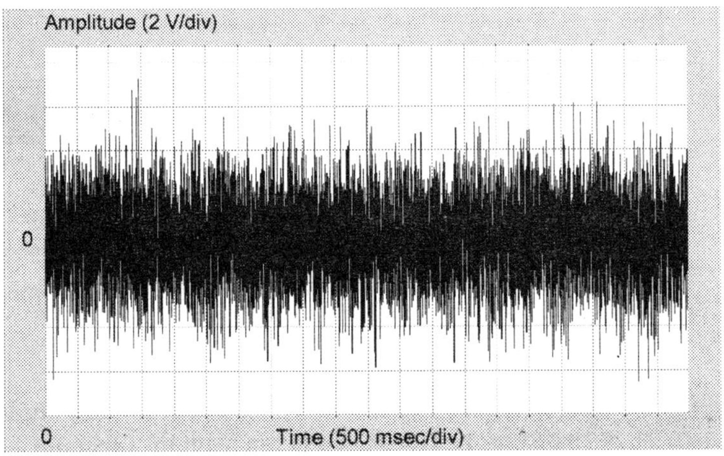

FIGURE 2.28 Gaussian noise source $\sigma = 12.53$ processed by the Integrate and Dump token.

The statistics for the Gaussian noise source processed by the integrate-and-dump σ_o^2 can be obtain for the Analysis window, as described in Chapter 1, and shown in Figure 2.29. The variance of the Gaussian noise source σ^2 was 12.53, but the processed noise has an approximate variance $\sigma_o^2 = (1.772)^2 = 3.139$.

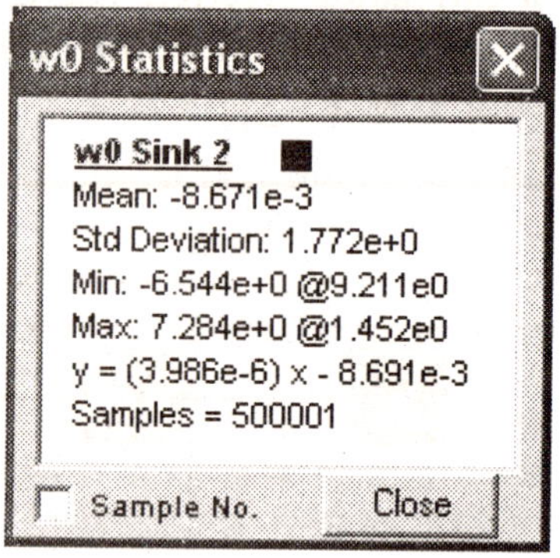

FIGURE 2.29 Statistics for the Gaussian noise source processed by the Integrate and Dump token.

The apriori probabilities P_1 and P_0 in Equation 2.29 for the optimum threshold τ_{opt} are not necessarily known exactly, or can change with the source of information. A test system can be devised to count the number of binary 1s and binary 0s in an arbitrary data set, as shown in Figure 2.30 (see Fig2-30.svu on the CD-ROM).

ON THE CD

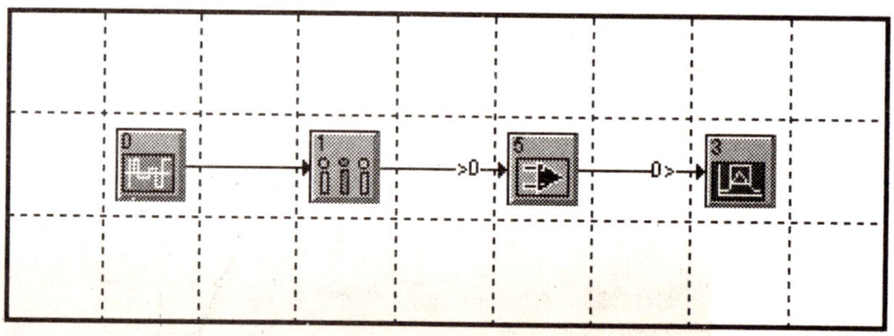

FIGURE 2.30 Test system for the determination of apriori probabilities P_1 and P_0 from a data source.

The PN Sequence token data source has an output amplitude of ± 5 V and a data rate r_b of 1 kb/sec. The Decimator token from the Operator Library has a decimation value of 50, because the System Time: Sampling Rate is 50 kHz, and provides one data sample per bit time T_b. The System Time: Stop Time was 100 seconds and the Analog Comparator token is set to provide a 1 or 0 V output to facilitate the determination of the apriori probabilities.

The resulting simulation of 10^5 data points provided a mean of 0.4994 V, which implies that $P_1 = 0.4994$ and $P_0 = 1 - P_1 = 0.5006$ approximately for the PN data source. These test systems for the determination of the Gaussian noise variance σ_o^2 in Figure 2.27 and for the apriori probabilities P_1 and P_0 in Figure 2.30 can be used to investigate the what-ifs of digital communication systems design.

ON THE CD

A binary rectangular PAM digital communication system with AWGN and a correlation receiver for asymmetrical signals is shown in Figure 2.31 (see Fig2-31.svu on the CD-ROM). This system is similar to the symmetrical PAM digital communication system shown in Figure 2.26. To facilitate the simulation in SystemVue, all of the Analysis tokens, except that for the BER token output as the Final Value token, are deleted in the simulation model Fig2-31DT. As described in Chapter 1, the SystemVue Textbook Edition does not permit tokens to be deleted.

The asymmetrical data source and PAM transmitter is the PN Sequence token from the Source Library, with parameters of a binary amplitude of 0 and +5 V, a data rate of 1 kb/sec, and 0 V voltage and 0° phase offset.

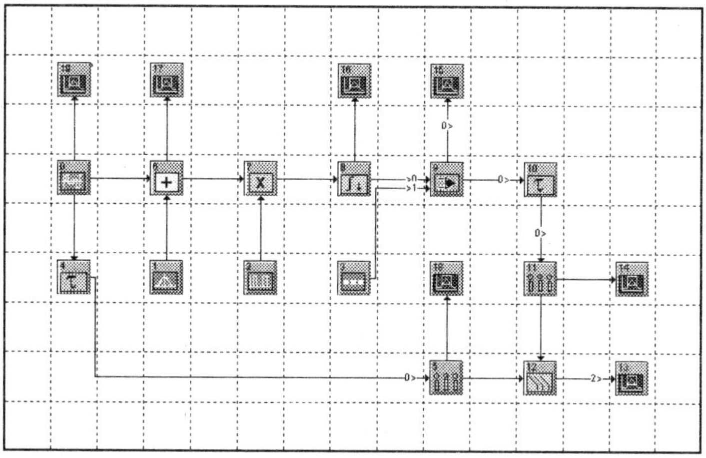

FIGURE 2.31 A binary rectangular asymmetrical PAM digital communication system with BER analysis.

The reference signal for the correlation receiver $s_{ref}(t)$ is the difference in the PAM pulses, given by Equation 2.30.

$$s_{ref}(t) = s_1(t) - s_0(t) \tag{2.30}$$

The Analog Comparator token comparison input is set to the optimum threshold τ_{opt} given by Equation 2.24, because the PAM signals even though asymmetrical are equally probable. The optimum threshold τ_{opt} is 12.5 V here, because $z_0(iT_b) = 0$ and $z_1(iT_b) = 25$ V, as given by Equation 2.23. The output of the integrate-and-dump process can also be observed in the Analysis window. The Custom token from the Source Library provides the constant comparator input, with parameters of one output with the algebraic simulation equation, that is $p(0) = 12.5$.

ON THE CD

An alternative structure for the correlation receiver for binary asymmetrical PAM signals is shown in Figure 2.32 (see Fig2-32.svu on the CD-ROM). The reference signal inputs to the two multipliers are $s_0(t)$ and $s_1(t)$ respectively. The optimum threshold τ_{opt} for the comparator is given by Equation 2.24 if the apriori probabilities P_0 and P_1 are equal, and by Equation 2.29 if not. The optimum threshold is set by the Custom token.

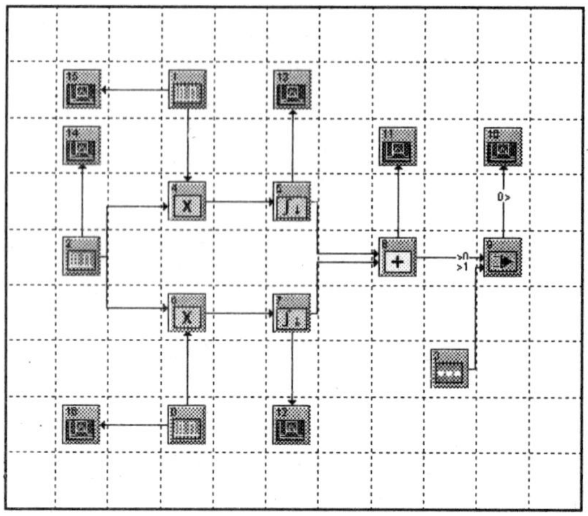

FIGURE 2.32 A binary asymmetrical PAM correlation receiver.

Probability of Bit Error for Baseband Asymmetrical Signals

For AWGN, the noise processed by the optimum receiver again has a Gaussian probability distribution. The asymmetrical received signal $s_j(t)$ is attenuated by the communication channel by an amount γ, and can be considered here to have a normalized *energy difference per bit* E_d, given by Equation 2.31.

$$E_d = \int_{(i-1)T_b}^{iT_b} \left\{\gamma\left[s_1(t) - s_0(t)\right]\right\}^2 dt \quad (i-1)T_b \le t \le iT_b \tag{2.31}$$

The probability of bit error P_b for baseband asymmetrical signals due to either a binary 1 being transmitted and a binary 0 being received, or a binary 0 being transmitted and a binary 1 being received, with equally likely apriori probability of binary 1 or binary 0 being transmitted ($P_0 = P_1 = 0.5$), is given by Equation 2.32.

$$P_b = Q\left(\sqrt{\frac{E_d}{2N_o}}\right) \tag{2.32}$$

If the binary PAM signals are symmetrical, then $E_d = 4E_b$ and the probability of bit error P_b in Equation 2.32 reduces to Equation 2.28. If the apriori probabilities P_0 and P_1 are not equal, then the optimum threshold τ_{opt} is given by Equation 2.29, and the probability of bit error is given by Equation 2.33.

$$P_b = P_1\, Q\left(\frac{z_1(iT_b) - \tau_{opt}}{\sigma_o}\right) + P_0\, Q\left(\frac{\tau_{opt} - z_0(iT_b)}{\sigma_o}\right) \tag{2.33}$$

The standard deviation σ_o of the Gaussian noise source processed by the integrate-and-dump process is used in Equation 2.33 rather than the AWGN power spectral density $N_o/2$, because a voltage noise margin is used here. The signal $z_j(t)$ is the output of the correlation receiver when there is no additive noise, given by Equation 2.23. Equation 2.33 is an extension of Equation 2.27 for the overall probability of bit error P_b, where the Q function is the complementary error function.

If the apriori probabilities are equal ($P_0 = P_1 = 0.5$), the optimum threshold τ_{opt} is given by Equation 2.24, and the probability of bit error from Equation 2.33 is given by Equation 2.34.

$$P_b = Q\left(\frac{z_1(iT_b) - z_0(iT_b)}{2\sigma_o}\right) \tag{2.34}$$

The probability of bit error P_b for binary asymmetrical signals in Equation 2.34 is equivalent to that given by Equation 2.32.

Performance of Asymmetrical PAM for the Optimum Receiver in AWGN

The binary asymmetrical rectangular PAM transmitter and optimum PAM receiver in Figure 2.31 can be evaluated for its performance in the presence of AWGN. The reference signal here for the correlation receiver is given by Equation 2.30. The asymmetrical PAM signals are $s_1(t) = +5$ V and $s_0(t) = 0$ V. The reference signal

$s_{ref}(t)$ then is a Pulse Train token from the Source Library, with a duration that is $T_b - T_{system}$, where T_{system} is the System Time interval. The bit time is $T_b = 1$ msec and $T_{system} = 20$ μsec, because the System Time: Sampling Rate is 50 kHz. A Pulse Train token cannot have a 100% duty cycle.

A BER analysis is conducted with the same parameters for the binary symmetrical rectangular PAM receiver, as given in Table 2.7. The Gaussian noise source is specified by its standard deviation σ and the resulting noise power is given by Equation 2.5. The Gaussian noise source is specified by the power spectral density N_o W/Hz. The energy difference per bit E_d for the asymmetrical rectangular PAM signal from Equation 2.31 is 2.5×10^{-2} V^2-sec, because the bit time T_b is 1 msec. The attenuation of the communication channel is assumed to be zero ($\gamma = 1$).

A bit error analysis for the asymmetrical binary rectangular PAM communication system with the optimum receiver can be obtained by varying N_o, and using the ratio of E_d/N_o as the SNR parameter. Table 2.8 is the result and compares the BER obtained for a small, single trial of 10 000 information bits to the theoretical P_b, given by Equation 2.32.

TABLE 2.8 Observed BER and Theoretical P_b as a Function of E_d/N_o in an Asymmetrical Binary Rectangular PAM Digital Communication System with Optimum Receiver

E_d/N_o dB	N_o V^2-sec	BER	P_b
∞	0	0	0
12	1.58×10^{-3}	2.8×10^{-3}	2.53×10^{-3}
10	2.50×10^{-3}	1.19×10^{-2}	1.25×10^{-2}
8	3.96×10^{-3}	3.86×10^{-2}	3.75×10^{-2}
6	6.28×10^{-3}	8.34×10^{-2}	7.93×10^{-2}
4	9.96×10^{-3}	1.291×10^{-1}	1.318×10^{-1}
2	1.58×10^{-2}	1.867×10^{-1}	1.872×10^{-1}
0	2.50×10^{-2}	2.322×10^{-1}	2.394×10^{-1}1

MULTILEVEL (M-ARY) PULSE AMPLITUDE MODULATION

Multilevel (M-ary) pulse amplitude modulation (PAM) transmits a symbol that represents $2^N = M$ ($N = \log_2 M$) bits of information. The SystemVue simulation presented here is for an $M = 4$, 4-level PAM signal, but can be easily extended to any 2^N-level rectangular PAM digital communication system.

SystemVue Simulation of M-ary Rectangular PAM

ON THE CD

An M-ary rectangular PAM digital communication system with AWGN and optimum receiver implemented as a correlator is shown in Figure 2.33 (see Fig2-33.svu on the CD-ROM). To facilitate the simulation in SystemVue, all of the Analysis tokens, except that for the BER token output as the Final Value token, are deleted in the simulation model Fig2-33DT. As described in Chapter 1, the SystemVue Textbook Edition does not permit tokens to be deleted.

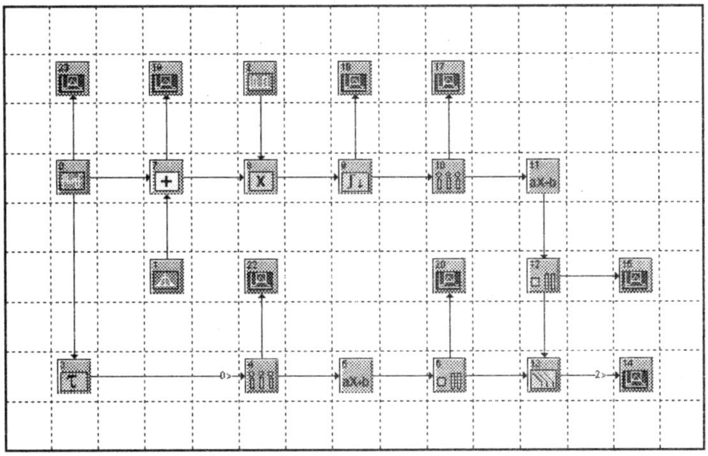

FIGURE 2.33 A 4-level straight-binary rectangular PAM digital communication system with BER analysis.

The 4-level rectangular PAM pulses have equal apriori probabilities of occurrence ($P_1 = P_2 = P_3 = P_4$, $P_j = 0.25$). The M-ary data source and PAM transmitter is a 4-level pseudonoise PN Sequence token from the Source Library, with parameters of amplitude ± 5 V and 0 V voltage and 0° phase offset. These parameters for the PN Sequence token result in output amplitudes of +5 V, +1.66 V, 1.66 V, and 5 V for one *symbol time* T_s, a *symbol rate* r_s of 500 Hz, or a resulting binary data rate r_b of 1 kb/sec (2 bits per symbol). This data rate is the same as that used for the binary rectangular, sinc, and raised cosine PAM transmitter, and facilitates the comparison of the power spectral densities.

The communication channel is represented by an Adder token and a Gaussian Noise token from the Source Library, with initial parameters of a standard deviation $\sigma = 0$ V and $\mu = 0$ V mean, or a power spectral density $N_o = 0$ W/Hz.

The System Time: Sample Rate is set to 50 kHz, well above the nominal 1 kb/sec data rate or 500 symbols/sec, with 2500 samples and a resulting simulation time of 50 msec. This number of samples is used to display the signal data at a scale that can conveniently verify the performance of the SystemVue simulation. The 4-level M-ary data source is shown in Figure 2.34.

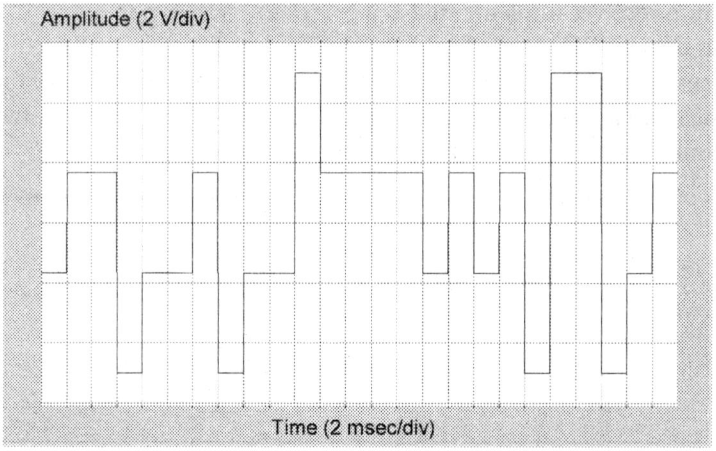

FIGURE 2.34 A 4-level M-ary rectangular PAM pseudonoise data source at 1 kb/sec or 500 symbols/sec.

The 4-level PAM communication system uses four rectangular pulses to signify four symbols, each of which are specified to represent (as a *data protocol*) the *di-bits* 11, 10, 01, and 00 respectively. These symbols are coded as *straight-binary* data. A symbol error interpreted as an adjacent symbol could produce two bits in error for the di-bits 10 and 01. However, because the di-bit 11 is represented by a pulse of amplitude +5 V, it would be difficult for it to be misinterpreted because of noise as the di-bit 00, which is represented as a pulse of amplitude –5 V.

M-ary Rectangular PAM Power Spectral Density

M-ary PAM transmits N bits per symbol ($N = \log_2 M$) but retains the same pulse shape as binary PAM. The bandwidth of a PAM transmission is not affected by the pulse amplitude, but only by the pulse shape. The bi-sided power spectral density (PSD) $PSD_M(f)$ for M-ary rectangular PAM then follows the analysis for binary rectangular PAM, and results in the PSD given by Equation 2.35.

$$PSD_M(f) = \frac{A_{avg}^2\, T_s}{R_L}\ \text{sinc}^2\left(2\pi\left(T_s/2\right)f\right) \tag{2.35}$$

This is the same PSD as given in Equation 2.3, with the substitution here of the symbol time T_s, rather than the bit time T_b. For the 4-level PAM transmitter here, $N = 2$, $M = 4$, $T_b = 0.5\ T_s$, and the resulting PSD bandwidth is reduced by a factor of two. The square of the average amplitude $A_{avg}{}^2$ in Equation 2.35 is given by Equation 2.36, where P_j is the probability of occurrence of an M-ary symbol.

$$A_{avg}^2 = \sum_{j=1}^{4} A_j^2 \, P_j \qquad (2.36)$$

Substituting the four equally likely ($P_1 = P_2 = P_3 = P_4$, $P_j = 0.25$) and equally spaced symbol amplitudes $A_j = -5$ V, -1.666 V, 1.666 V, and 5 V here results in an amplitude $A_{avg} = 3.727$ V. The symbol amplitudes can also be expressed as $A_j = -1.5A_v$, $-0.5A_v$, $0.5A_v$, and $1.5A_v$, where $A_v = 3.333$ V and the amplitude difference between symbols is $A_v = 3.333$ V.

Setting the System Time: Number of Samples specification to 262 144 (2^{18}) points results in a spectral resolution of approximately 0.19 Hz. The normalized ($R_L = 1\ \Omega$) PSD of the M-ary ($M = 4$) PAM transmitter is shown unscaled, and as a single-sided spectrum in Figure 2.35 in dBm/Hz. The maximum frequency displayed is 25 kHz, the Nyquist frequency or half of the system simulation sampling rate (50 kHz/2). Because the rectangular pulse width τ is the symbol time T_s, there are approximate nulls in the power spectral density that occur at the symbol rate $r_s = 1/T_s$ (500 symbols/sec or 500 Hz), and multiples of the symbol rate due to the sinc^2 term in Equation 2.35.

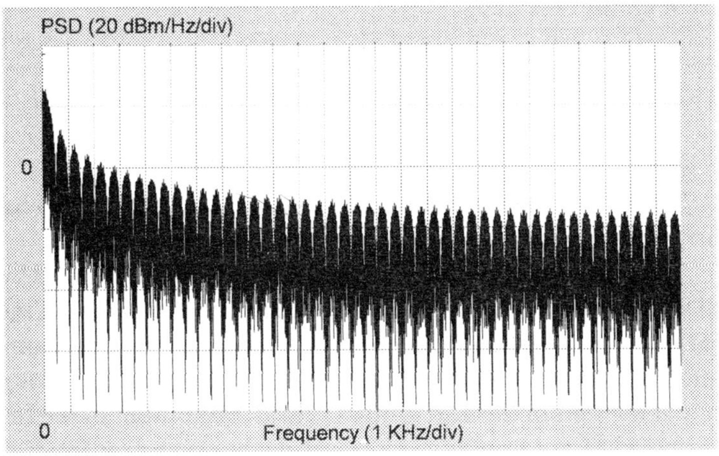

FIGURE 2.35 Power spectral density of a 1 kb/sec pseudonoise M-ary ($M = 4$) rectangular PAM signal.

The PSD is not smooth because, although there are a large number of data, the ensemble is finite, but the spectral envelope clearly demonstrates the sinc^2 behavior. Compare the 4-level rectangular PAM PSD to that of the binary rectangular PAM PSD shown in Figure 2.6.

The bandwidth of the 4-level rectangular PAM signal is somewhat problematical. Defining the bandwidth as a percentage of the total power of the 4-level rectangular PAM signal again is the usual course, because all of the available power would require an infinite bandwidth. The total power is easily calculated in the temporal domain. The power in the rectangular PAM signal is the square of the average amplitude A_{avg} here, divided by R_L, as given by Equation 2.4.

The equally probable 4-level rectangular PAM signal has an average amplitude $A_{avg} = 3.727$ V, and the normalized ($R_L = 1\ \Omega$) power is 13.84 W = 41.41 dBm (10 $\log_{10}[13840]$). Numerically integrating the power spectral density shown in Equation 2.35 from 0 to an arbitrary frequency, and comparing the resulting power to the expected total power, provides the same data as that in Table 2.1 with the substitution of the 4-level symbol time T_s for the bit time T_b ($T_b = T_s/2$). The 4-level PAM bandwidth here is one-half ($1/N$ with $M = 2^N$ and $M = 4$, $N = 2$) that of the binary PAM bandwidth as a percentage of the total power.

The Correlation Receiver for M-ary Baseband Signals

The optimum receiver for M-ary PAM signals is the matched receiver implemented as the correlator, as shown in Figures 2.25 and 2.33. The output of the correlator $z_j(t)$ ($j = 1, 2, \ldots, M$) is sampled at the end of each symbol period, $t = iT_s$, as given by Equation 2.37.

$$z_j(iT_s) = \frac{1}{T_s} \int_{(i-1)T_s}^{iT_s} \gamma s_j(\tau) s_M(\tau)\, d\tau \quad (i-1)T_s \le t \le iT_s \quad j = 1, 2, \ldots, M \tag{2.37}$$

The arbitrary correlator reference signal $s_M(t)$ is the M-ary baseband signal that corresponds to the maximum amplitude symbol.

The received signal $s_j(t)$ ($j = 1, 2, \ldots, M$) is first multiplied by the reference signal $s_M(t)$, synchronized to the start of the symbol period and then integrated over that one symbol period. The output of the Integrate and Dump token from the Communications Library is held for each symbol period, and is reset to zero internally at the start of the next symbol period. The output of the Integrate and Dump token is an M-ary amplitude representing the symbol for the N-bit PN data source.

The output is processed by the Sampler token from the Operator Library with the parameters of a sampling rate of 500 Hz, or the symbol rate $r_s = 1/T_s$, and a sampling aperture of $T_{system} = 20$ μsec, where T_{system} is the System Time interval. This sampling provides one amplitude per symbol time and facilitates the BER analysis.

The reference signal $s_4(t)$ here for the correlation receiver is the Pulse Train token from the Source Library, a 5 V rectangular pulse, with a duration that is $T_s - T_{system}$. A Pulse Train token cannot have a 100% duty cycle. The attenuation of the communication channel is taken to be zero ($\gamma = 1$) in Figure 2.33.

The output of the sampler is scaled by the Polynomial token from the Function Library to restore a 4-level amplitude of 0, 1, 2, and 3 V as required by the input of the Symbol-to-Bit token from the Communications Library. The maximum correlator output at the end of the symbol time T_s is $z_1(T_s) = 25$ V in Equation 2.27. The parameters of the Polynomial token are that the x^0 (offset) coefficient is 1.5, the x^1 (linear) coefficient is 0.06 ($0.06 \times 25 = 1.5$ V), and all other coefficients are zero. The parameters of the data source Polynomial token are that the x^0 (offset) coefficient is 1.5, but the x^1 (linear) coefficient is 0.3 ($0.3 \times 5 = 1.5$ V), and all other coefficients are zero. The Polynomial token can also provide non-linear output adjustments.

The output of the Polynomial token is processed by the Symbol-to-Bit decoder from the Communications Library, with parameters that are 2 bits per symbol with the most significant bit (MSB) produced first at a data rate r_b of 1 kb/sec. The output of the Symbol-to-Bit decoder is 0 or 1 V.

There are three optimum thresholds for the 4-level PAM digital communication system with equal apriori probabilities, as given by Equation 2.38.

$$\begin{aligned} \tau_{opt1} &= \frac{z_1(iT_s) + z_2(iT_s)}{2} \\ \tau_{opt2} &= \frac{z_2(iT_s) + z_3(iT_s)}{2} \\ \tau_{opt3} &= \frac{z_3(iT_s) + z_4(iT_s)}{2} \end{aligned} \tag{2.38}$$

The Symbol-to-Bit decoder is configured as the 4-level threshold comparator by specifying that there are 2 bits per symbol. The three thresholds for the 4-level signal of the Symbol-to-Bit token are set to 0.5 V, 1.5 V, and 2.5 V, which are the optimum thresholds τ_{opt} of Equation 2.38 after scaling by the Polynomial token in Figure 2.33. The Polynomial tokens provide scaling to verify that the PN source data has been recovered correctly, and to facilitate the cross-correlation of the recovered data with that of the original PN data source.

Probability of Bit Error for M-ary Baseband Signals

For AWGN, the noise processed by the optimum receiver again has a Gaussian probability distribution. The M-ary PAM received signal $s_j(t)$ is attenuated by the communication channel by an amount γ, and can be considered here to have a normalized *energy difference per symbol* $E_{d,symbol}$ given by Equation 2.39.

$$E_{d,symbol} = \int_{(i-1)\,T_s}^{i\,T_s} \left\{\gamma\left[s_j(t) - s_k(t)\right]\right\}^2 dt \quad (i-1)T_s \le t \le iT_s \quad j,k = 1, 2, \ldots, M \tag{2.39}$$

If the energy difference per symbol $E_{d,symbol}$ is equal for all $j, k = 1, 2, \dots, M$, then the probability of bit error for M-ary signals corresponds to the analysis for binary asymmetrical signals, as given by Equations 2.31 and 2.32. The probability of symbol error P_s due to a symbol being misinterpreted as an adjacent symbol, with equally likely apriori probability of the M-ary baseband symbols being transmitted ($P_j = 1/M, j = 1, 2, \dots, M$) is given by Equation 2.40.

$$P_s = \frac{2(M-1)}{M} Q\left(\sqrt{\frac{E_{d,symbol}}{2N_o}}\right) \tag{2.40}$$

In Equation 2.40, there are $2(M - 1)$ *error regions* due to only adjacent symbols being misinterpreted with M equally probable symbols. The probability of occurrence for a misinterpreted symbol is therefore also equally likely and given by Equation 2.41.

$$P_j = \frac{1}{2(M-1)} \tag{2.41}$$

The normalized ($R_L = 1\ \Omega$) energy difference per symbol for the 4-level rectangular PAM digital communication system in Figure 2.33, from Equation 2.39, is $E_{d,symbol} = A_v^2\, T_s = 2.222 \times 10^{-2}$ V^2-sec. The amplitude difference between symbols A_v here is 3.333 V, the symbol time $T_s = 2$ msec, and the attenuation of the communication channel is zero ($\gamma = 1$).

The average energy difference per symbol $E_{avg,symbol}$ can be expressed in terms of the average amplitude A_{avg} as $E_{avg,symbol} = A_{avg}^2\, T_s = 2.778 \times 10^{-2}$ V^2-sec. Noting that $E_{d,symbol} = 0.8\, E_{avg,symbol}$ and $M = 4$ in Equation 2.40 gives Equation 2.42.

$$P_s = \frac{3}{2} Q\left(\sqrt{\frac{0.4\, E_{avg,symbol}}{N_o}}\right) \tag{2.42}$$

Because each 4-level symbol represents two bits here, $E_{avg,symbol} = 2\, E_{b,4\text{-}level}$ and Equation 2.42 can be given as Equation 2.43.

$$P_s = \frac{3}{2} Q\left(\sqrt{\frac{0.8\, E_{b,4\text{-}level}}{N_o}}\right) \tag{2.43}$$

Although Equation 2.43 utilizes the energy per bit $E_{b,\,4\text{-}level}$, the probability of symbol error P_s is not the probability of bit error $P_{b,4\text{-}level}$. The data protocol representation for the 4-level PAM symbols is a straight-binary and there are equally probable six error regions for the 4-level PAM signal. An error due to a misrepresentation of an adjacent equally likely symbol can produce either one bit in error in four of the regions, or two bits in error in two of the regions, as shown in Table 2.9.

TABLE 2.9 Transmitted Di-Bit, Received Di-Bit, and the Number of Bits in Error for the Six Error Regions for a 4-Level PAM Digital Communications System

Transmitted Di-Bit	Received Di-Bit	Bits In Error
00	01	1
01	00	1
01	10	2
10	01	2
11	10	1
10	11	1

The equally likely probability of occurrence P_i for each of the error regions for M-ary PAM is given by Equation 2.41 and $M = 4$ here. From Table 2.9, the probability of bit error $P_{b,4\text{-}level}$ is given by Equation 2.44.

$$P_{b,4\text{-}level} = \frac{4}{6}P(1 \text{ of } 2 \text{ bits in error}) + \frac{2}{6}P(2 \text{ of } 2 \text{ bits in error})$$
$$P_{b,4\text{-}level} = \frac{4}{6}\frac{1}{2}P_s + \frac{2}{6}P_s = \frac{2}{3}P_s \tag{2.44}$$

From Equations 2.43 and 2.44, the probability of bit error $P_{b,4\text{-}level}$ can be given as Equation 2.45.

$$P_{b,4\text{-}level} = Q\left(\sqrt{\frac{0.8\,E_{b,4\text{-}level}}{N_o}}\right) \tag{2.45}$$

The probability of bit error for binary PAM is given by Equation 2.28, and a comparison with Equation 2.45 shows that 4-level PAM requires larger values of the E_b/N_o ratio for the same bit error. This is the performance tradeoff required for M-ary PAM, which also results in a lower transmission bandwidth when compared to binary PAM.

The M-ary PAM communication system as initially implemented used the four rectangular pulses –5 V, –1.666 V, 1.666 V, and 5 V to signify four symbols representing the straight-binary di-bits 00, 01, 10, and 11-respectively. However, these symbols can also be *Gray-coded* with the same amplitudes representing the di-bits

00, 01, 11, and 10. A symbol error then interpreted as an adjacent symbol would now only produce one bit in error, and the probability of bit error $P_{b,4\text{-level},Gray\text{-}coded}$ is given by Equation 2.46.

$$P_{b,4\text{-}level,Gray\text{-}coded} = \frac{6}{6} P(1 \text{ of } 2 \text{ bits in error}) = \frac{1}{2} P_s \quad (2.46)$$

From Equations 2.43 and 2.46 the probability of bit error $P_{b,4\text{-level},Gray\text{-}coded}$ can be given as Equation 2.47.

$$P_{b,4\text{-}level,Gray\text{-}coded} = \frac{3}{4} Q\left(\sqrt{\frac{0.8\,E_{b,4\text{-}level}}{N_o}}\right) \quad (2.47)$$

The probability of bit error for straight-binary coded 4-level PAM is given by Equation 2.45, and a comparison with Equation 2.47 shows that the 2-bit Gray-coded 4-level PAM has a lower bit error for the same values of the E_b/N_o ratio. This is the performance gain that results for merely Gray-coding an M-ary PAM communication system while still using the same transmission bandwidth and energy per bit.

Performance of M-ary PAM for the Optimum Receiver in AWGN

The M-ary ($M = 4$) rectangular PAM transmitter and optimum PAM receiver in Figure 2.33 can be evaluated for its performance in the presence of AWGN. The PN Sequence token data Sampler input is delayed by 2 msec to correlate the symbol data with the output of the correlator at the end of a symbol time T_s. The value of this delay can be observed by the cross-correlation of the two sampler signals as symbols before the Polynomial and Symbol-to-Bit in the BER analysis, as described in Chapter 1.

A BER analysis is conducted with the same parameters for the optimum binary symmetrical PAM receiver as given in Table 2.7. Table 2.10 is a tabular list of the observed BER in a single trial of 10 000 information bits as a function of E_b/N_o to the theoretical P_b for straight-binary coded source data, as given by Equation 2.45. The normalized energy per bit here is $E_{b,4\text{-level}} = E_{avg,symbol}/2 \approx 1.39 \times 10^{-2}$ V^2-sec.

A comparison of Tables 2.7 and 2.10 shows an approximate –4 dB E_b/N_o decrease in BER performance for 4-level PAM over that for rectangular PAM, as predicted by Fquations 2.28 and 2.45. The performance tradeoff is that 4-level PAM has half the transmission bandwidth of binary PAM.

TABLE 2.10 Observed BER and Theoretical P_b as a Function of E_b/N_o in a 4-Level Rectangular Straight-Binary Coded PAM Digital Communication System with Optimum Receiver

E_b/N_o dB	N_o V2-sec	BER	Pb
∞	0	0	0
10	1.39 10^{-3}	2.6×10^{-3}	2.4×10^{-3}
8	2.20×10^{-3}	1.42×10^{-2}	1.25×10^{-2}
6	3.49×10^{-3}	3.41×10^{-2}	3.75×10^{-2}
4	5.53×10^{-3}	7.83×10^{-2}	7.78×10^{-2}
2	8.77×10^{-3}	1.288×10^{-1}	1.320×10^{-1}
0	1.39×10^{-2}	1.881×10^{-1}	1.867×10^{-1}

ON THE CD

The M-ary ($M = 4$) straight-binary coded rectangular PAM transmitter and optimum PAM receiver in Figure 2.33 can be modified to transmit Gray-coded symbols evaluated for its performance in the presence of AWGN, as shown in Figure 2.36 (see Fig2-36.svu on the CD-ROM). To facilitate the simulation in SystemVue, all of the Analysis tokens, except that for the BER token output as the Final Value token, are deleted in the simulation model Fig2-36DT. As described in Chapter 1, the SystemVue Textbook Edition does not permit tokens to be deleted.

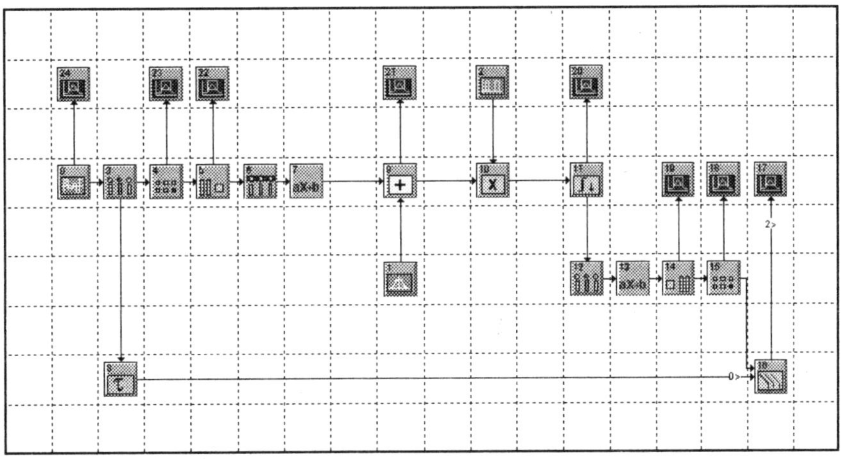

FIGURE 2.36 A 2-bit Gray-coded 4-level rectangular PAM digital communication system.

The PN Sequence token generates a unipolar binary data source at a rate $r_b = 1$ kb/sec of 0 and 1 V, with parameters of amplitude ± 0.5 V and 0.5 V offset. The System Time: Sample Rate is set to 50 kHz. The output of the PN Sequence token is downsampled by the Decimator token from the Operator Library by a factor of 50, producing one sample point per bit time $T_b = 1$ msec. The output of the Decimator token is inputted to the Gray Encode token from the Communications Library, with parameters of a threshold of 0.5 V, two bits per symbol, and the MSB outputted first, but only one system simulation point per symbol time $T_s = 2$ msec.

The Gray Encode token reassigns the input di-bits 00, 01, 10, and 11 as the output di-bits 00, 01, 11, and 10 respectively. The output of the Gray Encode token is processed by the Bit-to-Symbol token from the Communications Library, with parameters of a threshold of 0.5 V, two bits per symbol and the MSB inputted first. The output of the Bit-to-Symbol token is the 4-level PAM signal with amplitudes of 0, 1, 2, and 3 V at a data rate of 500 symbols/sec.

The output of the Bit-to-Symbol token is *upsampled* by the Resampler token from the Operator Library to restore the System Time: Sample Rate to 50 kHz. The output of the Resampler token is scaled by the Polynomial token from the Function Library to restore a 4-level amplitude of –5 V, –1.666 V, 1.666 V, and 5 V from the 0, 1, 2, and 3 V PAM input. The parameters of the Polynomial token are that the x^0 (offset) coefficient is –5, the x^1 (linear) coefficient is 3.3333 ($3.3333 \times 3 = 10$ V and with an offset of –5 V results in a +5 V output), and all other coefficients are zero.

The AWGN channel and the optimum PAM receiver are the same as that for the straight-binary PAM digital communication system, as shown in Figure 2.33. The output of the Integrate and Dump token of the optimum PAM receiver is downsampled by the Decimator token by a factor of 100, producing one sample point per symbol time $T_s = 2$ msec. The output of the Decimator token is scaled by the Polynomial token from the Function Library to produce a 4-level amplitude of 0 V, 1 V, 2 V, and 3 V.

The maximum correlator output at the end of the symbol time T_s is $z_4(T_s) = 25$ V in Equation 2.37. The parameters of the Polynomial token are that the x^0 (offset) coefficient is 1.5, the x^1 (linear) coefficient is 0.06 ($0.06 \times 25 = 1.5$ V and with an offset of 1.5 V results in a 3 V output), and all other coefficients are zero.

The output of the Polynomial token is processed by the Symbol-to-Bit token, with parameters of two bits per symbol and the MSB inputted first. The three thresholds for the 4-level signal of the Symbol-to-Bit token are set to 0.5 V, 1.5 V, and 2.5 V, which are the optimum thresholds τ_{opt} of Equation 2.38 after scaling by the Polynomial token in Figure 2.36. The output of the Symbol-to-Bit token is inputted to the Gray Decode token from the Communications Library, with parameters of a threshold of 0.5 V, two bits per symbol, the MSB outputted first, and one system simulation point per bit time $T_b = 1$ msec.

Complex SystemVue digital communication system simulations, such as the 2-bit Gray-coded 4-level rectangular PAM system in Figure 2.36, can be simplified by the creation of a MetaSystem token, as described in Chapter 1. The 2-bit Gray encoder and decoder of the 4-level rectangular PAM system each can be rendered as a MetaSystem token, as shown in Figure 2.37 (see Fig2-37.svu on the CD-ROM). To facilitate the simulation in SystemVue, all of the Analysis tokens, except that for the BER token output as the Final Value token, are deleted in the simulation model Fig2-37DT. As described in Chapter 1, the SystemVue Textbook Edition does not permit tokens to be deleted.

ON THE CD

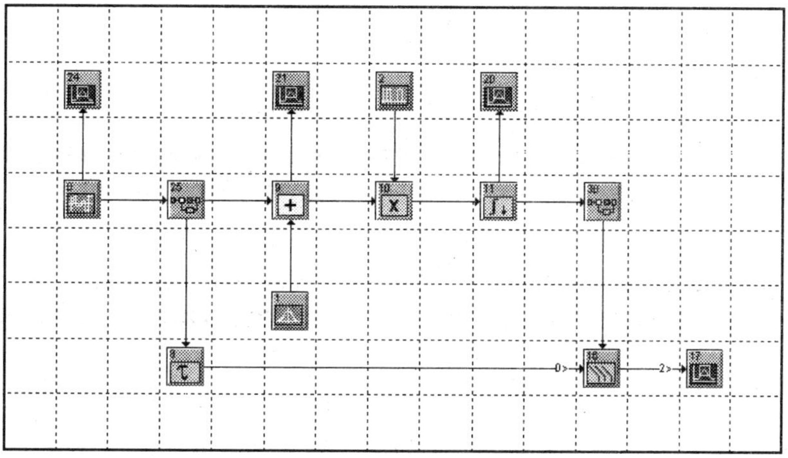

FIGURE 2.37 A 2-bit Gray-coded 4-level rectangular PAM digital communication system with MetaSystem tokens.

The Decimator token output in Figure 2.36, or within the Gray encoder MetaSystem shown expanded in Figure 2.38, is delayed by 8 msec to correlate the source binary data with the output of the Gray decoder at the end of each bit time T_b. The value of this delay can be observed by the cross-correlation of the two signals in the BER analysis, as described in Chapter 1.

A BER analysis is now conducted with the same parameters for the optimum straight-binary 4-level PAM receiver as given in Table 2.10. Table 2.11 is a tabular list of the observed BER in a single trial of 10 000 information bits a function of E_b/N_o to the theoretical P_b for Gray-coded source data, as given by Equation 2.47. The normalized energy per bit is the same, $E_{b,4\text{-}level} = E_{avg,symbol}/2 \approx 1.39 \times 10^{-2}$ V^2-sec. A comparison of Tables 2.10 and 2.11 shows an approximate –1 dB E_b/N_o increase in BER performance for Gray-coded over that for straight-binary 4-level rectangular PAM, as predicted by Equations 2.45 and 2.47.

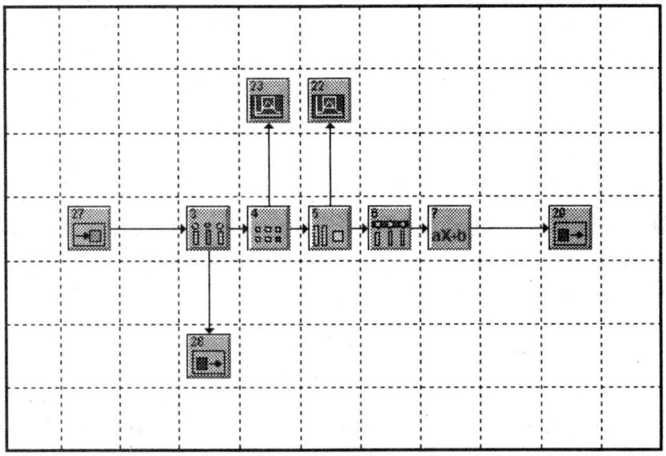

FIGURE 2.38 The 2-bit Gray encoder MetaSystem token of the 4-level rectangular PAM digital communication system.

TABLE 2.11 Observed BER and Theoretical P_b as a Function of E_b/N_o in a 4-Level Rectangular Gray-Coded PAM Digital Communication System with Optimum Receiver

E_b/N_o dB	N_o V2-sec	BER	Pb
∞	0	0	0
10	1.39×10^{-3}	1.5×10^{-3}	1.8×10^{-3}
8	2.20×10^{-3}	1.01×10^{-2}	9.38×10^{-3}
6	3.49×10^{-3}	2.49×10^{-2}	2.81×10^{-2}
4	5.53×10^{-3}	5.95×10^{-2}	5.84×10^{-2}
2	8.77×10^{-3}	9.96×10^{-2}	9.90×10^{-2}
0	1.39×10^{-2}	1.402×10^{-1}	1.404×10^{-1}

PARTIAL RESPONSE SIGNALING

Partial response signaling, also described as *correlative coding*, deliberately introduces intersymbol interference (ISI) into the transmitted signal. ISI is often described as an undesirable phenomenon, but the introduction of *controlled ISI* in a binary PAM digital communication system can achieve the theoretical maximum signaling rate of r_b b/sec in a bandwidth of $r_b/2$ Hz, as postulated by Nyquist.

Duobinary PAM Signaling

The binary sinc PAM digital communication system demonstrates no ISI and has the theoretical maximum signaling rate performance, as given by Equation 2.11 and shown in Figure 2.17. However, the frequency response of the communications channel must be *flat* from 0 to $r_b/2$ Hz to accommodate the transmission of the binary sinc PAM signal, and this is often problematical.

The *duobinary* pulse introduces a known ISI with an effect that can be interpreted in the receiver in a deterministic manner. The sinc pulse, given in Equation 2.9, has a peak amplitude of A at $t = 0$ and zero crossings at $\pm nT_b$ $(n \neq 0)$, satisfying Nyquist's First Criterion for no ISI. The duobinary pulse has an amplitude of A at $t = 0$ and $t = T_b$ and zero crossings at $\pm nT_b$ $(n \neq 0, 1)$. One form of the duobinary pulse $p_{db}(t)$ can be generated by the sum of two sinc pulses that are displaced in time by T_b seconds, as given by Equation 2.48. This duobinary pulse satisfies *Nyquist's Second Criterion* for no ISI.

$$p_{db}(t) = A\,\text{sinc}\left(\pi t/T_b\right) + A\,\text{sinc}\left[\pi\,(t - T_b)/T_b\right] \tag{2.48}$$

The magnitude of the voltage spectrum $P_{db}(f)$ of this duobinary pulse is given by Equation 2.49.

$$\begin{aligned} P_{db}(f) &= 2\,A T_b \cos\left(\pi f T_b\right) \quad -1/2T_b \le f \le 1/2T_b \\ P_{db}(f) &= 0 \quad \text{otherwise} \end{aligned} \tag{2.49}$$

The bi-sided energy spectral density $E_{db}(f)$ of n pseudonoise polar duobinary pulses is n times the square of the magnitude of the voltage spectrum divided by the load resistance R_L, as given by Equation 2.50.

$$\begin{aligned} E_{db}(f) &= \frac{2n\,A^2\,T_b^{\,2}}{R_L}\cos^2\left(\pi f T_b\right) \quad -1/2T_b \le f \le 1/2T_b \\ E_{db}(f) &= 0 \quad \text{otherwise} \end{aligned} \tag{2.50}$$

The bi-sided power spectral density (PSD) $PSD_{db}(f)$ is the energy spectral density divided by the time to transmit n duobinary PAM pulses or nT_b, as given by Equation 2.51.

$$PSD_{db}(f) = \frac{2\,A^2\,T_b}{R_L}\cos^2\left(\pi f T_b\right) \quad -1/2T_b \le f \le 1/2T_b \tag{2.51}$$

Unlike the sinc pulse, the duobinary pulse has a frequency response that is not flat, and simple filters can be realized that approximate this spectrum closely.

However, both the sinc and duobinary pulse require a communication channel that has a significant low frequency response to 0 Hz. The *modified duobinary* pulse $p_{mdb}(t)$ obviates this concern and is generated by the difference of two sinc pulses that are displaced in time by $2T_b$ seconds, as given by Equation 2.52. The modified duobinary pulse still satisfies Nyquist's Second Criterion for no ISI.

$$p_{mdb}(t) = A\,\mathrm{sinc}\left(\pi t/T_b\right) - A\,\mathrm{sinc}\left[\pi(t-2T_b)/T_b\right] \tag{2.52}$$

The magnitude of the voltage spectrum $P_{mdb}(f)$ of the modified duobinary pulse is given by Equation 2.53.

$$\begin{aligned} P_{mdb}(f) &= 2AT_b \sin\left(\pi f T_b\right) \quad -1/2T_b \le f \le 1/2T_b \\ P_{mdb}(f) &= 0 \quad \text{otherwise} \end{aligned} \tag{2.53}$$

With a development similar to that of the duobinary PAM pulse, the bi-sided $PSD_{mdb}(f)$ of the modified duobinary PAM pulse is given by Equation 2.54.

$$PSD_{mdb}(f) = \frac{2A^2 T_b}{R_L}\sin^2\left(\pi f T_b\right) \quad -1/2T_b \le f \le 1/2T_b \tag{2.54}$$

The modified duobinary pulse $PSD_{mdb}(f)$ has a frequency response null at 0 Hz. The spectrum is also not flat and can be realized approximately by even simpler filters, without utilizing a filter with a sinc impulse response.

SystemVue Simulation of Duobinary PAM

ON THE CD

A duobinary PAM transmitter, based on the sinc PAM transmitter of Figure 2.13, is shown in Figure 2.39 (see Fig2-39.svu on the CD-ROM). Simpler filters can be realized that approximate the duobinary pulse, but the sinc Communications Filter is used here. The configuration and parameters of the SystemVue simulation tokens are the same as for the sinc PAM transmitter with the addition of the Delay token from the Operator Library and the Adder token, which forms the sum of the sinc and delayed sinc PAM signals, as given by Equation 2.48. The delay is $T_b = 1$ msec.

Figure 2.40 shows the two initial polar duobinary PAM signals representing a binary 0 and binary 1, with a temporary data rate of 100 b/sec to adequately separate and display the duobinary pulses. The sinc Communications Filter has a width of $\pm 4T_b = \pm 4$ msec, centered at $t = 4$ msec, and the first duobinary PAM signal representing a binary 0 has a magnitude of $A = -5$ V at a time of 4 msec ($t = 0$) and 5 msec ($t = T_b$), and a zero value at relative times of $-T_b$, $\pm 2T_b$, $\pm 3T_b$, and $\pm 4T_b$, as given by Equation 2.48.

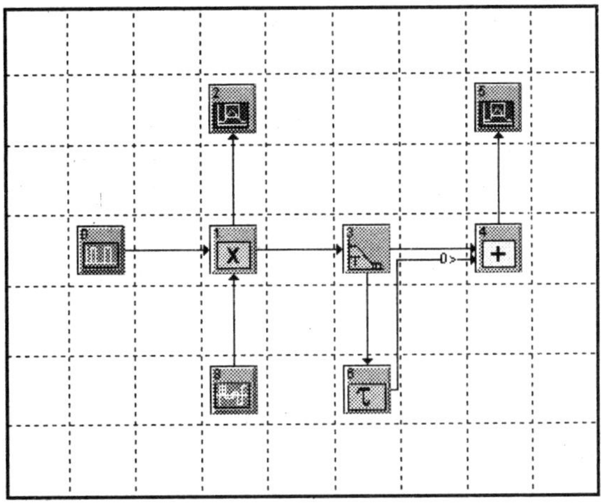

FIGURE 2.39 A duobinary sinc PAM transmitter.

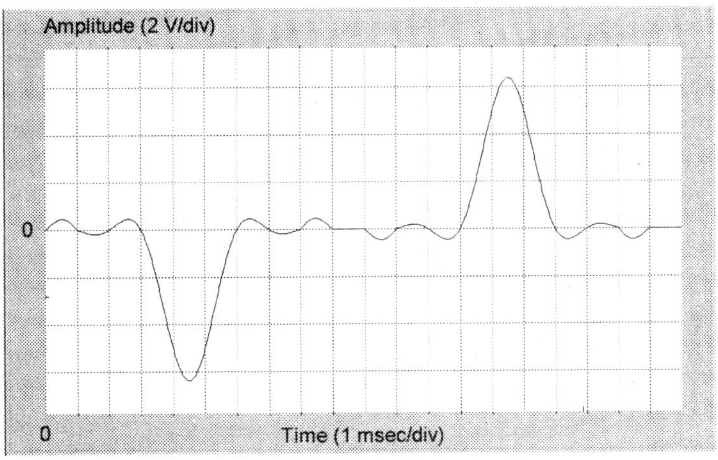

FIGURE 2.40 Two duobinary sinc PAM signals at a temporary data rate of 100 b/sec.

ON THE CD

The modified duobinary PAM transmitter based on Equation 2.50 is shown in Figure 2.41 (see Fig2-41.svu on the CD-ROM), which is similar to Figure 2.39 for the duobinary PAM transmitter. The Delay token has a delay parameter of $2T_b = 2$ msec. The output of the delayed duobinary PAM pulse then is inverted by the Negate token from the Operator Library before being inputted to the Adder token, as given by Equation 2.50.

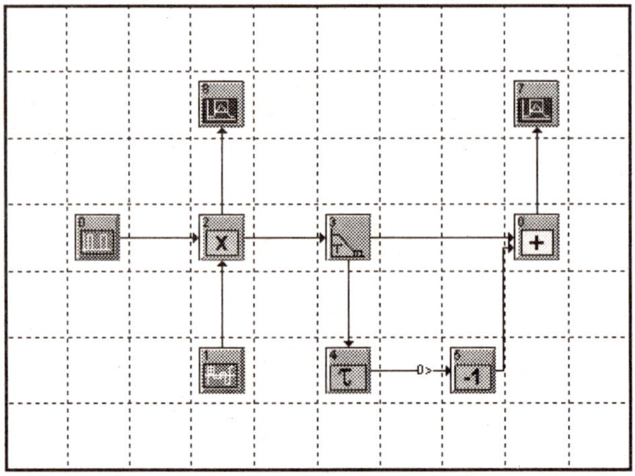

FIGURE 2.41 A modified duobinary sinc PAM transmitter.

Figure 2.42 shows the two initial polar-modified duobinary PAM signals representing a binary 0 and binary 1, with a temporary data rate of 100 b/sec to adequately separate and display the modified duobinary pulses. The sinc Communications Filter again has a width of $\pm\, 4T_b = \pm\, 4$ msec, centered at $t =$ 4 msec, and the first modified duobinary PAM signal representing a binary 0 has a magnitude of $A = -5$ V at $t = 4$ msec, but now at $t = 6$ msec ($t = 2T_b$). The zero values occur at $\pm\, T_b$, $-2T_b$, $\pm\, 3T_b$, and $\pm\, 4T_b$, as given by Equation 2.48.

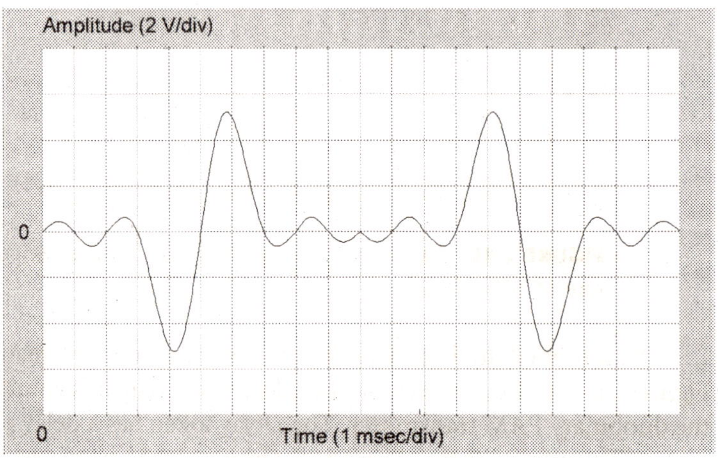

FIGURE 2.42 Two modified duobinary sinc PAM signals at a temporary data rate of 100 b/sec.

The Simple Receiver for Precoded Duobinary Signals

Resetting the data rate to 1 kb/sec for the duobinary sinc PAM transmitter of Figure 2.39 results in the 3-level PAM signal shown in Figure 2.43. Unlike the rectangular and sinc PAM signals, the duobinary sinc PAM signal has the amplitudes of $-2A$, 0, and $+2A$ at the bit time iT_b and $A = 5$ V here.

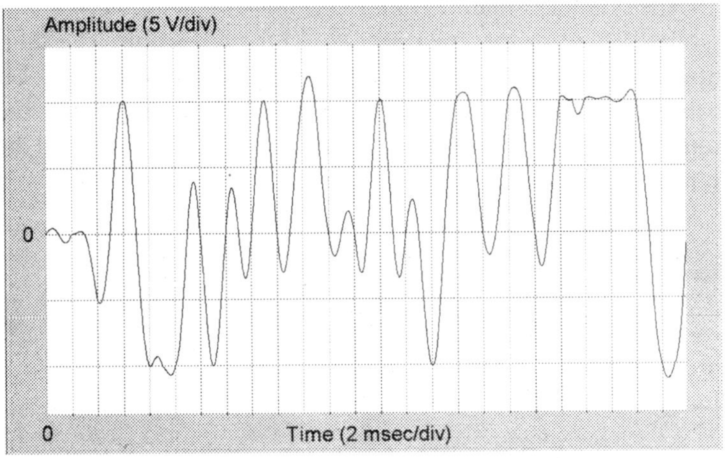

FIGURE 2.43 Summated duobinary sinc PAM signals at a data rate of 1 kb/sec.

The output amplitude $C(iT_b)$ of the duobinary PAM transmitter at the bit time iT_b then is given by Equation 2.55.

$$C(iT_b) = A\,b_i + A\,b_{i-1} \tag{2.55}$$

The binary polar ($\pm$ 1) data input at the current bit time iT_b is b_i and that at the immediate past bit time $(i-1)T_b$ is b_{i-1}. From Equation 2.55, $C(iT_b) = -10$ V if $b_i = -1$ and $b_{i-1} = -1$, $C(iT_b) = 10$ V if $b_i = 1$ and $b_{i-1} = 1$, but $C(iT_b) = 0$ V if $b_i = -1$ and $b_{i-1} = 1$ or if $b_i = 1$ and $b_{i-1} = -1$. The value of the detected binary polar ($\pm$ 1) data is given by Equation 2.56.

$$c_i = d_i - c_{i-1} \tag{2.56}$$

The detected 3-level PAM signal is $d_i = \pm\,2, 0$. However, a correct value for the detected binary polar data c_i requires that both the current detected value d_i and immediate past estimate c_{i-1} be error free. The process for the received polar data in Equation 2.56 is called *decision feedback*. However, a disadvantage of decision feedback is that once a detection error for d_i is made, errors *propagate.* A technique for avoiding such error propagation is the use of *precoding* in the duobinary sinc PAM transmitter, as given by Equation 2.57.

$$f_i = a_i \oplus f_{i-1} \tag{2.57}$$

The symbol ⊕ denotes binary *modulo-two addition* (0 + 1 = 1 + 0 = 1 and 0 + 0 = 1 + 1 = 0) or the *exclusive OR* (XOR) operation. The unipolar binary (0, 1) input data is a_i and the unipolar binary precoded output is f_i. The precoded data f_i is then converted to the polar binary data b_i for processing by the duobinary transmitter in Figure 2.39, as given by Equation 2.55. The precoded duobinary sinc PAM transmitter is shown as a MetaSystem in Figure 2.44.

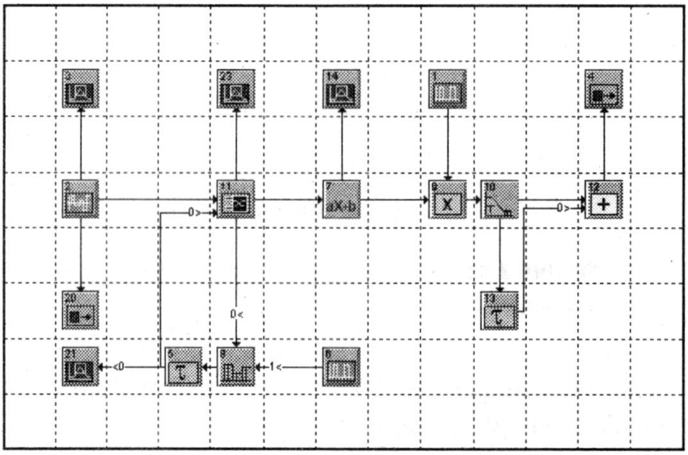

FIGURE 2.44 MetaSystem of the precoded duobinary sinc PAM transmitter.

The data source is the PN Sequence token from the Source Library with a binary amplitude of 0 and 1 V, a data rate of 1 kb/sec, and 0 V voltage and 0° phase offset. The PN Sequence data source is outputted from this MetaSystem to the duobinary sinc PAM digital communication system with AWGN channel, as shown in Figure 2.45 (see Fig2-45.svu on the CD-ROM). To facilitate the simulation in SystemVue, all of the Analysis tokens, except that for the BER token output as the Final Value token, are deleted in the simulation model Fig2-45DT. As described in Chapter 1, the SystemVue Textbook Edition does not permit tokens to be deleted.

ON THE CD

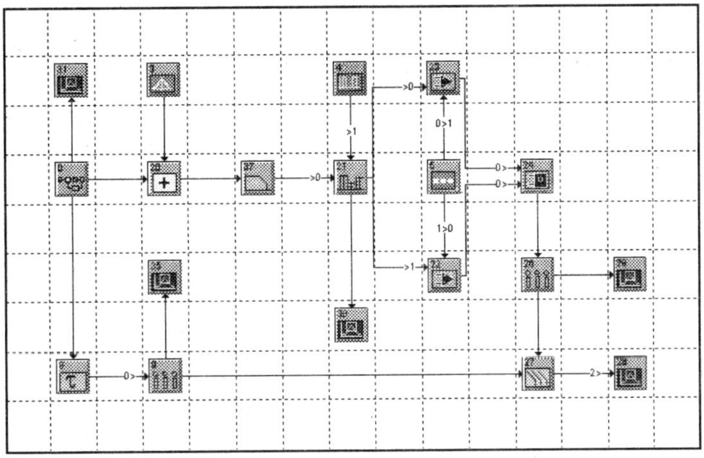

FIGURE 2.45 Precoded duobinary sinc PAM digital communication system with BER analysis.

The precoder, as given by Equation 2.57 and shown in Figure 2.44, is implemented as an XOR (exclusive OR) token from the Logic Library, with the parameters of an output of 0 and 1 V, a logical threshold of 0.5 V, and rise, fall, and gate delay times of 0 seconds, a Sample and Hold token from the Operator Library, with a control threshold of 0.5 V, and a Delay token from the Operator Library. The Delay token has a delay parameter of 0.98 msec, or $T_b - T_{system}$ sec, to avoid glitches due to the logic transitions that occur at $T_b = 1$ msec in the feedback system here. The SystemVue simulation System Time T_{system} here is 20 μsec or a Sample Rate of 50 kHz.

The output of the precoder in the MetaSystem of Figure 2.44 is applied to the Polynomial token from the Function Library to produce a polar binary (±5 V) pulse from the unipolar (0, 1) data source. The parameters of the Polynomial token are that the x^0 (offset) coefficient is –5, the x^1 (linear) coefficient is 10, and all other coefficients are zero. The polar binary pulse from the Polynomial token is applied to the Multiplier token, with another input derived from a Pulse Train token. The Pulse Train token is a periodic impulse source with a frequency of 1 kHz (bit time $T_b = 1$ msec), 0 V voltage and 0° phase offset, a pulse width of 20 μsec and a unipolar amplitude of 0 to 47.6 V peak.

The periodic impulse source excites a Linear System Filter token sinc Communications Filter to the nominal 5 V peak output voltage, as in the binary sinc PAM transmitter of Figure 2.13. The sinc Communications Filter is also delayed by $T_b = 1$ msec by a Delay token. The output of the non-delayed and delayed sinc Communication Filter PAM signal is then summed by the Adder token, as in Equation 2.48.

The communication channel is represented by an Adder token and a Gaussian Noise token from the Source Library, with initial parameters of a standard deviation $\sigma = 0$ V (no noise) in Figure 2.45.

The Linear System Filter token from the Operator Library provides the LPF for the simple PAM receiver here. The LPF is a 9-pole Chebyshev filter, with a cutoff frequency $f_{cutoff} = 600$ Hz and an inband ripple of 0.1 dB ($\approx 1\%$ ripple). An LPF with a cutoff frequency of 600 Hz would pass nearly all of the total signal power for a binary data rate of 1 kb/sec ($1/2T_b = 500$ Hz) for the duobinary and modified duobinary sinc PAM signals, as in Equations 2.51 and 2.54.

The duobinary sinc PAM receiver Pulse Train token provides the Sample and Hold token control voltage with a single point sample ($T_{system} = 20$ μsec) at 0° phase offset, which places the data sampling exactly at the start of the bit time T_b. The noise-free 3-level output of the Sample and Hold token at the bit time iT_b is $y(iT_b) = \pm 10$ V, 0 V, which corresponds to the detected 3-level PAM signal $d_i = \pm 2, 0$.

The detection of the original unipolar binary data a_i is accomplished with the *decision rule* derived from Equation 2.58 and the exclusive OR operation, as given by Equation 2.57.

$$\begin{aligned} &\text{if } \left| y(iT_b) \right| < 5 \text{ V then } a_i = 1 \\ &\text{if } \left| y(iT_b) \right| > 5 \text{ V then } a_i = 0 \end{aligned} \qquad (2.58)$$

The decision rule is implemented by two Analog Comparator tokens from the Logic Library, with a unipolar binary (0, 1) output. The comparator input signals are derived from the Custom token from the Source Library, with parameters that specify two output signals that are +5 and –5 V. The output of the Analog Comparators are processed by the OR token from the Logic Library with inverted logical outputs.

Two Sampler tokens are used to downsample the 50 kHz System Time rate to the 1 kb/sec binary data rate, or 1 kHz. The Sampler token parameters are set to a sample rate of 1 kHz, an aperture time of 20 μsec (T_{system}, a single point sample). The PN Sequence token data from the MetaSystem of Figure 2.44 is delayed by 6 msec to correlate the data with the received duobinary sinc PAM signal. The value of this delay can be observed by the cross-correlation of the two sampler signals at the input to the BER token, as described in Chapter 1. The nominal 6 msec delay consists of the 4 msec sinc PAM signal delay and 2 data samples or 2 msec of additional processing delay.

The Simple Receiver for Precoded Modified Duobinary Signals

Resetting the data rate to 1 kb/sec for the modified duobinary sinc PAM transmitter of Figure 2.41 results in the 3-level PAM signal with amplitudes of $-2A$, 0, and $+2A$ at the bit time iT_b and $A = 5$ V here, as shown in Figure 2.46.

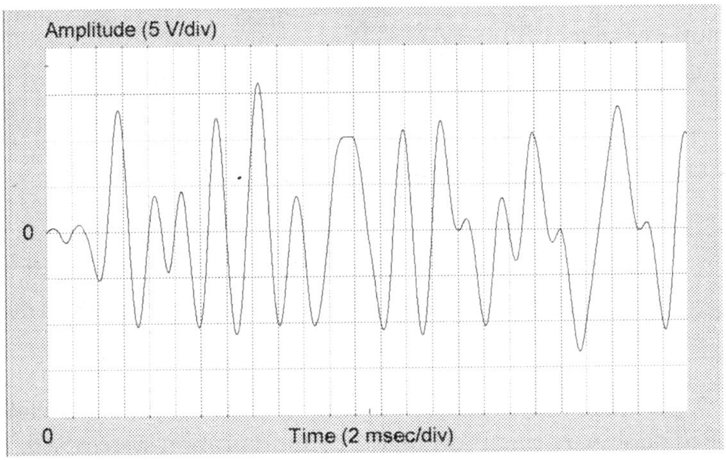

FIGURE 2.46 Summated modified duobinary sinc PAM signals at a data rate of 1 kb/sec.

The output amplitude $C(iT_b)$ of the modified duobinary PAM transmitter at the bit time iT_b then is given by Equation 2.59.

$$C(iT_b) = A\,b_i - A\,b_{i-2} \tag{2.59}$$

The binary polar ($\pm$ 1) data input at the current bit time iT_b is b_i and that at the second past bit time $(i-2)T_b$ is b_{i-2}. From Equation 2.59, $C(iT_b) = 10$ V if $b_i = 1$ and $b_{i-2} = -1$, $C(iT_b) = -10$ V if $b_i = -1$ and $b_{i-2} = 1$, but $C(iT_b) = 0$ V if $b_i = 1$ and $b_{i-1} = 1$ or if $b_i = -1$ and $b_{i-1} = 1$. The value of the detected binary polar ($\pm$ 1) data is given by Equation 2.53. The detected 3-level PAM signal is $d_i = \pm$ 2, 0, and precoding is again used to obviate error propagation with decision feedback, as given by Equation 2.57.

The unipolar binary (0, 1) input data is a_i and the unipolar binary precoded output is f_i. The precoded data f_i is then converted to the polar binary data b_i for processing by the duobinary transmitter in Figure 2.41, as given by Equation 2.59. The precoded modified duobinary sinc PAM transmitter is shown as a MetaSystem in Figure 2.47.

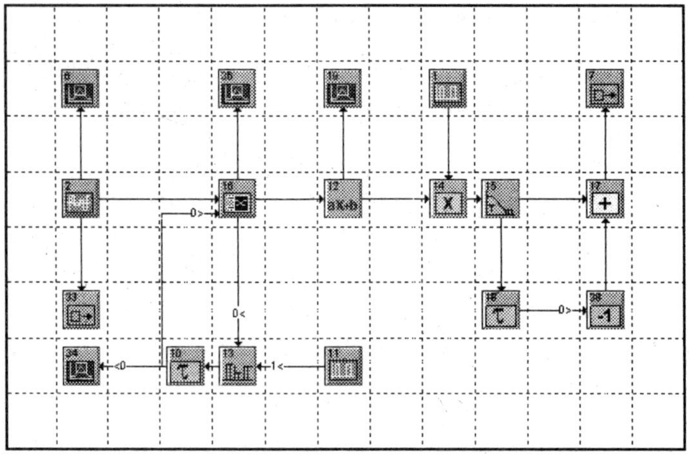

FIGURE 2.47 MetaSystem of the precoded modified duobinary sinc PAM transmitter.

The PN Sequence data source is outputted from this MetaSystem to the modified duobinary sinc PAM digital communication system with AWGN channel, as shown in Figure 2.48 (see Fig2-48.svu on the CD-ROM). To facilitate the simulation in SystemVue, all of the Analysis tokens, except that for the BER token output as the Final Value token, are deleted in the simulation model Fig2-48DT. As described in Chapter 1, the SystemVue Textbook Edition does not permit tokens to be deleted.

ON THE CD

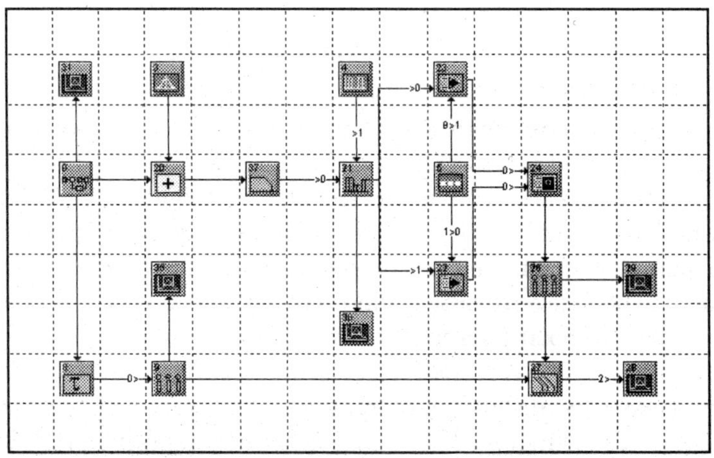

FIGURE 2.48 Precoded modified duobinary digital communication system with BER analysis.

The precoder, modified duobinary sinc PAM transmitter, communication channel, and simple PAM receiver are similar to that for the duobinary sinc PAM communication system, as shown in Figures 2.44 and 2.45. The precoder, as given by Equation 2.58 and shown in Figure 2.47, is again implemented as an XOR token from the Logic Library, a Sample and Hold token from the Operator Library, and a Delay token from the Operator Library. The Delay token here has a delay parameter of 1.98 msec, or $2T_b - T_{system}$ sec, to avoid glitches due to the logic transitions that occur at $T_b = 2$ msec in the feedback system. The SystemVue simulation System Time T_{system} here is 20 µsec or a Sample Rate of 50 kHz.

The output of the precoder in the MetaSystem of Figure 2.47 is applied to the Polynomial token from the Function Library to produce a polar binary (± 5 V) pulse from the unipolar (0, 1) data source. The parameters of the Polynomial token are that the x^0 (offset) coefficient is –5, the x^1 (linear) coefficient is 10, and all other coefficients are zero. The sinc Communications Filter is delayed here by $T_b = 2$ msec by a Delay token and negated by the Negate token from the Operator Library. The output of the non-delayed and the delayed and negated sinc Communication Filter PAM signal is then summed by the Adder token, as in Equation 2.50.

The detection of the original unipolar binary data a_i is accomplished with the decision rule derived from Equation 2.60 and the exclusive OR operation, as given by Equation 2.57.

$$\begin{aligned} &\text{if } \left| y(iT_b) \right| < 5\text{ V then } a_i = 0 \\ &\text{if } \left| y(iT_b) \right| > 5\text{ V then } a_i = 1 \end{aligned} \tag{2.60}$$

The decision rule is again implemented by two Analog Comparator tokens from the Logic Library, with a unipolar binary (0, 1) output. The output of the Analog Comparators are processed by the OR token from the Logic Library, but here with *non-inverted* logical outputs.

The PN Sequence token data from the MetaSystem of Figure 2.47 is delayed by 6 msec to correlate the data with the received modified duobinary sinc PAM signal. The value of this delay can be observed by the cross-correlation of the two sampler signals at the input to the BER token, as described in Chapter 1. The nominal 6 msec group delay consists of the 4 msec sinc PAM signal delay and 2 data samples, or 2 msec of additional processing delay.

Duobinary PAM Power Spectral Density

Setting the System Time: Number of Samples specification to 262 144 (2^{18}) points in the SystemVue simulation of duobinary and modified duobinary sinc PAM transmitters results in a spectral resolution of approximately 0.19 Hz. The normalized PSD of the duobinary sinc PAM signal from Figure 2.39 is shown unscaled in

Figure 2.49 in dBm/Hz. The normalized PSD of the modified duobinary sinc PAM in Figure 2.41 is shown unscaled in Figure 2.50 in dBm/Hz. The maximum frequency displayed is 25 kHz, the Nyquist frequency or half of the system simulation sampling rate (50 kHz/2).

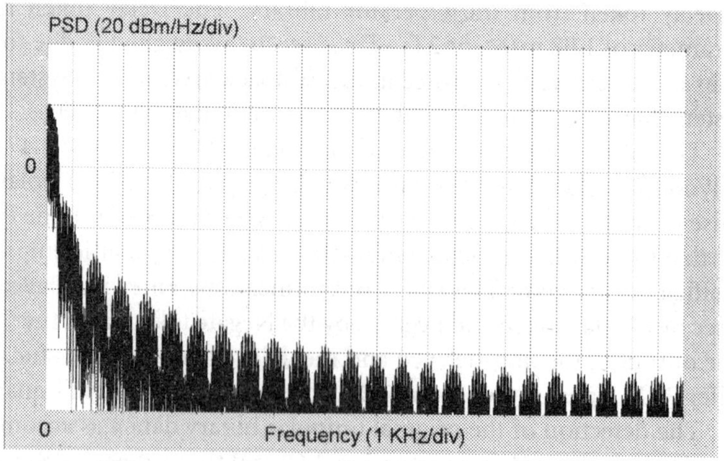

FIGURE 2.49 Power spectral density of a 1 kb/sec pseudonoise polar duobinary sinc PAM signal.

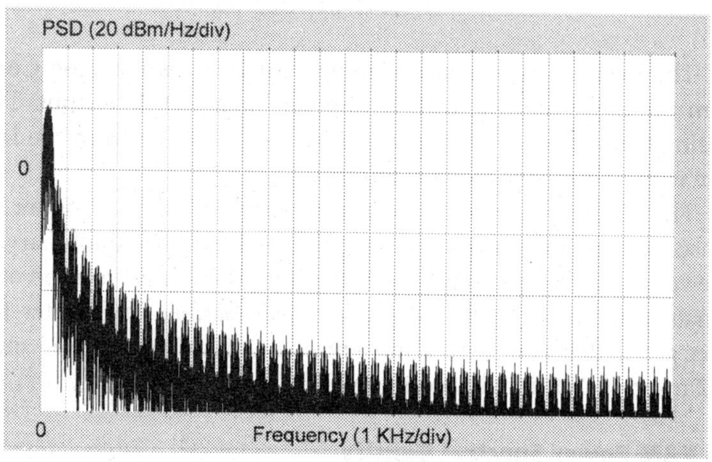

FIGURE 2.50 Power spectral density of a 1 kb/sec pseudonoise polar-modified duobinary sinc PAM signal.

A comparison of the PSDs of the sinc PAM signal in Figure 2.17 to those of the duobinary and modified duobinary PAM signals in Figures 2.46 and 2.44 demonstrates the expected power spectral differences. Figure 2.51 is an expanded representation of these PSDs to a maximum frequency of 1 kHz. The polar sinc PAM PSD is nearly flat to $r_b/2 = 500$ Hz, as in Equation 2.13, the duobinary sinc PAM PSD rolls off significantly and displays a null at $r_b/2$ Hz, as in Equation 2.49, and the modified duobinary sinc PAM PSD displays a null at both 0 and $r_b/2$ Hz, as in Equation 2.51. The PSD of either the duobinary or modified duobinary sinc PAM signal is not affected by the source data precoder, as given by Equations 2.54 and 2.57.

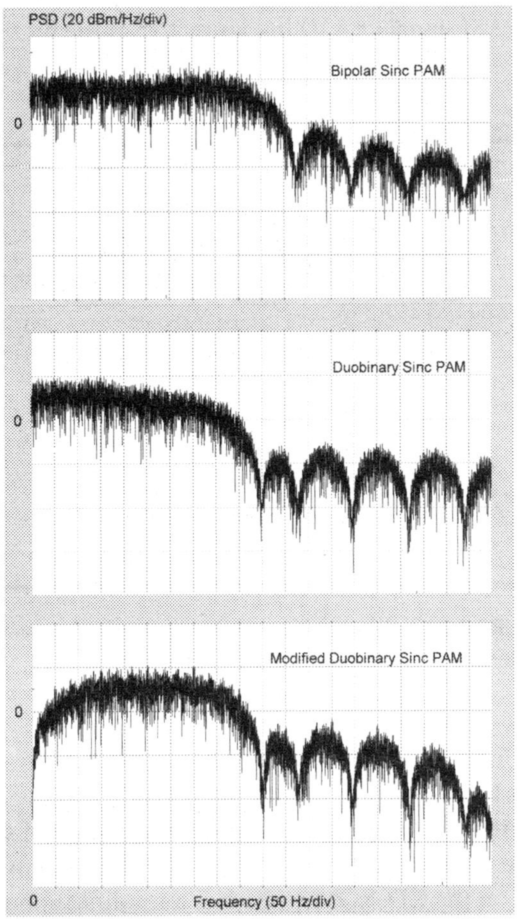

FIGURE 2.51 Power spectral densities of a 1 kb/sec pseudonoise polar rectangular, duobinary and modified duobinary sinc PAM signals.

Performance of Duobinary PAM in a Simple Receiver in AWGN

The precoded duobinary sinc PAM transmitter and simple PAM receiver in Figure 2.45 can be evaluated for its performance in the presence of AWGN. A BER analysis is conducted with the same parameters as the filtered sinc PAM communication system, as given in Table 2.4.

ON THE CD

The signal power, which contributes to the SNR, is problematical because the duobinary sinc PAM signal extends beyond one bit time. Although there is no ISI at the data sampling interval T_b in Figure 2.43, adjacent duobinary sinc PAM signals contribute or remove signal power everywhere else over the bit time. The duobinary sinc PAM signal in Figure 2.43 does not have the same voltage (± 10 V, 0 V) as the sinc PAM signal (± 5 V) in Figure 2.15 at the bit time T_b. However, both the

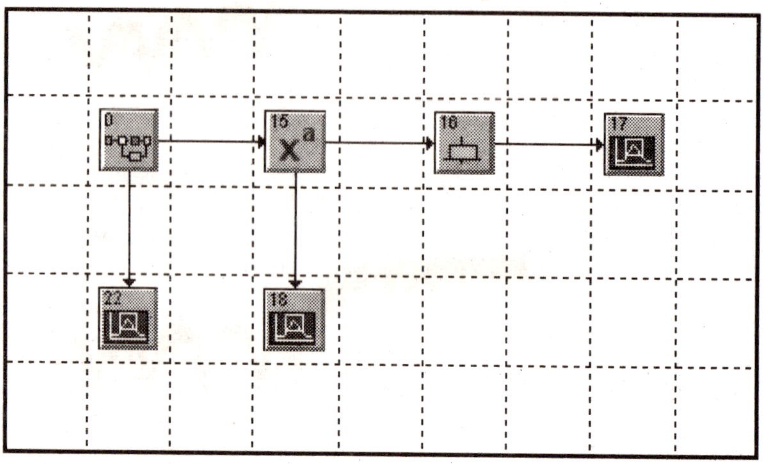

FIGURE 2.52 Computation of the normalized power in the precoded duobinary sinc Communications Filter output.

duobinary sinc and sinc PAM signals have a 10 V difference between their respective symbols by design.

The normalized power in the duobinary sinc PAM signal is the average of the square of the output and can be computed by the system shown in Figure 2.52 (see Fig2-52.svu on the CD-ROM), which is a modification of Figure 2.19. The MetaSystem of the precoded duobinary sinc PAM transmitter from Figure 2.44 is used here. The computed average normalized power in the duobinary sinc PAM signal is approximately 50.2 W, because the amplitudes at the bit times iT_b are ± 10 V and

0 V. The power in the sinc PAM signal in Figure 2.19 is approximately 24.4 W, because the amplitudes at the bit times iT_b are $\pm$ 5 V.

Table 2.12 is a tabular list of the observed BER in a single trial of 10 000 information bits as a function of the SNR for the precoded duobinary sinc PAM digital communication system in Figure 2.45. SNR is used here, rather than the E_b/N_o ratio, as for the performance of the sinc PAM signal. Tables 2.4 and 2.12 show an approximately comparable BER performance for sinc PAM and precoded duobinary sinc PAM.

A comparison of Figures 2.18 and 2.45 demonstrates that the design tradeoff for precoded duobinary sinc PAM is a slight increase in the complexity of the PAM digital communication system for the same transmission bandwidth, with only comparable BER performance. However, simpler filters can be realized that approximate the duobinary sinc pulse with its *relaxed* LPF response, rather than the *ideal* LPF response of the sinc pulse.

TABLE 2.12 Observed BER as a Function of SNR in an LPF (9-pole Chebyshev, 0.1 dB ripple, f_{cutoff} = 600 Hz) Precoded Duobinary Sinc PAM Digital Communication System, Normalized Signal Power ≈ 50.2 W

SNR dB	AWGN σ V	BER
∞	0	0
9.04	2.5	0
3.03	5	1×10^{-4}
−0.49	7.5	9×10^{-4}
−2.99	10	6.8×10^{-3}
−4.93	12.5	1.93×10^{-2}
−6.51	15	4.33×10^{-2}
−7.85	17.5	6.73×10^{-2}

ON THE CD

The precoded modified duobinary sinc PAM transmitter and simple PAM receiver in Figure 2.48 can also be evaluated for its performance in the presence of AWGN. The normalized power in the modified duobinary sinc PAM signal is the average of the square of the output and can be computed by the system shown in Figure 2.53 (see Fig2-53.svu on the CD-ROM). The MetaSystem of the precoded modified duobinary sinc PAM transmitter from Figure 2.47 is used here.

The Final Value token from the Sink Library is used in Figure 2.53, rather than an Analysis token to directly display the average normalized power. The computed average normalized power in the modified duobinary sinc PAM signal then is approximately 50.6 W, and that for the duobinary sinc PAM signal is nearly the same at 50.2 W.

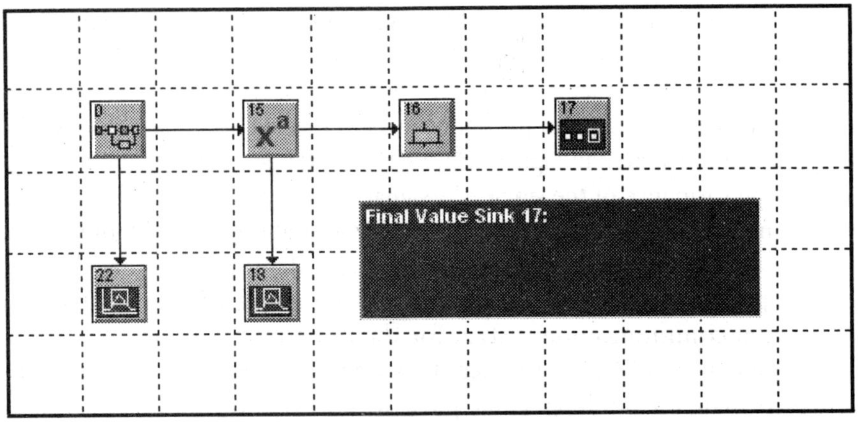

FIGURE 2.53 Computation of the normalized power in the precoded modified duobinary sinc Communications Filter output with the Final Value sink token.

Table 2.13 is a tabular list of the observed BER in a single trial of 10 000 information bits as a function of the SNR for the precoded modified duobinary sinc PAM digital communication system in Figure 2.48.

TABLE 2.13 Observed BER as a Function of SNR in an LPF (9-pole Chebyshev, 0.1 dB ripple, $f_{cutoff} = 600$ Hz) Precoded Duobinary Sinc PAM Digital Communication System, Normalized Signal Power ≈ 50.6 W

SNR dB	AWGN σ V	BER
∞	0	0
9.08	2.5	0
3.06	5	0
−0.46	7.5	8×10^{-3}
−2.96	10	6.4×10^{-3}
−4.90	12.5	2.08×10^{-2}
−6.48	15	3.82×10^{-2}
−7.82	17.5	7.07×10^{-2}

DELTA MODULATION

Delta modulation (DM) uses simple equipment to transmit an originally analog message signal by a digital technique. This is accomplished by deliberately oversampling the analog signal to increase the correlation between adjacent samples and then employing a quantizer with minimal complexity. In its simplest form, DM uses a *staircase* approximation for the oversampled message signal. The difference between the analog message signal input and the approximation is then quantized into only two levels $\pm \Delta$ V, or a positive or negative voltage difference.

If the staircase approximation retained in the transmitter *accumulator* falls below the analog input signal, a true binary signal is transmitted, representing $\pm \Delta$V. Alternatively, if the approximation is above the input signal, a false binary signal is transmitted, representing $-\Delta$V. The binary signal is a symbol, and transmitted with a symbol time T_s. The receiver also uses an accumulator to sum the binary symbol received representing $\pm \Delta$ V, and generate the staircase approximation to the original analog message signal.

DM is subject to two types of quantization error, described as *granular noise* and *slope overload* distortion. If the analog message signal is a constant voltage level, the delta modulator still sends a sequence of binary signals that cause the accumulator output in the receiver to *clock* around the constant level, because there is no symbol for 0 V in DM. Slope overload distortion occurs when the analog message signal has a steep segment that the staircase approximation cannot follow. Slope overload distortion can be avoided if Equation 2.61 is satisfied.

$$\frac{\Delta}{T_s} \geq \max \left| \frac{dm(t)}{dt} \right| \tag{2.61}$$

The ratio of the staircase level Δ to the symbol time T_s must be greater than or at least equal to the maximum absolute instantaneous slope (derivative) of the analog message signal *m(t)*. The *tradeoff* is that a large value of Δ would increase the DM granular noise and the symbol time T_s must be decreased (symbol rate r_s increased).

ON THE CD

A delta modulation system (DM) using the staircase approximation is shown in Figure 2.54 (see Fig2-54.svu on the CD-ROM). The analog message signal here is provided by a Sinusoid token from the Source Library, with the parameters of an amplitude $A = 1$ V, an initial frequency $f_o = 2$ Hz, and 0° phase. The DM transmitter consists of a feedback system with two Adder tokens, an Analog Comparator token from the Logic Library, two Sample and Hold tokens, a Negate token from the Operator Library, and a Pulse Train token from the Source Library. The System Time: Sample Rate is set to 100 kHz.

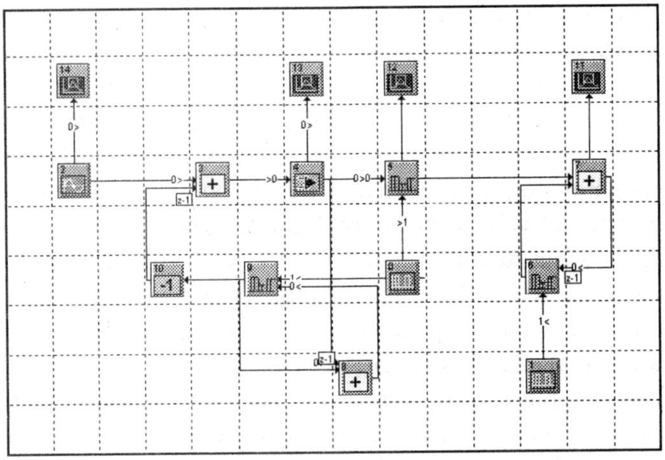

FIGURE 2.54 Delta modulation system using the staircase approximation.

The DM transmitter processes samples of the analog message signal m_i, as shown in Figure 2.54 and given by Equation 2.62.

$$
\begin{aligned}
e_i &= m_i - q_{i-1} \\
q_i &= q_{i-1} + \Delta\,\mathrm{sgn}(e_i)
\end{aligned}
\tag{2.62}
$$

Here, e_i is the error signal, which is the difference between the sample of the analog signal m_i and the immediate past output of the DM transmitter accumulator q_{i-1}. The DM transmitter accumulator output $q-_i$ is updated with the sum of the immediate past output q_{i-1} and the staircase levels $\pm\,\Delta$, where the sign function of the error signal $\mathrm{sgn}(e_i) = \pm\,1$.

The Analog Comparator token output is the staircase levels $\pm\,\Delta$ V $= \pm\,1$ mV, with a comparator input that is set to 0 V (unconnected). One Sample and Hold token is the transmitter accumulator output. The other Sample and Hold token assures that the DM transmitted signal is updated only each symbol time $T_s = 50$ μsec, and not the SystemVue simulation System Time $T_{system} = 10$ μsec here. The Pulse Train token has the parameters of an amplitude $A = 1$ V, a frequency $f_o = 20$ kHz, a pulse width $\tau = 10$ μsec, and 0° phase and 0 V offset. Setting the pulse width equal to the SystemVue simulation System Time T_{system} effectively produces an impulse sampler.

There is no communication channel shown in Figure 2.54. An additive white Gaussian noise (AWGN) channel could be implemented and the DM system performance assessed, but arbitrary bit errors only affect the accumulator by an amount $\pm\,\Delta$ V.

The DM receiver consists of an Adder token, a Pulse Train token, and a Sample and Hold token as the accumulator, which reconstructs the analog message signal, as given by Equation 2.63.

$$o_i = o_{i-1} \pm \Delta \tag{2.63}$$

Here o_{i-1} is the immediate past output of the DM receiver accumulator and $\pm\,\Delta$ is the staircase level. The DM receiver Pulse Train token has the same parameters as that for the DM transmitter.

The maximum instantaneous slope or derivative of a 1 V peak, 2-Hz sinusoid is $4\pi = 12.56$ V/sec. From Equation 2.61, the ratio of the staircase level Δ to the symbol time T_s here is 1 mV/50 μsec = 20 V/sec, and slope overload is avoided here. The 1 mV stair level implies that granular noise is ± 1 mV peak-to-peak, or only 0.1% of the 1 V peak sinusoidal signal here.

Figure 2.55 shows the difference between the input analog message signal and the DM reconstructed signal as 0.5 mV/div for a 2 sec SystemVue simulation. The DM reconstruction error here is never greater than ± 1.5 mV.

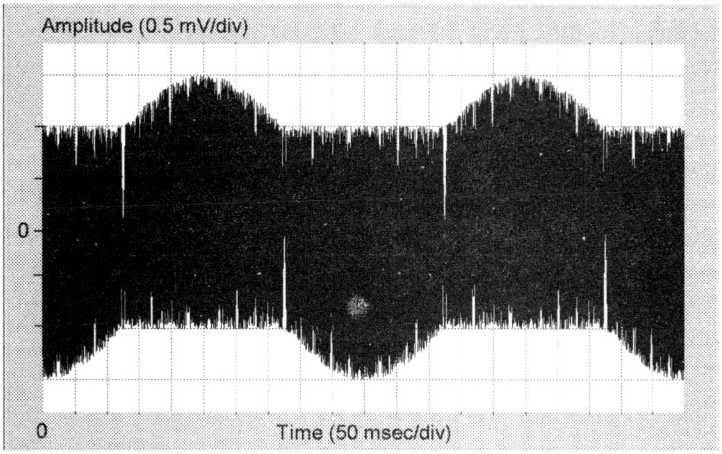

FIGURE 2.55 Error signal between a 2-Hz sinusoid input signal and the staircase approximation of a delta modulation system.

Slope overload can be shown by setting the frequency of the analog message signal to 5 Hz, with a maximum instantaneous slope of $10\pi = 31.41$ V/sec, which is greater than the DM system parameter of 20 V/sec. Figure 2.56 shows the 5-Hz sinusoidal input signal, overlaid with the DM staircase approximation.

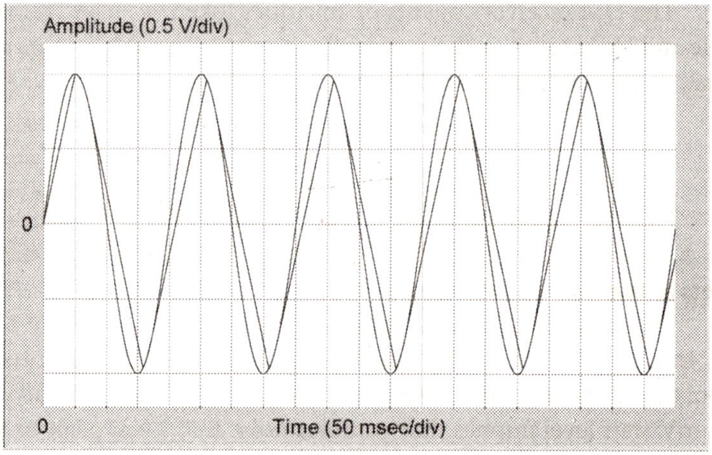

FIGURE 2.56 Sinusoidal input signal overlaid with the DM staircase approximation.

Granular noise can be shown by setting the voltage of the analog message signal to 0 V. Figure 2.57 shows the granular noise as the staircase levels ± Δ V = ± 1 mV, clocking every 50 msec (20 kHz rate).

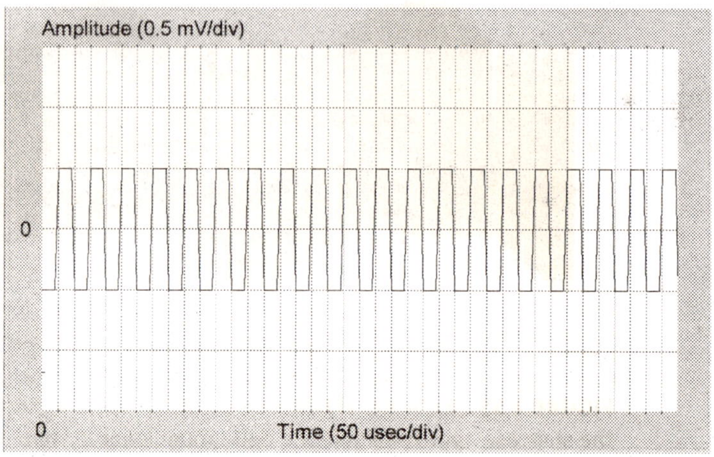

FIGURE 2.57 DM granular noise as the staircase levels (± 1 mV) clocking at the symbol rate (20 kHz).

The DM system bit rate here is 1 bit transmitted at 20 kHz or 20 kb/sec. An equivalent source bit rate for a baseband digital communication system can be assessed from the parameters of the DM system. The DM staircase level is ± 1 mV, and the analog message signal has an amplitude of ± 1 V (a range of 2 V), which implies that 11 bits of resolution is required ($2^{11} = 2048 \approx 2$ V/1 mV = 2000). A reasonable sampling rate to accurately define the 2-Hz sinusoid here would be 200 Hz, and the resulting data rate at 11 bits per sample is only 2.2 kb/sec. The tradeoff is that DM requires only very simple equipment and not the *analog-to-digital* and *digital-to-analog* converters required for analog signal source conversion.

EYE DIAGRAMS

The effect of additive white Gaussian noise (AWGN), instability in the bit rate (jitter), communication channel bandwidth and distortion, and intersymbol interference (ISI) can be qualitatively assessed by the eye diagram. An *eye diagram* is an overlay temporal display of the baseband binary or M-ary PAM signal synchronized to the bit rate r_b. Figure 2.58 shows the eye pattern of the binary rectangular symmetrical PAM digital communication system at the channel in Figure 2.26 with no noise.

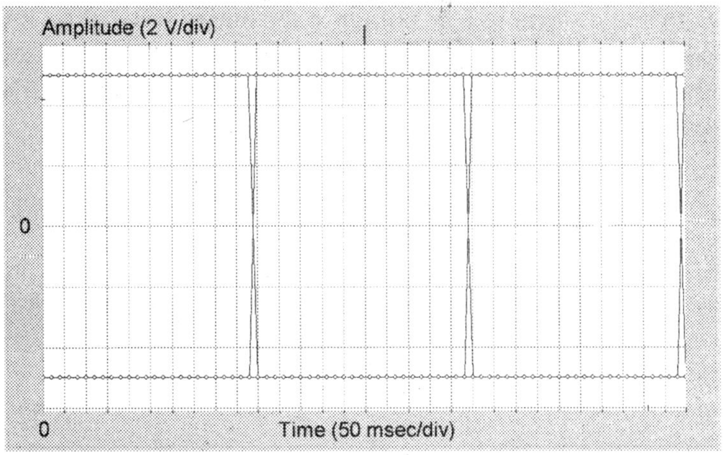

FIGURE 2.58 Eye diagram of the binary rectangular PAM signal with no noise at the input to the correlation receiver.

The eye diagram Analysis window of the communication channel is a repeating time slice, starting at 0.5 msec, with a length of 1.5 msec. The System Time: Sample Rate here is 50 kHz and the bit rate is $r_b = 1$ kb/sec. The sample points are shown in the eye diagrams as small circles and the System Stop Time is 100 msec. The eye is the opening framed by the multiple binary bit transition in the overlay temporal display, and is entirely open with no noise.

For the binary rectangular PAM signal in the simple receiver in Figure 2.1, the data is sampled at the middle of the bit time T_b (as indicated by |). Figure 2.59 is the binary rectangular PAM signal with AWGN $\sigma = 1$ V. The normalized signal to noise power ratio (SNR) is 13.98 dB here. No errors are produced in the bit error response (BER), as the binary rectangular PAM signal with AWGN at the input to either the simple or optimum correlation receiver does not cross the symmetrical signal optimum threshold $\tau_{opt} = 0$ V.

Decreasing the SNR to –6.02 dB, AWGN $\sigma = 10$ V, produces the eye diagram shown in Figure 2.60. The binary rectangular PAM signal with AWGN at the input to the correlation receiver in Figure 2.26 does cross the threshold, as shown in Figure 2.60. However, both the eye diagram and the BER measurement of the output of the optimum correlation receiver show no discernable errors. The eye diagram at the output of the correlation receiver here is open, as shown in Figure 2.58.

The eye diagram does not display bit errors with AWGN directly, but does show the effect of channel bandwidth and distortion and jitter. The transmitter of the binary rectangular PAM digital communication system in Figure 2.26 is shown in Figure 2.61, where the communication channel has jitter, and decreased bandwidth and distortion. One output of the binary rectangular PAM transmitter PN Sequence token from the Source Library is inputted to an Adder token, and the other input to the Adder token is a Gaussian Noise token to assess the effect on noise on the eye diagram, as shown in Figures 2.59 and 2.60.

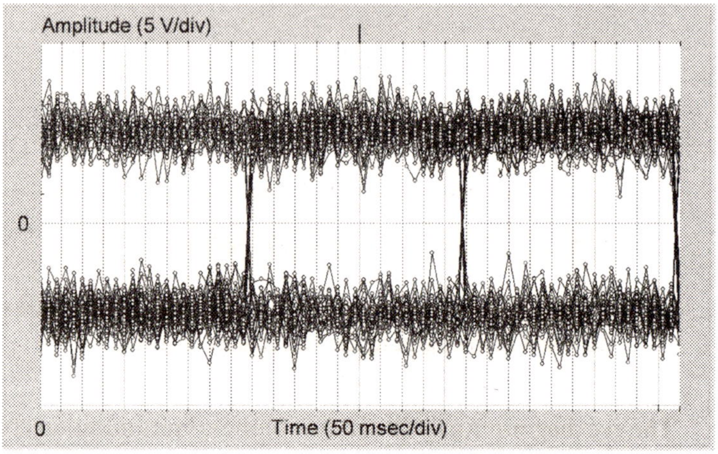

FIGURE 2.59 Eye diagram of the binary rectangular PAM signal with SNR = 13.98 dB at the input to the correlation receiver.

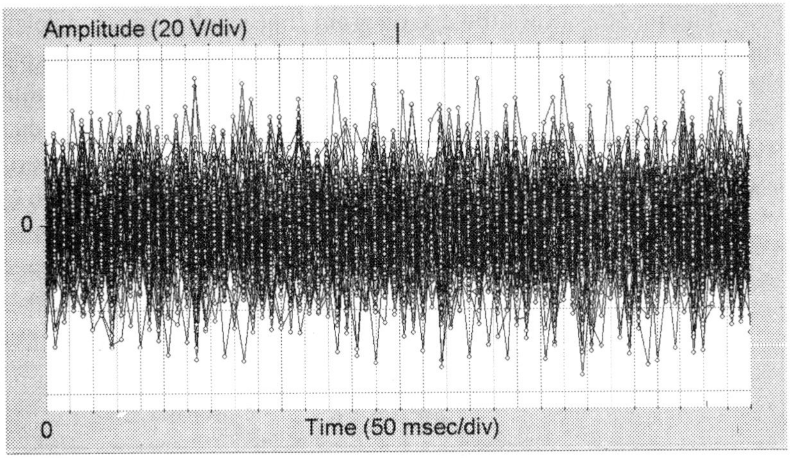

FIGURE 2.60 Eye diagram of the binary rectangular PAM signal with SNR = –6.02 dB at the input to the correlation receiver.

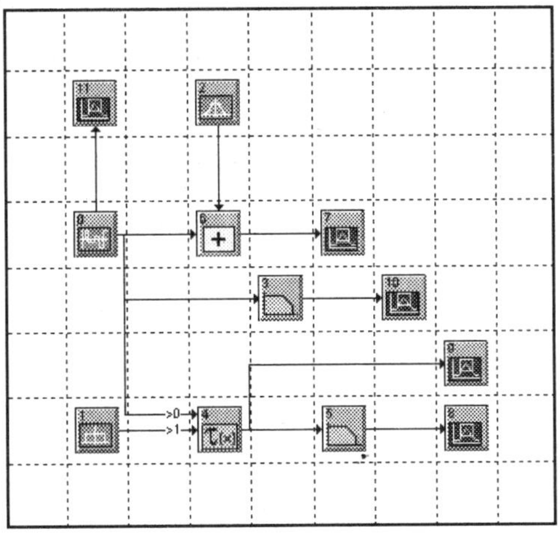

FIGURE 2.61 Test system to display the eye diagram of a binary rectangular PAM signal with AWGN, channel bandwidth and distortion, and jitter.

Another output of the binary rectangular PAM transmitter in Figure 2.61 is inputted to a Linear System Filter token from the Operator Library, which simulates the distortion of a communication channel with limited bandwidth. The Linear System Filter token is selected to be a 9-pole Chebyshev filter, with a cutoff frequency f_{cutoff} = 1.2 kHz ($1.2 \times r_b$), and an inband ripple of 0.1 dB ($\approx$ 1% ripple), as shown in Figure 2.10.

Figure 2.62 shows the eye diagram that results, which displays an amplitude and phase distortion due to a limited channel bandwidth (sample points are not shown as small circles here). The opening of the eye, the optimum data sampling point, is smaller and displaced from the nominal position (as indicated by |) by 50 msec because of the limited bandwidth of the communication channel. With AWGN, the smaller eye opening induces additional bit errors to occur. The compensation for this distortion in a digital communication system is channel equalization, which is described in Chapter 4.

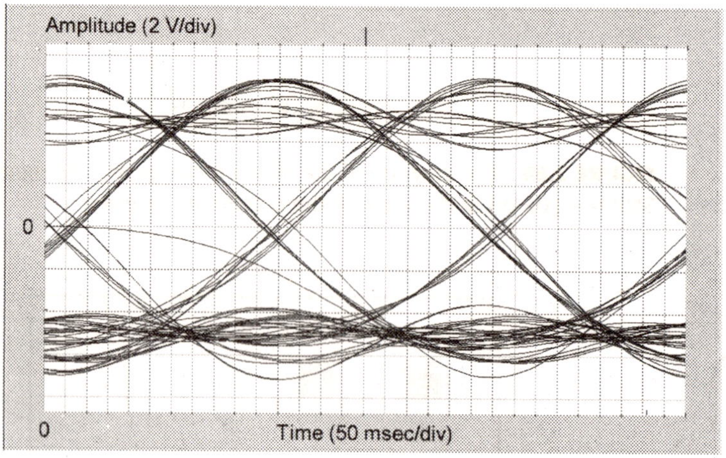

FIGURE 2.62 Eye diagram of the binary rectangular PAM signal with a limited channel bandwidth and no noise.

Another output of the binary rectangular PAM transmitter in Figure 2.61 is inputted to a Variable Delay token from the Operator Library. The control input of the Variable Delay token is derived from a Uniform Noise token from the Source Library with a uniform distributed probability density function (pdf) of amplitudes from 0 to 1 V. The Variable Delay token has parameters of a minimum delay of 0 and a maximum delay of 0.1 msec, which is 10% of the bit time $T_b = 1$ msec here.

Figure 2.63 shows the eye diagram with jitter that results (sample points are again shown as small circles here). The jitter here in the rectangular PAM signal simulates a variable delay in a communication channel, but not the direct variation in bit time T_b. The recovery of the bit rate for optimum sampling in the receiver can be compromised by jitter in either case. The compensation for the jitter is synchronization, which is discussed in Chapter 4.

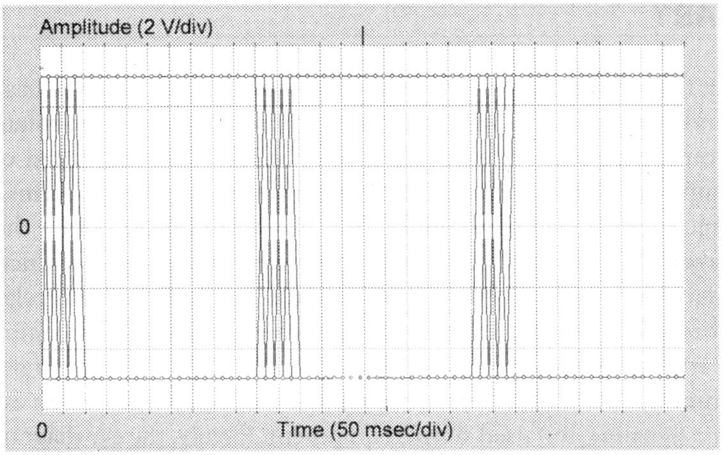

FIGURE 2.63 Eye diagram of the binary rectangular PAM signal with jitter and no noise.

The final output of the binary rectangular PAM transmitter in Figure 2.61 is inputted to a Variable Delay token and a Linear System Filter token as a lowpass filter (LPF), with the same parameters as in Figure 2.62. Figure 2.64 shows the eye diagram with both jitter and a limited channel bandwidth that results (sample points are again not shown as small circles here). The opening of the eye, the optimum data sampling point, is smaller and displaced from the nominal position (as indicated by |), now by 150 msec because of the jitter (100 msec) and limited bandwidth of the communication channel (50 msec).

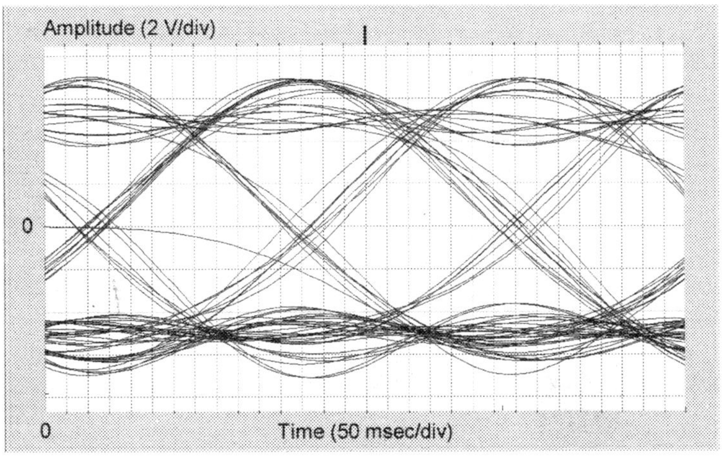

FIGURE 2.64 Eye diagram of the binary rectangular PAM signal with jitter, limited channel bandwidth and no noise.

SUMMARY

In this chapter, SystemVue was used to design and analyze baseband modulation and demodulation systems. Binary pulse amplitude modulation (PAM) digital communication systems, using rectangular, sinc, and raised cosine pulses, were implemented. Multilevel PAM digital communication systems, using rectangular pulses, introduced the concept of source encoding to reduce the probability of bit error (P_b). Bit error performance was analyzed in the presence of additive white Gaussian noise (AWGN) in the simple and optimum correlation receiver. The theoretical *Pb* was determined for each baseband digital communication system. Partial response signaling was shown to attain the maximum rate of signaling in the minimum bandwidth. Delta modulation actually required an excessive bandwidth for signaling, but used simple equipment. Finally, the eye diagram was presented to qualitatively assess the performance of these digital communication systems with AWGN, limited channel bandwidth, and jitter. In the next chapter, bandpass modulation and demodulation digital communication systems, which use a carrier, are presented.

REFERENCES

[Lathi98] Lathi, B.P., *Modern Digital and Analog Communication Systems.* Oxford University Press, 1998.

[Stern04] Stern, Harold, et al., *Communication Systems Analysis and Design.* Prentice Hall, 2004.

3 Bandpass Modulation and Demodulation

In This Chapter

- Optimum Bandpass Receiver: The Correlation Receiver
- Binary Amplitude Shift Keying
- Binary Frequency Shift Keying
- Binary Phase Shift Keying
- Multilevel (M-ary) Amplitude Shift Keying
- Multilevel (M-ary) Frequency Shift Keying
- Multilevel (M-ary) Phase Shift Keying
- Quadrature Amplitude Modulation
- Differential Phase Shift Keying
- Differential Quaternary Phase Shift Keying
- Noncoherent Demodulation of Binary Frequency Shift Keying
- Noncoherent Demodulation of Binary Amplitude Shift Keying
- Constellation Plots

Bandpass modulation techniques encode information as the amplitude, frequency, phase, or phase and amplitude of a sinusoidal carrier. These bandpass modulation schemes are known by their acronyms ASK (amplitude shift keying), FSK (frequency shift keying), PSK (phase shift keying), and QAM (quaternary amplitude modulation), where *keying* or *modulation* is used to indicate that a carrier signal is modified in some manner.

The carrier is a sinusoidal signal that is initially devoid of any information. The purpose of the carrier is to translate essentially a baseband information signal to a frequency and wavelength that can be sent with a guided or propagating electromagnetic (EM) wave.

Bandpass ASK is similar to baseband pulse amplitude modulation (PAM) in Chapter 2, "Baseband Modulation and Demodulation," but FSK, PSK, and QAM are new non-linear modulation techniques. ASK, FSK, and PSK can be readily extended to multiple level (M-ary) signaling and demodulated *coherently* or *noncoherently*. The optimum receiver for bandpass symmetrical or asymmetrical signals is the correlation receiver, which is developed for baseband signals in Chapter 2. Coherent demodulation uses a reference signal with the same frequency and phase as the received signal. Noncoherent demodulation of bandpass signaling may use differential encoding of the information to derive the reference signal in the correlation receiver.

The observed bit error rate (BER) for a single, limited trial in a SystemVue simulation for several bandpass digital communication systems with coherent and noncoherent correlation receivers is compared to the theoretical probability of bit error (P_b). Digital communication systems are subject to performance degradations with additive white Gaussian noise (AWGN). SystemVue simulations of bandpass communication systems are used to investigate the effect upon BER of the performance of the correlation receiver, the reduction in BER with Gray-coding of M-ary data, and binary and quaternary differential signaling.

SystemVue simulations of such bandpass digital communication systems and investigations of their characteristics and performance are provided here. These simulations confirm the theoretical expectation for P_b and are the starting point for the what-ifs of bandpass digital communication system design.

Finally, the constellation plot depicts the demodulated in-phase and quadrature signals of complex modulation schemes in the presence of AWGN. The optimum decision regions are shown, and the observed BER performance of the bandpass digital communication system can be qualitatively assessed.

ON THE CD

The source files for these SystemVue simulations in bandpass modulation and demodulation are located in the Chapter 3 folder on the CD-ROM, and are identified by the figure number (such as Fig3-1.svu). Appendix A includes a complete description of the contents of the CD-ROM.

OPTIMUM BANDPASS RECEIVER: THE CORRELATION RECEIVER

The correlation receiver was shown to be the optimum receiver for baseband signals in Chapter 2. The development of the optimal receiver for coherent bandpass signals now follows a similar path and also results in the correlation receiver. The optimum threshold τ_{opt} and probability of bit error P_b for symmetrical and asymmetrical bandpass signals is found to be the same as that for baseband signals. Although these analytical results are somewhat reassuring, the SystemVue simulation of the modulation and demodulation of the bandpass signals with the resulting BER measurement provides an insight into the process and a verification of the expected performance.

The Correlation Receiver for Bandpass Symmetrical Signals

Demodulation of bandpass digital communication systems can be accomplished either coherently, where the receiver internally produces a reference signal of the same frequency and phase as the input signal, or noncoherently, where the receiver does not use a reference signal, but employs either differential encoding of the information or even simpler means. The optimum coherent receiver for bandpass signals can be shown to be the matched filter, with an impulse response that is given for baseband signals by Equation 2.20 in Chapter 2 [Stern04].

The optimum receiver for bandpass signals can be implemented as a correlation receiver for either symmetrical or asymmetrical signals. The bit time is T_b and symmetrical signals are defined by Equation 3.1.

$$s_1(t) = -s_0(t) \quad (i-1)\,T_b \le t \le i\,T_b \tag{3.1}$$

The bandpass signal $s_1(t)$ represents a binary 1 and $s_0(t)$ represents a binary 0. The signals are defined over one bit time (i is an integer) and initially assumed to be equally likely to occur (equal *apriori* probability). The optimum receiver for bandpass symmetrical signals is the correlation receiver, as shown in Figure 3.1 (see Fig3-1.svu on the CD-ROM).

ON THE CD

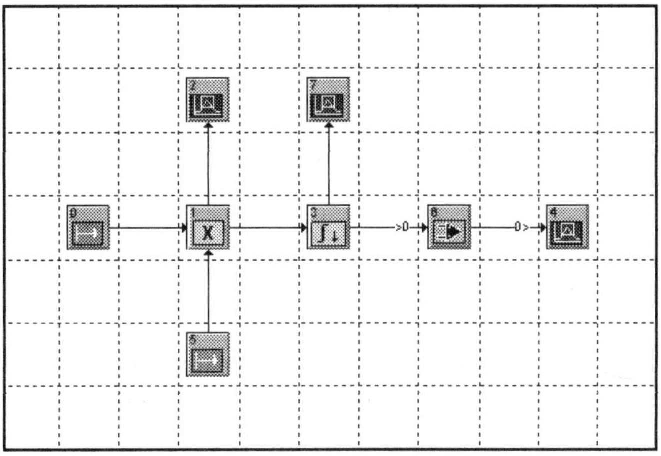

FIGURE 3.1 The matched filter implemented as a correlation receiver for bandpass symmetrical signals.

The received signal $s_j(t)$ ($j = 0, 1$) is first multiplied by the coherent reference signal $s_{ref}(t) = s_1(t)$ and synchronized to the start of the bit time and integrated over that one bit time T_b. The output of the integrate-and-dump process is compared to the optimum threshold τ_{opt} to decide if the received signal is a binary 1 or binary 0 then reset to zero. The output of the integrate-and-dump operation $z_j(iT_b)$ is given by Equation 3.2.

$$z_j(iT_b)=\frac{1}{T_b}\int_{(i-1)T_b}^{iT_b} \gamma\, s_j(\tau)s_1(\tau)\, d\tau \quad (i-1)T_b \le t \le iT_b \quad j=0,1 \tag{3.2}$$

Here it is assumed that the received signal is attenuated by the communication channel by an amount γ and there is no additive noise. The optimum threshold for equally probable binary bandpass signals is given by Equation 3.3.

$$\tau_{opt} = \frac{z_0(iT_b)+z_1(iT_b)}{2} \tag{3.3}$$

This threshold is the same as Equation 2.24 in Chapter 2 for binary baseband signals. Because the signals are symmetrical and in the presence of no noise, the optimum threshold is $\tau_{opt} = 0$ V.

Probability of Bit Error for Bandpass Symmetrical Signals

For additive white Gaussian noise (AWGN), the noise processed by the bandpass optimum receiver also has a Gaussian probability distribution. The bandpass received signal $s_j(t)$ is attenuated by the communication channel by an amount γ and has a normalized *energy per bit* E_b^j given by Equation 3.4, which is the same as Equation 2.25 in Chapter 2 for baseband signals.

$$E_b^j = \int_{(i-1)T_b}^{iT_b} \gamma^2 s_j^2(t)\, dt \quad (i-1)T_b \le t \le iT_b \quad j=0,1 \tag{3.4}$$

Because the signals are symmetrical, from Equation 3.1, the E_b^j are equal ($E_b^0 = E_b^1 = E_b$). The probability of bit error P_b^j for bandpass signals due to either a binary 1 being transmitted and a binary 0 being received (P_b^1), and a binary 0 being transmitted and a binary 1 being received (P_b^0) is given by Equation 3.5, which is the same as Equation 2.26 in Chapter 2 for baseband signals.

$$P_b^j = Q\left(\sqrt{\frac{2E_b}{N_o}}\right) \tag{3.5}$$

$N_o/2$ is the nearly uniform amplitude of the AWGN bi-sided normalized ($R_L = 1$ Ω) power spectral density [Stern04]. The *Q function* is also called the *complementary error function* or *co-error function*. The Q function cannot be evaluated in closed form and is usually presented in tabular form for zero mean μ and unit standard deviation σ Gaussian distributions [Haykin01].

Again as for baseband signals in Chapter 2, the overall probability of bit error P_b includes the apriori probability of a binary 1 (P_1) or binary 0 (P_0) being transmitted, as given by Equation 3.6 for the binary transmission of data ($j = 0, 1$).

$$P_b = \sum_j P_j P_b^j = P_0 P_b^0 + P_1 P_b^1 \tag{3.6}$$

Because the normalized energy per bit are equal ($E_b^0 = E_b^1$), the probabilities of bit error are equal ($P_b^1 = P_b^0$) and, regardless of the value of the apriori probabilities, the overall probability of bit error for bandpass symmetrical signals is given by Equation 3.7. This is the same as Equation 2.28 in Chapter 2 for baseband symmetrical signals.

$$P_b = \mathrm{Q}\left(\sqrt{\frac{2E_b}{N_o}}\right) \tag{3.7}$$

The Correlation Receiver for Bandpass Asymmetrical Signals

The bandpass signals $s_0(t)$ and $s_1(t)$ that represent a binary 0 and binary 1 need not be symmetrical. If the binary bandpass signals remain equally probable, then Equation 3.3 still determines the optimum threshold value. If the bandpass signals are not equally probable ($P_1 \neq P_0$), then the optimum threshold is given by Equation 3.8, which is the same as Equation 2.29 in Chapter 2 for baseband asymmetrical signals.

$$\tau_{opt} = \frac{2\sigma_o^2 \ln\dfrac{P_1}{P_0} + z_0(iT_b)^2 - z_1(iT_b)^2}{2\left[z_0(iT_b) - z_1(iT_b)\right]} \tag{3.8}$$

If $P_1 = P_0$, then the optimum threshold τ_{opt} in Equation 3.8 reduces to Equation 3.3. The output of the correlation receiver $z_j(iT_b)$ ($j = 0, 1$) is assumed to have no additive noise. The additive Gaussian noise is zero mean with a variance of σ^2, but σ_o^2 is the variance of the noise processed by the integrate-and-dump implementation of the matched filter as a correlation receiver. The Gaussian noise variance σ_o^2 can be determined by the test system shown in Figure 2.27 in Chapter 2.

The apriori probabilities P_1 and P_0 in Equation 3.8 for the optimum threshold τ_{opt} are not necessarily known exactly or can change with the source of information.

A test system can be devised to count the number of binary 1s and binary 0s in an arbitrary data set, as shown in Figure 2.30 in Chapter 2.

If the transmitted bandpass signals are not symmetrical, the optimum coherent receiver is a modified correlation receiver, as shown in Figure 3.2 (see Fig3-2.svu on the CD-ROM). The received signal $s_j(t)$ ($j = 0, 1$) is processed by two multipliers, with inputs that are the two reference signal $s_1(t)$ and $s_0(t)$ synchronized to the start of the bit time. The outputs of the multipliers are then separately integrated over that one bit time T_b by two Integrate and Dump tokens from the Communications Library synchronized to the input signal.

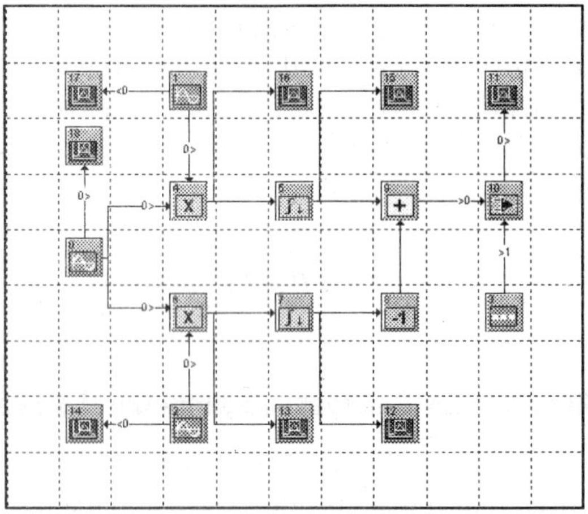

FIGURE 3.2 The matched filter implemented as a correlation receiver for bandpass asymmetrical signals.

The outputs of the two integrate-and-dump operations $z_j(iT_b)$ ($j = 0, 1$) are given by Equation 3.2. The Negate token from the Operator Library negates the output of one of the integrate-and-dump operations and the Adder token then effectively subtracts them. For equally probable binary asymmetrical signals, the optimum threshold τ_{opt} is given by Equation 3.3. The threshold is set by the Custom token from the Source Library, which provides the Analog Comparator token comparison input from one output with an algebraic simulation equation that is $p(0) = \tau_{opt}$ V.

The correlation receiver for asymmetrical bandpass signals can also be implemented by an alternative representation, as shown in Figure 3.3 (see Fig3-3.svu on the CD-ROM). This configuration is similar to the correlation receiver for symmetrical bandpass signals, as shown in Figure 3.1. The reference signal $s_{ref}(t)$ here is the difference signal, as given by Equation 3.9.

$$s_{ref}(t) = s_1(t) - s_0(t) \tag{3.9}$$

The reference signal is synchronized to the start of the bit time, multiplied by the input signal, and then integrated over that one bit time T_b. The threshold is again set by the Custom token from the Source Library, as in Figure 3.2.

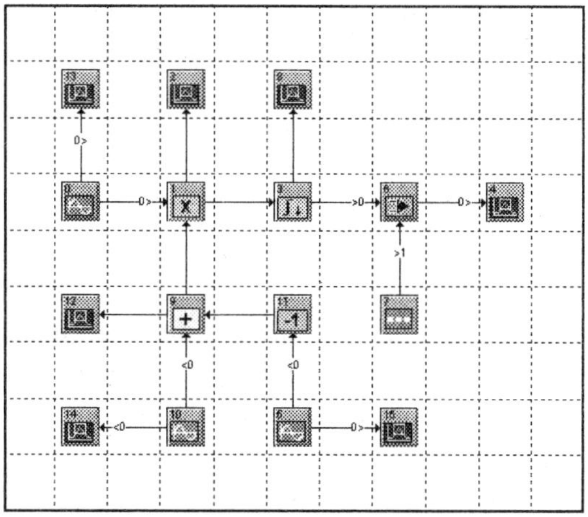

FIGURE 3.3 An alternative structure for the correlation receiver for bandpass asymmetrical signals.

Probability of Bit Error for Bandpass Asymmetrical Signals

For AWGN, the noise processed by the optimum receiver has a Gaussian probability distribution. The asymmetrical received signal $s_j(t)$ is attenuated by the communication channel by an amount γ and can be considered here to have a normalized energy difference per bit E_d given by Equation 3.10, which is the same as Equation 2.31 in Chapter 2 for baseband signals.

$$E_d = \int_{(i-1)T_b}^{iT_b} \left\{\gamma\left[s_1(t) - s_0(t)\right]\right\}^2 dt \quad (i-1)T_b \le t \le iT_b \tag{3.10}$$

The probability of bit error P_b for bandpass asymmetrical signals due to either a binary 1 being transmitted and a binary 0 being received or a binary 0 being transmitted and a binary 1 being received, with equally likely apriori probability of binary 1 or binary 0 being transmitted ($P_0 = P_1 = 0.5$) is given by Equation 3.11, which is the same as Equation 2.32 in Chapter 2 for baseband asymmetrical signals.

$$P_b = Q\left(\sqrt{\frac{E_d}{2\,N_o}}\right) \tag{3.11}$$

If the binary bandpass signals are symmetrical, then $E_d = 4E_b$ and the probability of bit error P_b in Equation 3.11 reduces to Equation 3.7. If the apriori probabilities P_0 and P_1 are not equal, then the optimum threshold τ_{opt} is given by Equation 3.8 and the probability of bit error is given by Equation 3.12.

$$P_b = P_1\,Q\left(\frac{z_1(iT_b) - \tau_{opt}}{\sigma_o}\right) + P_0\,Q\left(\frac{\tau_{opt} - z_0(iT_b)}{\sigma_o}\right) \tag{3.12}$$

The standard deviation σ_o of the Gaussian noise source processed by the integrate-and-dump operation is used in Equation 3.12, rather than the AWGN power spectral density $N_o/2$, because a voltage noise margin is used here. The signal $z_j(t)$ is the output of the correlation receiver when there is no additive noise, given by Equation 3.2. Equation 3.12 is an extension of Equation 3.6 for the overall probability of bit error P_b, where the Q function is the complementary error function.

If the apriori probabilities are equal ($P_0 = P_1 = 0.5$), the optimum threshold τ_{opt} is given by Equation 3.3, and the probability of bit error from Equation 3.12 is given by Equation 3.13.

$$P_b = Q\left(\frac{z_1(iT_b) - z_0(iT_b)}{2\,\sigma_o}\right) \tag{3.13}$$

The probability of bit error P_b for binary bandpass asymmetrical signals in Equation 3.13 is equivalent to that given by Equation 3.11.

BINARY AMPLITUDE SHIFT KEYING

Binary amplitude shift keying (ASK) keys the carrier frequency as one of two discrete levels during the bit time T_b for the representation of binary logic signals for the transmission of information. A convenient set of amplitudes for binary ASK is 1 V for binary 1 and 0 V for binary 0, which is also known as on-off keying (OOK). The modulated sinusoidal carrier signal has an amplitude of A V, a carrier frequency of f_c Hz, and a 0° reference phase angle, as given by the analytical expression in Equation 3.14.

$$s_j(t) = A\,m_j(t)\sin(2\pi f_c t) \quad (i-1)T_b \le t \le iT_b \quad j = 0, 1 \tag{3.14}$$

The information signal or data source is $m_j(t)$ ($j = 0, 1$) and for on-off keying $m_j(t) = 0$ V or 1 V for one bit time T_b.

SystemVue Simulation of Binary ASK

ON THE CD

A binary amplitude shift keying (ASK) coherent digital communication system with AWGN is shown in Figure 3.4 (see Fig3-4.svu on the CD-ROM). To facilitate the simulation in SystemVue, all of the Analysis tokens, except that for the BER token output as the Final Value token, are deleted in the simulation model Fig3-4DT. As described in Chapter 1, "Communication Simulation Techniques," the SystemVue Textbook Edition does not permit tokens to be deleted.

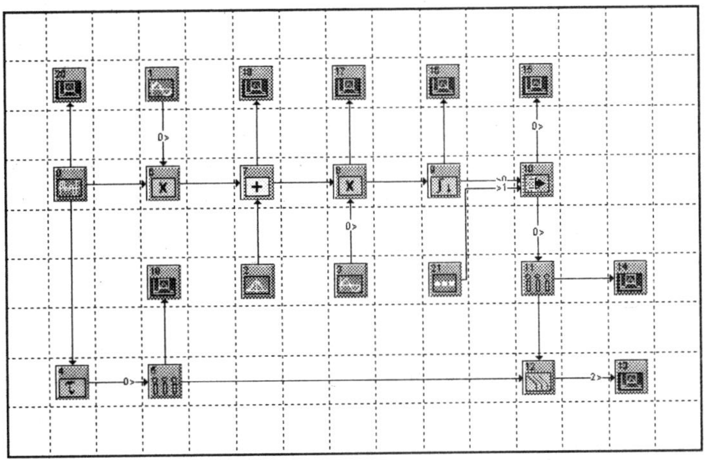

FIGURE 3.4 A binary ASK (OOK) coherent digital communication system with BER analysis.

The data source is the polar pseudonoise PN Sequence token from the Source Library, with parameters of a binary amplitude $A = \pm 0.5$ V, a bit rate $r_b = 1$ kb/sec, and 0.5 V voltage and 0° phase offset. The bit rate is the same as that used for baseband modulation and demodulation systems in Chapter 2, which facilitates the comparison of power spectral densities. This PN Sequence token will transmit unipolar rectangular pulses for $m_j(t)$ with an amplitude of either 1 V or 0 V.

The carrier generator is the Sinusoid token from the Source Library, with parameters of an amplitude $A = 5$ V peak, a carrier frequency $f_c = 20$ kHz, and 0° phase offset.

The non-linear binary ASK modulator is the Multiplier token.

An ASK signal is asymmetrical, because the bandpass signals $s_i(t)$ ($i = 0, 1$) do not satisfy Equation 3.1, and for on-off keying are given by Equation 3.15.

$$
\begin{aligned}
s_0(t) &= 0 && (i-1)T_b \le t \le iT_b \quad b_i = 0 \\
s_1(t) &= 5\sin(2\pi f_c t) && (i-1)T_b \le t \le iT_b \quad b_i = 1 \\
s_j(t) &= s_{PAM}(t)\sin(2\pi f_c t) && j = 0, 1
\end{aligned}
\tag{3.15}
$$

The ASK signal can be decomposed into a baseband PAM signal $s_{PAM}(t)$ and the sinusoidal carrier. The carrier frequency f_c here is 20 kHz, the information bit is b_i, and the bit time is T_b. The optimum receiver for ASK is the correlation receiver for asymmetrical signals, as described in Chapter 2 and shown in Figure 2.31 for binary baseband PAM, but modified for bandpass demodulation, as shown in Figure 3.4.

The communication channel is represented by the Adder token and the Gaussian Noise token from the Source Library, with initial parameters of a standard deviation $\sigma = 0$ V, and $\mu = 0$ V mean, or a power spectral density $N_o = 0$ W/Hz.

The Multiplier token multiplies the input signal $s_j(t)$ with the coherent reference signal $s_{ref}(t)$, the difference signal $s_1(t) - s_0(t)$, as given by Equations 3.9 and 3.14. The reference signal then is the Sinusoid token from the Source Library, with parameters of an amplitude $A = 5$ V peak, a frequency $f_c = 20$ kHz, and 0° phase offset. The output of the Multiplier token is processed by the Integrate and Dump token from the Communications Library synchronized to the input signal. The parameters of the Integrate and Dump token are specified as an integration time equal to the bit time T_b of 1 msec and a continuous output.

The Analog Comparator token comparison input is set to the optimum threshold τ_{opt} given by Equation 3.3, because the binary ASK signals, though asymmetrical, are assumed equally probable. The optimum threshold τ_{opt} is 6.25 V here, because $z_0(iT_b) = 0$ and $z_1(iT_b) = 12.5$ V, as given by Equation 3.2. The output of the integrate-and-dump process can also be observed in the Analysis window. The Custom token from the Source Library provides the constant comparator input, with parameters of one output with an algebraic simulation equation that is $p(0) = 6.25$.

The System Time: Sample Rate is set to 500 kHz, well above the nominal 1 kb/sec bit rate and the 20 kHz carrier frequency. Two Sampler tokens are used to downsample the 500 kHz System Time rate to the 1 kb/sec bit rate or 1 kHz. The Sampler token parameters are set to a sample rate of 1 kHz, and an aperture time of 2 μsec (T_{system}, a single point sample).

The PN Sequence token data Sampler input is delayed by 1 msec to correlate the data. The value of this delay can be observed by the cross-correlation of the two sampler signals at the input to the BER token, as described in Chapter 1. The BER token parameters are set to output on every trial (number of trials = 1), with a threshold of 0.5 V for binary ASK, and a 1 msec offset.

The BER Analysis window displays the cumulative average error, but the BER averaging is performed over 1 bit and the output is the total observed errors. The two Analysis windows, which display the two comparison inputs to the BER token, are to ensure that the error rate determination is proceeding correctly.

The System Stop Time is set to 10 sec for convenience, which would process 10,000 information bits in possible error. The Gaussian Noise token is specified by the power spectral density (PSD) N_o W/Hz. Setting the Gaussian noise source PSD to 0 should produce 0 observed errors.

Binary ASK Power Spectral Density

The magnitude of the frequency domain representation $P_{BASK\ (OOK)}(f)$ (derived from the Fourier transform) of a binary ASK (OOK) signal of width T_b seconds, peak amplitude A volts, and carrier frequency f_c Hz, is given by Equation 3.16.

$$P_{BASK\ (OOK)}(f) = \frac{1}{2}\left[\frac{AT_b}{2}\ \mathrm{sinc}\left(2\pi\left(T_b/2\right)\left(f-f_c\right)\right)+\frac{A}{2}\delta\left(f-f_c\right)\right] - \frac{1}{2}\left[\frac{AT_b}{2}\ \mathrm{sinc}\left(2\pi\left(T_b/2\right)\left(f+f_c\right)\right)-\frac{A}{2}\delta\left(f+f_c\right)\right] \quad (3.16)$$

This is the bi-sided voltage spectrum, extending from all negative to all positive frequencies ($-\infty < f < +\infty$). The analysis for the voltage spectrum utilizes that for the rectangular PAM pulse, as given by Equation 2.1. The *unipolar* PAM pulse $s_{PAM}(t)$ in Equation 3.15 has an amplitude 0 V and $A = 5$ V, but can be decomposed into a *polar* PAM pulse of $\pm A/2 = \pm 2.5$ V and a constant (DC) level of $A/2 = 2.5$ V in Equation 3.16.

The polar PAM pulse is represented by the sinc ($\sin x/x$) function and the constant term by the impulse ($\delta(f)$) in the frequency domain. The *modulation property* of the Fourier transform shifts the sinc and impulse representations in the frequency domain by replacing the argument f by $f - f_c$ and $f + f_c$ and scaling by 1/2 in Equation 3.16.

The bi-sided energy spectral density $E_{BASK\ (OOK)}(f)$ of n pseudonoise binary ASK (OOK) signals is n times the square of the magnitude of the voltage spectrum divided by the load resistance R_L, as given by Equation 3.17.

$$E_{BASK\ (OOK)}(f) = \frac{n}{4R_L}\left[\frac{A^2T_b^2}{4}\ \mathrm{sinc}^2\left(2\pi\left(T_b/2\right)\left(f-f_c\right)\right)+\frac{A^2T_b}{4}\delta\left(f-f_c\right)\right] + \frac{n}{4R_L}\left[\frac{A^2T_b^2}{4}\ \mathrm{sinc}^2\left(2\pi\left(T_b/2\right)\left(f+f_c\right)\right)+\frac{A^2T_b}{4}\delta\left(f+f_c\right)\right] \quad (3.17)$$

The bi-sided power spectral density (PSD) $PSD_{BASK\,(OOK)}(f)$ is the energy spectral density divided by the time to transmit the n binary ASK (OOK) signals or nT_b, as given by Equation 3.18.

$$PSD_{BASK\,(OOK)}(f) = \frac{1}{4R_L}\left[\frac{A^2T_b}{4}\,\text{sinc}^2\left(2\pi\left(T_b/2\right)\left(f-f_c\right)\right)+\frac{A^2}{4}\,\delta\left(f-f_c\right)\right] + \frac{1}{4R_L}\left[\frac{A^2T_b}{4}\,\text{sinc}^2\left(2\pi\left(T_b/2\right)\left(f+f_c\right)\right)+\frac{A^2}{4}\,\delta\left(f+f_c\right)\right] \tag{3.18}$$

The impulse terms are discrete components and not affected by the PSD transformation, the energy spectral density averaged over multiple bit times nT_b.

Setting the Gaussian noise PSD $N_o = 0$ W/Hz and increasing the System Time: Number of Samples specification to 524 288 (2^{19}) points results in a spectral resolution of approximately 0.95 Hz. The normalized ($R_L = 1\ \Omega$) PSD of the Adder token channel output without AWGN is shown unscaled in Figure 3.5 and as a single-sided spectrum in Figure 3.6 in dBm/Hz (decibels referenced to 1 milliwatt per Hertz). The maximum frequency displayed is 250 kHz, the Nyquist frequency or half of the system simulation sampling rate (500 kHz/2). The carrier component is shown as a spectral impulse ($\delta(f)$) at $f_c = 20$ kHz, as given by Equation 3.18 and shown in Figure 3.5. However details of the PSD are difficult to discern.

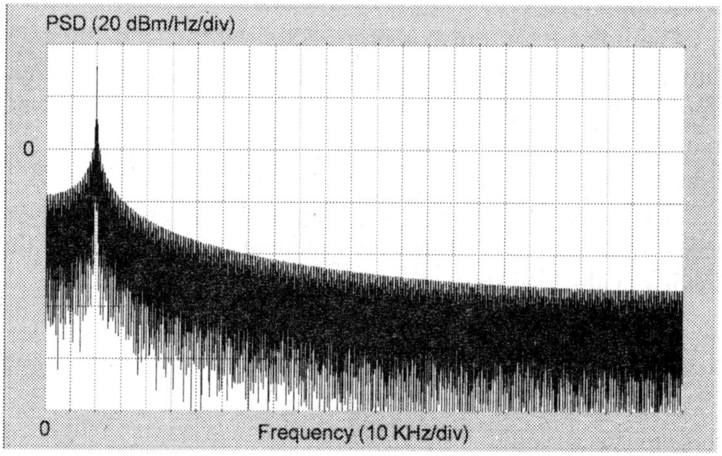

FIGURE 3.5 Power spectral density of a 1 kb/sec pseudonoise binary ASK (OOK) signal.

The normalized PSD can be scaled to resolve greater spectral detail, as shown in Figure 3.6. Because the rectangular PAM pulse width τ is the bit time T_b, there are approximate nulls in the PSD that occur at the bit rate $r_b = 1/T_b$ (1 kb/sec or 1 kHz) and multiples of the bit rate centered about the carrier frequency $f_c = 20$ kHz due to the sinc^2 term in Equation 3.18. The PSD is not smooth because, although there are a large number of data, the ensemble is finite, but the spectral envelope clearly demonstrates the sinc^2 behavior.

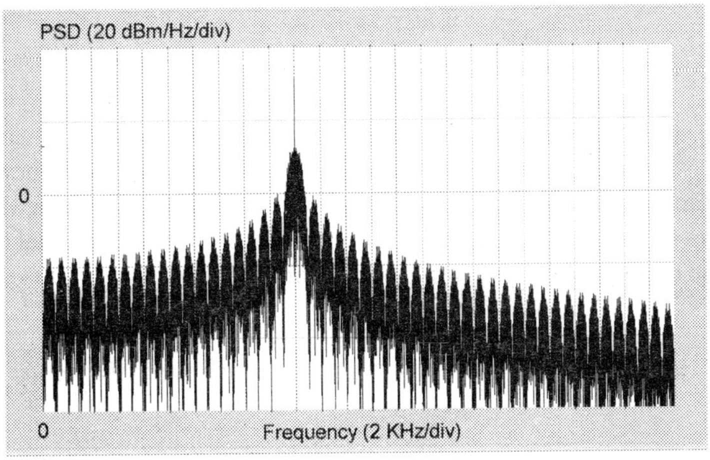

FIGURE 3.6 Scaled power spectral density of a 1 kb/sec pseudonoise binary ASK (OOK) signal.

The PSD of an ASK signal is often referred to as a *double sideband spectrum* (DSB), with an *upper sideband* (USB) and *lower sideband* (LSB) centered about the carrier, as shown in Figure 3.6. Because the bandwidth of a bandpass ASK signal is twice that of the baseband PAM signal, the transmission bandwidth to the first nulls in the PSD of data at a rate r_b b/sec requires $2r_b$ Hz. The theoretical minimum for the binary bandpass transmission of data at a rate r_b b/sec is a bandwidth of r_b Hz as postulated by Nyquist [Stern04].

The carrier frequency is deliberately chosen to be relatively low to facilitate the SystemVue simulation, but this causes *aliasing* to occur. The amplitude of the PSD centered about the carrier frequency should be symmetrical, as given by Equation 3.18. However, the amplitude of the PSD below the carrier frequency is elevated due to this aliasing. The alias affect, although noticeable, can be alleviated to a degree by a bandpass filter centered at the carrier frequency and with a bandwidth sufficient to pass the information.

The bandwidth of the binary ASK signal is somewhat problematical, as it is for binary PAM in Chapter 2. Defining the bandwidth as a percentage of the total power of the binary ASK signal again is the usual course, because all of the available power would require an infinite bandwidth. The total power is easily calculated in the temporal domain. The power P in the binary ASK signal per bit time T_b, derived from Equation 3.14 is given by Equation 3.19.

$$
\begin{aligned}
P &= \frac{A^2}{T_b R_L} \int_{(i-1)T_b}^{iT_b} m_j^2(t) \sin^2(2\pi f_c t)\,dt \quad (i-1)T_b \le t \le iT_b \\
P &= \frac{A^2}{2R_L} \quad \text{if } m_j(t) = 1 \\
P &= 0 \quad \text{if } m_j(t) = 0
\end{aligned}
\tag{3.19}
$$

Equation 3.19 assumes that the ratio of the carrier frequency to the bit rate f_c/r_b is an integer. The integration then over one bit time T_b is for a whole number of cycles of the carrier sinusoid. If the apriori probabilities are equal ($P_0 = P_1 = 0.5$) then the total power P_T is given by Equation 3.20.

$$P_T = \frac{A^2}{2R_L} P_1 + 0\,P_0 = \frac{A^2}{4R_L} \tag{3.20}$$

The binary ASK signal has an amplitude $A = 5$ V and the normalized ($R_L = 1\ \Omega$) power is 6.25 W = 37.95 dBm ($10 \log_{10}[6250]$). Numerically integrating the power spectral density, Equation 3.18, from the carrier frequency to an arbitrary, but symmetrical frequency on either side of the carrier frequency and comparing the resulting power to the expected total power provides the data in Table 3.1. Table 3.1 for binary bandpass ASK is similar to Table 2.1 for binary baseband rectangular PAM. For the same percentage of the total power the bandwidth of the bandpass ASK signal is double that for the baseband PAM signal.

TABLE 3.1 Bandwidth of a Binary ASK (OOK) Signal as a Percentage of the Total Power

Bandwidth (Hz)	Percentage of Total Power
$2/T_b$	90%
$3/T_b$	93%
$4/T_b$	95%
$6/T_b$	96.5%
$8/T_b$	97.5%
$10/T_b$	98%

Performance of Binary ASK for the Optimum Receiver in AWGN

The binary ASK (OOK) transmitter and optimum bandpass receiver in Figure 3.4 can be evaluated for its performance in the presence of AWGN. A bandpass BER analysis is conducted with the same parameters as the baseband BER analysis in Chapter 2. The System Stop Time is set to 10 sec for convenience, which would process 10 000 information bits in possible error. However, the 500 kHz System Time rate results in 5×10^6 simulation sample points.

To facilitate the simulation in SystemVue, all of the Analysis tokens, except that for the BER token are deleted from the binary ASK (OOK) digital communication system with BER analysis in Figure 3.4, as shown in Figure 3.7. The Final Value token from the Sink Library records the total number of errors in 10 000 trials. Figure 3.7 is the same as the simulation model Fig3-4DT. As described in Chapter 1 the SystemVue Textbook Edition does not permit tokens to be deleted.

ON THE CD

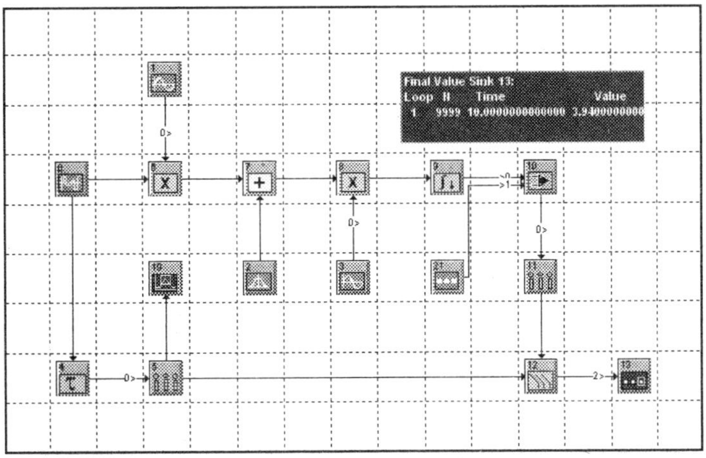

FIGURE 3.7 A binary ASK (OOK) digital communication system with BER analysis recorded by the Final Value token.

Table 3.2 is a tabular list of the observed BER in a single trial of 10 000 information bits as a function of E_d/N_o to the theoretical P_b for binary ASK (OOK), as given by Equation 3.11 for asymmetrical signals. Binary ASK does not satisfy Equation 3.1 and is asymmetric. The optimum receiver is implemented as the correlation receiver for asymmetrical signals, as shown in Figure 3.3.

The normalized energy difference per bit here is $E_d \approx 1.25 \times 10^{-2}$ V^2-sec, as given by Equations 3.10 and 3.15. The attenuation of the communication channel is taken to be zero ($\gamma = 1$). Binary ASK (OOK) provides the largest energy difference per bit E_d for a given power in the sinusoidal carrier and therefore the lowest P_b for any binary ASK. A comparison of Table 3.2 and Table 2.8 in Chapter 2 shows the same BER performance for binary ASK (OOK) using the bandpass optimum receiver as that for rectangular PAM using the baseband optimum receiver.

TABLE 3.2 Observed BER and Theoretical P_b as a Function of E_d/N_o in a Binary ASK (OOK) Digital Communication System with Optimum Receiver

E_d/N_o dB	N_o V2-sec	BER	P_b
∞	0	0	0
12	7.89×10^{-4}	1.9×10^{-3}	2.5×10^{-3}
10	1.25×10^{-3}	1.23×10^{-2}	1.25×10^{-2}
8	1.98×10^{-3}	3.61×10^{-2}	3.75×10^{-2}
6	3.14×10^{-3}	8.15×10^{-2}	7.93×10^{-2}
4	4.98×10^{-3}	1.325×10^{-1}	1.314×10^{-1}
2	7.89×10^{-3}	1.884 v 10^{-1}	1.872×10^{-1}
0	1.25×10^{-2}	2.387×10^{-1}	2.393×10^{-1}

BINARY FREQUENCY SHIFT KEYING

Binary frequency shift keying (FSK) shifts the carrier frequency to one of two discrete frequencies during the bit time T_b for the representation of binary logic signals for the transmission of information. The modulated sinusoidal carrier signal has an amplitude of A V, a frequency of f_c Hz, and a 0° reference phase angle, as given by the analytical expression in Equation 3.21.

$$s_j(t) = A\sin(2\pi(f_c + k_f m_j(t))\,t) \quad (i-1)T_b \le t \le iT_b \quad j = 0,1 \tag{3.21}$$

The information signal or data source is $m_j(t)$ ($j = 0$, 1) and for binary FSK $m_j(t) = \pm 1$ V for one bit time T_b. The factor k_f, with units that are Hz/V, is the frequency deviation factor (or the modulation gain) and the frequency deviation Δf is given by Equation 3.22.

$$|\Delta f| = |k_f m_j(t)| \tag{3.22}$$

Because $m_j(t) = \pm 1$ V the magnitude of the frequency deviation Δf is equal on either side of the carrier frequency f_c.

SystemVue Simulation of Binary FSK

A binary frequency shift keying (FSK) coherent digital communication system with AWGN is shown in Figure 3.8 (see Fig3-8.svu on the CD-ROM). To facilitate the simulation in SystemVue, all of the Analysis tokens, except that for the BER token output as the Final Value token, are deleted in the simulation model Fig3-8DT. As described in Chapter 1 the SystemVue Textbook Edition does not permit tokens to be deleted.

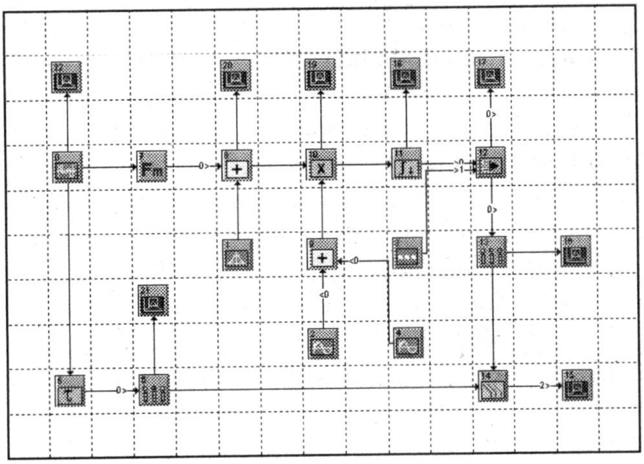

FIGURE 3.8 A binary FSK coherent digital communication system with BER analysis.

The SystemVue simulation is similar to that of the binary ASK communication system in Figure 3.4. The data source is the polar pseudonoise PN Sequence token from the Source Library, with parameters of a binary amplitude $A = \pm 1$ V, a bit rate $r_b = 1$ kb/sec, and 0 V voltage and 0° phase offset. The data rate is the same as that used for bandpass ASK and the baseband modulation and demodulation systems in Chapter 2, which facilitates the comparison of power spectral densities. This PN Sequence token will transmit polar rectangular pulses for $m_j(t)$ with an amplitude $A = \pm 1$ V.

The non-linear FSK modulator is the Frequency Modulator token from the Function Library, with parameters of an amplitude $A = 5$ V, a frequency $f_c = 20$ kHz, 0° phase offset, and an initial modulation gain (or frequency deviation factor k_f) of 2000 Hz/V. The resulting initial frequency deviation Δf here is ± 1 V × 2000 = ± 2 kHz, as given by Equation 3.22.

A binary FSK signal is asymmetrical, because the bandpass signals $s_j(t)$ ($j = 0$, 1) do not satisfy Equation 3.1 and are given by Equation 3.23.

$$\begin{aligned} s_0(t) &= 5\sin(2\pi(f_c - \Delta f)t) \quad (i-1)T_b \le t \le iT_b \quad b_i = 0 \\ s_1(t) &= 5\sin(2\pi(f_c + \Delta f)t) \quad (i-1)T_b \le t \le iT_b \quad b_i = 1 \end{aligned} \tag{3.23}$$

The carrier frequency f_c here is 20 kHz, the frequency deviation Δf initially is 2 kHz, the information bit is b_i and the bit time is T_b. The optimum receiver for FSK is the correlation receiver for asymmetrical signals, as described in Chapter 2 and shown in Figure 2.31 for binary baseband PAM, but modified for passband demodulation, as shown in Figure 3.8.

The communication channel is represented by the Adder token and the Gaussian Noise token from the Source Library, with initial parameters of a standard deviation $\sigma = 0$ V and $\mu = 0$ V mean or a power spectral density $N_o = 0$ W/Hz.

The Multiplier token multiplies the input signal with the coherent reference signal $s_{ref}(t)$, the difference signal $s_1(t) - s_0(t)$, as given by Equations 3.9 and 3.23. The reference signal then is derived from two Sinusoid tokens from the Source Library, with parameters of an amplitude of 5 V peak, frequencies of $f_c \pm \Delta f$ Hz, and phase offsets of 0° and 180° (to provide the difference). The two sinusoids are then summed by the Adder token to provide the coherent reference signal $s_{ref}(t)$.

The output of the Multiplier token is processed by the Integrate and Dump token from the Communications Library synchronized to the input signal. The parameters of the Integrate and Dump token are specified as an integration time equal to the bit time T_b of 1 msec and a continuous output.

The Analog Comparator token comparison input is set to the optimum threshold τ_{opt} given by Equation 3.3, because the binary FSK signals, although asymmetrical, are assumed equally probable. The optimum threshold τ_{opt} is 0 V here, because $z_0(iT_b) = -12.5$ V and $z_1(iT_b) = 12.5$ V, as given by Equation 3.2. The Custom token from the Source Library provides the constant comparator input, with parameters of one output with an algebraic simulation equation that is $p(0) = 0$.

The System Time: Sample Rate is set to 500 kHz. Two Sampler tokens are used to downsample the 500 kHz System Time rate to the 1 kb/sec bit rate or 1 kHz. The Sampler token parameters are set to a sample rate of 1 kHz, and an aperture time of 2 µsec (T_{system}, a single point sample).

The PN Sequence token data Sampler input is delayed by 1 msec to correlate the data. The BER token parameters are set to output on every trial (number of trials = 1), with a threshold of 0.5 V for binary FSK, and a 1 msec offset. The System Stop Time is set to 10 sec for convenience, which would process 10 000 information bits in possible error. The Gaussian Noise token is specified by the power spectral density (PSD) N_o W/Hz. Setting the Gaussian noise source PSD to 0 should produce 0 observed errors.

Binary FSK Power Spectral Density

The power spectral density (PSD) $PSD_{BFSK}(f)$ of a binary FSK signal of width T_b seconds, peak amplitude A volts, and carrier frequency f_c Hz is somewhat difficult to develop. An approximate method is to decompose the binary FSK signal as a sum of two ASK signals with two data sources, as given by Equation 3.24.

$$s_j(t) = A\, m_{1,j}(t)\ \sin(2\pi(f_c + \Delta f)\,t) + A\, m_{2,j}(t)\ \sin(2\pi(f_c - \Delta f)\,t) \qquad (i-1)T_b \le t \le iT_b \quad j = 0, 1 \tag{3.24}$$

The data sources or information signals $m_{1,j}(t)$ and $m_{2,j}(t)$ ($j = 0, 1$) are given in Equation 3.25.

$$\begin{aligned} m_{1,j}(t) &= 1 \quad (i-1)T_b \le t \le iT_b \quad b_i = 1 \\ m_{1,j}(t) &= 0 \quad (i-1)T_b \le t \le iT_b \quad b_i = 0 \\ \\ m_{2,j}(t) &= 1 \quad (i-1)T_b \le t \le iT_b \quad b_i = 0 \\ m_{2,j}(t) &= 0 \quad (i-1)T_b \le t \le iT_b \quad b_i = 1 \end{aligned} \tag{3.25}$$

The decomposed FSK signal is then two ASK signals with carrier frequencies of $f_c \pm \Delta f$ and the pulse modulation signals $s_{PAM}(t) = m_{k,j}(t)$ ($k, j = 0, 1$), as given by Equation 3.15. The bi-sided PSD for a binary FSK signal $PSD_{BFSK}(f)$ can be approximated as two PSDs from the two ASK signals from Equation 3.18, as given by Equation 3.26.

$$\begin{aligned} &PSD_{BFSK}(f) = \\ &\frac{A^2T_b}{16R_L}\text{sinc}^2\big(2\pi(T_b/2)(f - f_c - \Delta f)\big) + \frac{A^2}{16R_L}\delta(f - f_c - \Delta f) + \\ &\frac{A^2T_b}{16R_L}\text{sinc}^2\big(2\pi(T_b/2)(f - f_c + \Delta f)\big) + \frac{A^2}{16R_L}\delta(f - f_c + \Delta f) + \\ &\frac{A^2T_b}{16R_L}\text{sinc}^2\big(2\pi(T_b/2)(f + f_c + \Delta f)\big) + \frac{A^2}{16R_L}\delta(f + f_c + \Delta f) + \\ &\frac{A^2T_b}{16R_L}\text{sinc}^2\big(2\pi(T_b/2)(f + f_c - \Delta f)\big) + \frac{A^2}{16R_L}\delta(f + f_c - \Delta f) \end{aligned} \tag{3.26}$$

This approach is a useful approximation, but is not rigorous. A rigorous analysis for the binary FSK PSD would require the addition of the voltage spectral density using complex numbers prior to squaring to obtain the energy and power spectral densities. However, the approximation for the binary FSK PSD in Equation 3.26 is adequate for the analysis here [Lathi98].

The minimum frequency deviation Δf is $1/2T_b = r_b/2$ Hz. With this value of Δf the peak of the PSD of each of the decomposed ASK signals in binary FSK would be at a null of the PSD of the other signal, as given by Equation 3.26. FSK using this value of Δf is known as *minimum frequency shift keying* (MFSK).

In the temporal domain, binary FSK signals with $\Delta f = nr_b/2$ (multiples of the minimum frequency deviation, n is an integer) are *orthogonal* to each other over one bit time T_b. An orthogonal signal $o(t)$ is described by Equation 3.27, which is similar to the correlation receiver process in Equation 3.2.

$$o(t) = \int_{(i-1)T_b}^{iT_b} s_j(t)\, s_k(t)\, \mathrm{d}t = 0 \quad j \neq k \quad (i-1)T_b \leq t \leq iT_b$$
$$o(t) = \int_{(i-1)T_b}^{iT_b} s_j(t)\, s_k(t)\, \mathrm{d}t = A_o \quad j = k \quad (i-1)T_b \leq t \leq iT_b \tag{3.27}$$

The initial frequency deviation Δf here is $2/T_b = 2r_b$ Hz. Although not the minimum value, this Δf is a multiple (4) of the minimum and the peak of one PSD still would be at a null of the other PSD. This allows the resulting PSD to clearly demonstrate that of the decomposed ASK signals in binary FSK.

Setting the Gaussian noise PSD $N_o = 0$ W/Hz and increasing the System Time: Number of Samples specification to 524 288 (2^{19}) points results in a spectral resolution of approximately 0.95 Hz. The frequency deviation Δf is 2 kHz here. The normalized ($R_L = 1\ \Omega$) PSD of the Adder token channel output in Figure 3.8 without AWGN is shown scaled and as a single-sided spectrum in Figure 3.9 in dBm/Hz. The carrier components are shown as spectral impulses at $f_c \pm \Delta f = 20 \pm 2$ kHz, as given by Equation 3.26 and shown in Figure 3.9.

There are approximate nulls in the PSD that occur at the bit rate $r_b = 1/T_b$ (1 kb/sec or 1 kHz) and multiples of the bit rate centered about the two decomposed ASK carrier frequencies of $f_c = 20 \pm 2$ kHz due to the sinc^2 term in Equation 3.26. The PSD is not smooth because, although there are a large number of data, the ensemble is still finite, however the spectral envelope clearly demonstrates the sinc^2 behavior. The frequency deviation $\Delta f = 2$ kHz is also a multiple of the bit rate $r_b = 1$ kHz, so that the PSD of the decomposed ASK signals can be demonstrated.

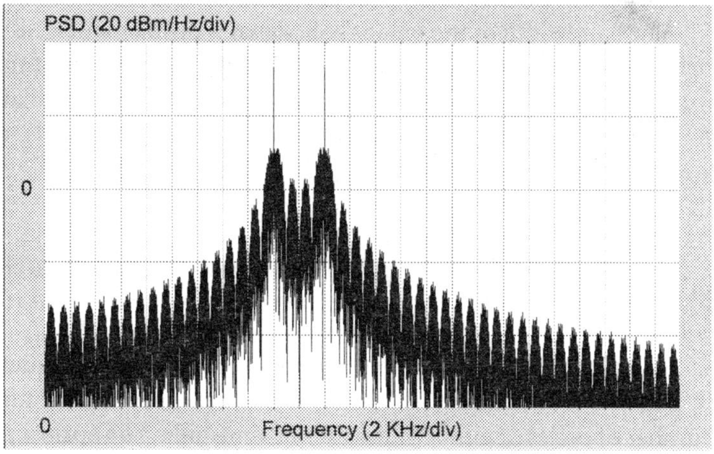

FIGURE 3.9 Power spectral density of a 1 kb/sec pseudonoise binary FSK signal, $f_c = 20$ kHz, $\Delta f = 2$ kHz.

Minimum frequency shift keying (MFSK) occurs when $\Delta f = 1/2T_b = r_b/2$ Hz. Setting the modulation gain $k_f = 500$ Hz/V results in $\Delta f = \pm 1 \text{ V} \times 500 = \pm 500$ Hz, because $r_b/2 = 500$ Hz. The resulting normalized PSD is shown scaled in Figure 3.10. The approximate nulls in the PSD remain at the bit rate r_b and multiples of the bit rate centered about the two decomposed ASK carrier frequencies due to the sinc^2 term in Equation 3.26. The carrier components are shown as a spectral impulses $(\delta(f))$ at $f_c \pm \Delta f = 20 \pm 0.5$ kHz, as given by Equation 3.26 and shown in Figure 3.10. This carrier frequency separation is the minimum possible, because each carrier spectral impulse is at the null of the PSD of the other decomposed ASK signal.

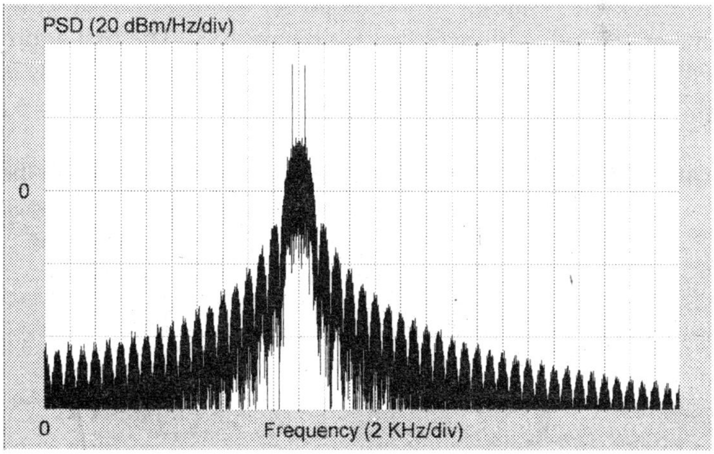

FIGURE 3.10 Power spectral density of a 1 kb/sec pseudonoise binary FSK (MFSK) signal, $f_c = 20$ kHz, $\Delta f = 500$ Hz.

The bandwidth of the binary FSK signal is problematical, as it is for binary ASK. Defining the bandwidth as a percentage of the total power of the binary FSK signal again is against usual course. The total power is easily calculated in the temporal domain. The power P in the binary FSK signal per bit time T_b, derived from Equations 3.21 and 3.22, is given by Equation 3.28.

$$P = \frac{A^2}{T_b R_L} \int_{(i-1)T_b}^{iT_b} \sin^2(2\pi(f_c \pm \Delta f)t)\,dt = \frac{A^2}{2R_L} \quad (i-1)T_b \le t \le iT_b \tag{3.28}$$

Equation 3.28 assumes that the ratio of the carrier frequencies to the bit rate $(f_c \pm \Delta f)/r_b$ is an integer. The integration then over one bit time T_b is for a whole number of cycles of either of the carrier sinusoids. The apriori probabilities (P_0, P_1) do not affect the total power because $P_0 + P_1 = 1$, as for binary ASK in Equation 3.20, and the total power P_T for binary FSK is given by Equation 3.29.

$$P_T = \frac{A^2}{2R_L} P_1 + \frac{A^2}{2R_L} P_0 = \frac{A^2}{2R_L} \tag{3.29}$$

The binary FSK signal has an amplitude $A = 5$ V and the normalized ($R_L = 1\ \Omega$) power is 12.5 W = 40.97 dBm (10 $\log_{10}$[12500]).

The decomposition of a binary FSK signal into two binary ASK signals separated by $2\Delta f$, with the bandwidth of a binary ASK signal as a percentage of the total power given by Table 3.1, results in Table 3.3. For the same percentage of the total power the bandwidth of the binary FSK signal is greater than that for the binary ASK (OOK) signal by $2\Delta f$ Hz. Because the minimum value of Δf is $1/2T_b$ Hz in FSK (MFSK), the resulting 90% total power bandwidth is at least $3/T_b$ Hz, a 50% increase in bandwidth over that for binary ASK. The first approximate nulls in the PSD for binary FSK (MFSK) in Figure 3.10 occur at 18.5 kHz and 21.5 kHz. This is the 90% total power bandwidth here of $2\Delta f + 2/T_b = 3/T_b = 3$ kHz.

TABLE 3.3 Bandwidth of a Binary FSK Signal as a Percentage of the Total Power

Bandwidth (Hz)	Percentage of Total Power
$2\Delta f + 2/T_b$	90%
$2\Delta f + 3/T_b$	93%
$2\Delta f + 4/T_b$	95%
$2\Delta f + 6/T_b$	96.5%

→

Bandwidth (Hz)	Percentage of Total Power
$2\Delta f + 8/T_b$	97.5%
$2\Delta f + 10/T_b$	98%

Performance of Binary FSK for the Optimum Receiver in AWGN

The binary FSK (MFSK) transmitter and optimum bandpass receiver in Figure 3.8 can be evaluated for its performance in the presence of AWGN. A BER analysis is conducted with the same parameters as that for binary ASK. The System Stop Time is set to 10 sec for convenience, which would process 10 000 information bits in possible error. To facilitate the simulation in SystemVue, all of the Analysis tokens, except that for the BER token are deleted from the binary FSK digital communication system with BER analysis in Figure 3.8. The Final Value token from the Sink Library records the total number of errors in 10 000 trials, as shown in Figure 3.7 for binary ASK.

Table 3.4 is a tabular list of the observed BER in a single trial of 10 000 information bits as a function of E_d/N_o to the theoretical P_b for binary FSK (MFSK), as given by Equation 3.11 for asymmetrical signals. Binary FSK does not satisfy Equation 3.1 and is asymmetric. The optimum receiver is implemented as the correlation receiver for asymmetrical signals, as shown in Figure 3.3.

TABLE 3.4 Observed BER and Theoretical P_b as a Function of E_d/N_o in a Binary FSK (MFSK) Digital Communication System with Optimum Receiver

E_d/N_o dB	N_o V²-sec	BER	P_b
∞	0	0	0
12	1.58×10^{-3}	1.7×10^{-3}	2.5×10^{-3}
10	2.50×10^{-3}	1.19×10^{-2}	1.25×10^{-2}
8	3.96×10^{-3}	3.82×10^{-2}	3.75×10^{-2}
6	6.28×10^{-3}	8.19×10^{-2}	7.93×10^{-2}
4	9.95×10^{-3}	1.382×10^{-1}	1.314×10^{-1}
2	1.58×10^{-2}	1.854×10^{-1}	1.872×10^{-1}
0	2.50×10^{-2}	2.363×10^{-1}	2.393×10^{-1}

The normalized energy difference per bit here is $E_d \approx 2.5 \times 10^{-2}$ V^2-sec, as given by Equations 3.10 and 3.23. The attenuation of the communication channel is taken to be zero ($\gamma = 1$). A comparison of Tables 3.4 and 3.2 shows the same BER performance for binary FSK (MFSK) as that for binary ASK (OOK) using the optimum receiver.

BINARY PHASE SHIFT KEYING

Binary phase shift keying (PSK) shifts the phase angle of the carrier frequency to one of two discrete phases during the bit time T_b for the representation of binary logic signals for the transmission of information. The modulated sinusoidal carrier signal has an amplitude of A V, a frequency of f_c Hz, as given by the analytical expression in Equation 3.30.

$$s_j(t) = A\sin(2\pi f_c t + k_p m_j(t) + \theta) \quad (i-1)T_b \le t \le iT_b \quad j = 0, 1 \tag{3.30}$$

The phase angle θ is the fixed initial phase of the PSK signal. The information signal or data source is $m_j(t)$ ($j = 0, 1$) and for PSK $m_j(t) =$ 0V and 1 V for one bit time T_b. The factor k_p, with units that are 2π (radians)/V, is the *phase deviation factor* (or the modulation gain) and the *phase deviation* $\Delta\varphi$ is given by Equation 3.31.

$$\Delta\varphi = k_p m_j(t) \tag{3.31}$$

The *phase shift* φ is the sum of the phase deviation and the fixed initial phase, as given by Equation 3.32.

$$\varphi = \Delta\varphi + \theta = k_p m_j(t) + \theta \tag{3.32}$$

SystemVue Simulation of Binary PSK

A binary phase shift keying (PSK) coherent digital communication system with AWGN is shown in Figure 3.11 (see Fig3-11.svu on the CD-ROM). To facilitate the simulation in SystemVue, all of the Analysis tokens, except that for the BER token output as the Final Value token, are deleted in the simulation model Fig3-11DT. As described in Chapter 1 the SystemVue Textbook Edition does not permit tokens to be deleted.

The SystemVue simulation is similar to that of the binary FSK communication system in Figure 3.8. The data source is the polar pseudonoise PN Sequence token from the Source Library, with parameters of a binary amplitude of ± 0.5 V, a bit rate

$r_b = 1$ kb/sec, and 0.5 V voltage and 0° phase offset. The initial phase θ provides the PSK phase shift for binary 0 when $m_j(t) = 0$ V. The data rate is the same as that used for binary ASK and FSK, which facilitates the comparison of power spectral densities. This PN Sequence token will transmit binary unipolar rectangular pulses for $m_j(t)$ with an amplitude of 0 V or 1 V.

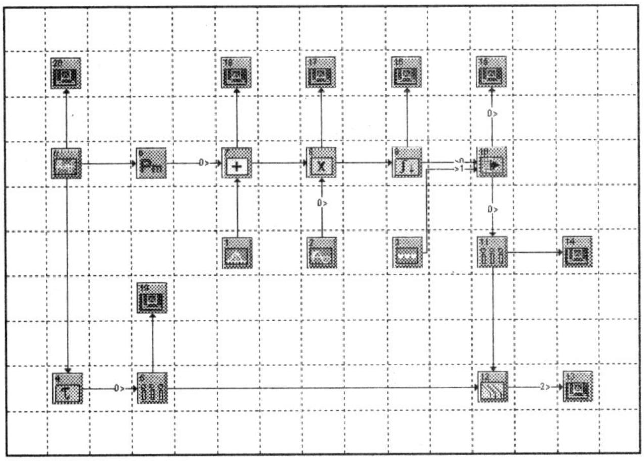

FIGURE 3.11 A binary PSK coherent digital communication system with BER analysis.

The non-linear binary PSK modulator is the Phase Modulator token from the Function Library, with parameters of an amplitude $A = 5$ V, a carrier frequency $f_c = 20$ kHz, a modulation gain (or phase deviation factor k_p) of 0.5 2π (radians)/V, and 180° phase offset. This phase offset provides the initial phase $\theta = 180° = \pi$ radians in Equations 3.30 and 3.31. Here the requisite PSK phase shift φ for binary 0 when $m_j(t) = 0$ V is $(0\text{ V} \times 0.5\ 2\pi/\text{V}) + \pi = \pi$ radians. The phase shift φ for binary 1 when $m_j(t) = 1$ V is $\Delta\varphi + \theta = (1\text{ V} \times 0.5\ 2\pi/\text{V}) + \pi = 0$ radians.

The binary PSK signals are symmetrical, because the bandpass signals $s_i(t)$ $(i = 0, 1)$ do satisfy Equation 3.1 and are given by Equation 3.33.

$$
\begin{aligned}
s_0(t) &= 5\sin(2\pi f_c t + \pi) \quad (i-1)T_b \le t \le iT_b \quad b_i = 0 \\
s_0(t) &= -5\sin(2\pi f_c t) \\
s_1(t) &= 5\sin(2\pi f_c t) \qquad\quad (i-1)T_b \le t \le iT_b \quad b_i = 1 \\
s_1(t) &= -s_0(t) \\
s_j(t) &= s_{PAM}(t)\sin(2\pi f_c t) \quad j = 0, 1
\end{aligned}
\tag{3.33}
$$

The PSK signal can be decomposed into a baseband PAM signal $s_{PAM}(t)$ and the sinusoidal carrier. The carrier frequency f_c here is 20 kHz, the information bit is b_i and the bit time is T_b. The optimum receiver for PSK is the correlation receiver, as described in Chapter 2 and shown in Figure 2.25, but modified for passband demodulation, as shown in Figure 3.11.

The communication channel is represented by the Adder token and the Gaussian Noise token from the Source Library, with initial parameters of a standard deviation $\sigma = 0$ V and $\mu = 0$ V mean or a power spectral density $N_o = 0$ W/Hz.

The Multiplier token multiplies the input signal of the receiver with the coherent reference signal $s_{ref}(t) = s_1(t)$, as given by Equation 3.33. The reference signal is provided by the Sinusoid token from the Source Library, with parameters of an amplitude $A = 5$ V peak, a frequency of $f_c = 20$ kHz, and 0° phase offset. The output of the Multiplier token is processed by the Integrate and Dump token from the Communications Library synchronized to the input signal. The parameters of the Integrate and Dump token are specified as an integration time equal to the bit time T_b of 1 msec and a continuous output.

The Analog Comparator token comparison input is set to the optimum threshold τ_{opt} given by Equation 3.3, because the binary PSK signals are symmetrical. The optimum threshold τ_{opt} is 0 V here, because $z_0(iT_b) = -12.5$ V and $z_1(iT_b) = 12.5$ V, as given by Equation 3.2. The Custom token from the Source Library provides the constant comparator input, with parameters of one output with an algebraic simulation equation that is $p(0) = 0$. The comparator input could also have been left unconnected.

The System Time: Sample Rate is set to 500 kHz. Two Sampler tokens are used to downsample the 500 kHz System Time rate to the 1 kb/sec bit rate r_b or 1 kHz. The Sampler token parameters are set to a sample rate of 1 kHz, and an aperture time of 2 µsec (T_{system}, a single point sample).

The PN Sequence token data Sampler input is delayed by 1 msec to correlate the data. The BER token parameters are set to output on every trial (number of trials = 1), with a threshold of 0.5 V for binary FSK, and a 1 msec offset. The System Stop Time is set to 10 sec for convenience, which would process 10 000 information bits in possible error. The Gaussian Noise token is specified by the power spectral density (PSD) N_o W/Hz. Setting the Gaussian noise source PSD to 0 should produce 0 observed errors.

Binary PSK Power Spectral Density

The magnitude of the frequency domain representation $P_{BPSK}(f)$ (derived from the Fourier transform) of a binary PSK signal of width T_b seconds, peak amplitude A volts, and carrier frequency f_c Hz is given by Equation 3.34.

$$P_{BPSK}(f) = \frac{1}{2}\left[AT_b \operatorname{sinc}\left(2\pi\left(T_b/2\right)\left(f - f_c\right)\right) + AT_b \operatorname{sinc}\left(2\pi\left(T_b/2\right)\left(f + f_c\right)\right)\right] \tag{3.34}$$

This bi-sided voltage spectrum is similar to Equation 3.16 for binary ASK (OOK) and extends from all negative to all positive frequencies ($-\infty < f < +\infty$). The analysis for the voltage spectrum utilizes that for the rectangular PAM pulse, as given by Equation 2.1. The polar PAM pulse $s_{PAM}(t)$ in Equation 3.33 has an amplitude $A = \pm 5$ V and is represented by the sinc ($\sin x/x$) function in the frequency domain. The modulation property of the Fourier transform shifts the sinc representations in the frequency domain by replacing the argument f by $f - f_c$ and $f + f_c$ and scaling by 1/2 in Equation 3.34.

The bi-sided energy spectral density $E_{BPSK}(f)$ of n pseudonoise PSK signals is n times the square of the magnitude of the voltage spectrum divided by the load resistance R_L, as given by Equation 3.35.

$$E_{BPSK}(f) = \frac{n}{4R_L}\left[A^2 T_b^2 \operatorname{sinc}^2\left(2\pi\left(T_b/2\right)\left(f - f_c\right)\right) + A^2 T_b^2 \operatorname{sinc}^2\left(2\pi\left(T_b/2\right)\left(f + f_c\right)\right)\right] \tag{3.35}$$

The bi-sided power spectral density $PSD_{BPSK}(f)$ is the energy spectral density divided by the time to transmit the n PSK signals or nT_b, as given by Equation 3.36.

$$PSD_{BPSK}(f) = \frac{1}{4R_L}\left[\frac{A^2 T_b}{4} \operatorname{sinc}^2\left(2\pi\left(T_b/2\right)\left(f - f_c\right)\right) + \frac{A^2 T_b}{4} \operatorname{sinc}^2\left(2\pi\left(T_b/2\right)\left(f + f_c\right)\right)\right] \tag{3.36}$$

Setting the Gaussian noise PSD $N_o = 0$ W/Hz and increasing the System Time: Number of Samples specification to 524 288 (2^{19}) points results in a spectral resolution of approximately 0.95 Hz. The normalized ($R_L = 1\ \Omega$) PSD of the Adder token channel output in Figure 3.11 without AWGN is shown scaled and as a single-sided spectrum in Figure 3.12 in dBm/Hz.

The PSD of a binary PSK signal is similar to that for a binary ASK signal in Figure 3.6, but does not display the carrier component as a spectral impulse ($\delta(f)$). There are approximate nulls in the PSD that occur at the bit rate $r_b = 1/T_b$ (1 kb/sec or 1 kHz) and multiples of the bit rate centered about the carrier frequency $f_c = 20$ kHz due to the sinc^2 term in Equation 3.36. The bandwidth of a bandpass PSK signal is the same as that for an ASK signal as a percentage of the total power in the signal, as given by Table 3.1. The power P in the binary PSK signal per bit time T_b, derived from Equation 3.30, is given by Equation 3.37.

$$P = \frac{A^2}{T_b R_L} \int_{(i-1)T_b}^{iT_b} \sin^2(2\pi f_c t + k_p m_j(t) + \theta)\, dt = \frac{A^2}{2R_L} \qquad (3.37)$$

$$(i-1)T_b \le t \le iT_b \quad j = 0, 1$$

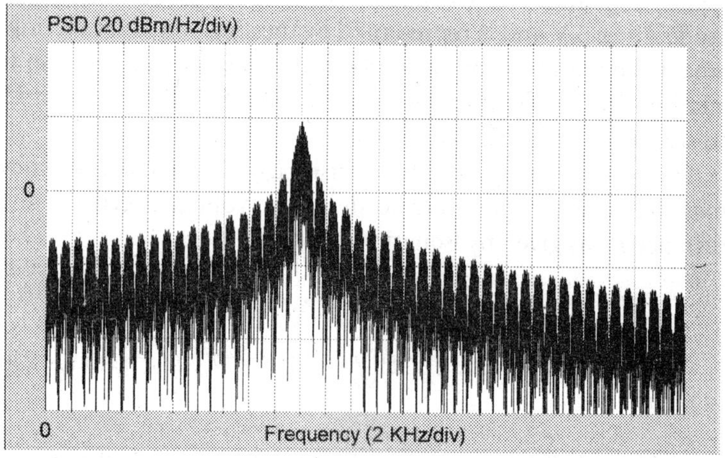

FIGURE 3.12 Power spectral density of a 1 kb/sec pseudonoise binary PSK signal.

Equation 3.37 assumes that the ratio of the carrier frequency to the bit rate f_c/r_b is an integer. The integration then over one bit time T_b is for a whole number of cycles of the carrier sinusoid. The binary PSK signal has an amplitude $A = 5$ V and the normalized ($R_L = 1\ \Omega$) power is 12.5 W = 40.97 dBm (10 $\log_{10}$[12500]).

Performance of Binary PSK for the Optimum Receiver in AWGN

The binary PSK transmitter and optimum bandpass receiver in Figure 3.11 can be evaluated for its performance in the presence of AWGN. A BER analysis is conducted with the same parameters as that for binary ASK (OOK) and FSK (MFSK). The System Stop Time is set to 10 sec for convenience, which would process 10 000 information bits in possible error.

Table 3.5 is a tabular list of the observed BER in a single trial of 10 000 information bits as a function of E_b/N_o to the theoretical P_b for binary PSK, as given by Equation 3.7 for symmetrical signals. Binary PSK satisfies Equation 3.1 and is symmetric. The optimum receiver is implemented as the correlation receiver for symmetrical signals, as shown in Figure 3.1.

The normalized energy per bit here is $E_b = 1.25 \times 10^{-2}$ V^2-sec, as given by Equations 3.4 and 3.33. The attenuation of the communication channel is taken to be zero ($\gamma = 1$). A comparison of Table 3.5 with Tables 3.2 and 3.4 shows a + 6 dB increase in BER performance for binary PSK over that for binary ASK (OOK) and FSK (MFSK) using the optimum receiver.

TABLE 3.5 Observed BER and Theoretical P_b as a Function of E_b/N_o in a Binary PSK Digital Communication System with Optimum Receiver

E_b/N_o dB	N_o V^2-sec	BER	P_b
∞	0	0	0
10	1.25×10^{-3}	0	4.05×10^{-6}
8	1.98×10^{-3}	2×10^{-4}	2.06×10^{-4}
6	3.14×10^{-3}	2.9×10^{-3}	2.41×10^{-3}
4	4.98×10^{-3}	1.24×10^{-2}	1.25×10^{-2}
2	7.89×10^{-3}	3.77×10^{-2}	3.75×10^{-2}
0	1.25×10^{-2}	8.02×10^{-2}	7.93×10^{-2}

MULTILEVEL (M-ARY) AMPLITUDE SHIFT KEYING

Multilevel (M-ary) amplitude shift keying (ASK) transmits a symbol that represents $N = \log_2 M$ bits of information. An M-ary ASK signal *keys* the carrier frequency as to one of M discrete levels during the symbol time T_s for the representation of $N = \log_2 M$ binary logic signals for the transmission of information. The modulated sinusoidal carrier signal has an amplitude of A V, a carrier frequency of f_c Hz, and a 0° reference phase angle, as given by the analytical expression in Equation 3.38.

$$s_j(t) = A\, m_j(t) \sin(2\pi f_c t) \quad (i-1)T_s \le t \le iT_s \quad j = 1, 2, \ldots, M \tag{3.38}$$

SystemVue Simulation of M-ary ASK

ON THE CD

An M-ary ASK coherent digital communication system with AWGN and optimum receiver implemented as a correlator is shown in Figure 3.13 (see Fig3-13.svu on the CD-ROM). To facilitate the simulation in SystemVue, all of the Analysis tokens, except that for the BER token output as the Final Value token, are deleted in the simulation model Fig3-13DT. As described in Chapter 1 the SystemVue Textbook Edition does not permit tokens to be deleted.

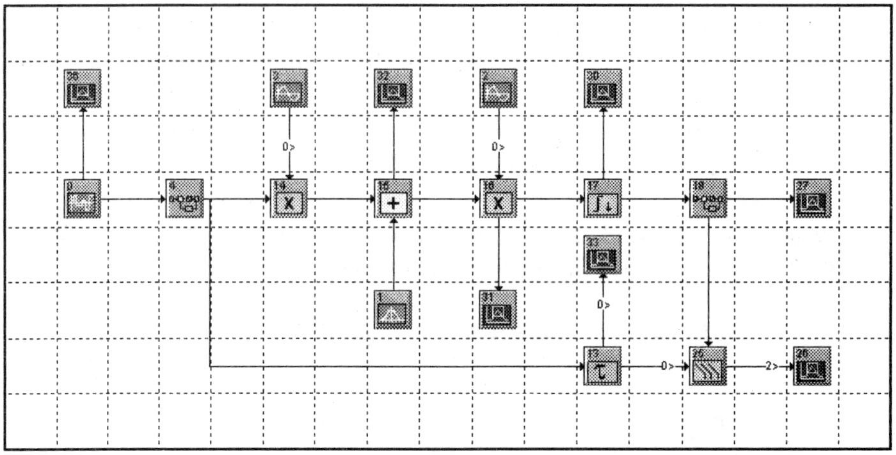

FIGURE 3.13 A 4-ASK coherent digital communication system with BER analysis.

The SystemVue simulation presented here is for an $M = 4$, 4-ASK signal, but can be easily extended to any 2^N-ASK ($N = \log_2 M$) digital communication system. A convenient set of positive, equally spaced amplitudes $m_j(t)$ for 4-level ASK is 0 V, 0.333 V, 0.666 V and 1 V for one symbol time T_s, as given by Equation 3.38.

This 4-ASK system is similar to the 2-bit Gray-coded 4-level PAM system, as shown in Figure 2.37 in Chapter 2. The 4-ASK signals are assumed to have equal apriori probabilities of occurrence ($P_1 = P_2 = P_3 = P_4$, $P_j = 0.25$).

The data source is the pseudonoise PN Sequence token from the Source Library, with parameters of a bit rate $r_b = 1$ kb/sec, amplitude ± 0.5 V and 0.5 V voltage and 0° phase offset. The PN Sequence token results in binary output amplitudes of 0 and 1 V for one bit time T_b. This data rate is the same as that used for the binary ASK, FSK, and PSK signals and facilitates the comparison of the power spectral densities.

The System Time: Sample Rate is set to 500 kHz. The output of the PN Sequence token is inputted to an encoder MetaSystem, where it is downsampled by the Decimator token from the Operator Library by a factor of 500, producing one sample point per bit time $T_b = 1$ msec. The 2-bit Gray encoder MetaSystem is shown in Figure 3.14. The output of the Decimator token is inputted to the Gray Encode token from the Communications Library, with parameters of a threshold of 0.5 V, two bits per symbol and the MSB outputted first, but only one system simulation point per symbol time $T_s = 2$ msec.

The Gray code improves the BER performance of the M-ary digital communication system, as described in Chapter 2. A symbol error interpreted as an adjacent symbol would now only produce one bit in error, as given by Equation 2.46.

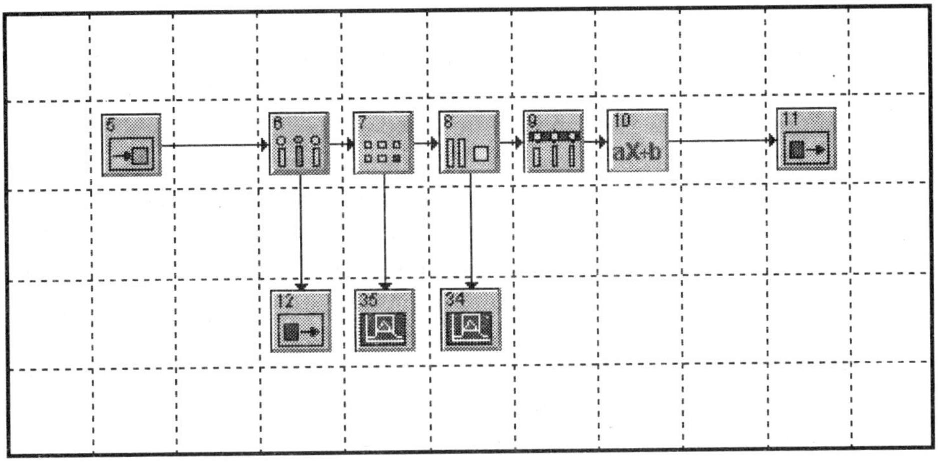

FIGURE 3.14 The 2-bit Gray encoder MetaSystem token of the 4-ASK digital communication system in Figure 3.13.

The Gray Encode token reassigns the input *di-bits* 11, 10, 01, and 00 as the output di-bits 10, 11, 01, and 00, respectively. The output of the Gray Encode token is processed by the Bit-to-Symbol token from the Communications Library, with parameters of a threshold of 0.5 V, two bits per symbol and the MSB inputted first. The output of the Bit-to-Symbol token is the 4-level signal with amplitudes of 0 V, 1 V, 2 V, and 3 V at a data rate of 500 symbols/sec.

The output of the Bit-to-Symbol token is upsampled by the Resampler token from the Operator Library to restore the System Time: Sample Rate to 500 kHz. The output of the Resampler token is scaled by the Polynomial token from the Function Library to restore a 4-level amplitude of 0 V, 0.333 V, 0.666 V, and 1 V from the 0 V, 1 V, 2 V, and 3 V input. The parameters of the Polynomial token are that the x^0 (offset) coefficient is 0, the x^1 (linear) coefficient is 0.33333 (0.3333 × 3 V = 1 V) and all other coefficients are zero.

The carrier generator is the Sinusoid token from the Source Library, with parameters of an amplitude A = 5 V peak, a carrier frequency f_c = 20 kHz, and 0° phase offset. The non-linear M-ary ASK modulator is the Multiplier token, which multiples the output of the Polynomial token and the Sinusoid token. The output of the Multiplier token is a 4-ASK signal with peak amplitudes of 0 V, 1.666 V, 3.333 V, and 5 V at the carrier frequency f_c.

The AWGN communication channel is the same and the optimum M-ary ASK receiver is similar to that for the binary ASK digital communication system, as shown in Figure 3.4. The communication channel is represented by the Adder token and the Gaussian Noise token from the Source Library, with initial parameters of a standard deviation σ = 0 V and a mean μ = 0 V or a power spectral density N_o = 0 W/Hz.

The reference signal $s_{ref}(t) = s_4(t)$ input to the Multiplier token of the correlation receiver is the Sinusoid token from the Source Library, with parameters of an amplitude $A = 5$ V peak, a carrier frequency $f_c = 20$ kHz, and 0° phase offset. The output of the Integrate and Dump token at the end of the symbol time T_s (rather than bit time T_b here) is given by Equation 3.42 with $M = 4$.

The four resulting integrate-and-dump output amplitudes of 0 V, 4.166 V, 8.333 V, and 12.5 V is inputted to the 2-bit Gray decoder MetaSystem, as shown in Figure 3.15. Here the signal is downsampled by the Decimator token by a factor of 1000, producing one sample point per symbol time $T_s = 2$ msec.

The output of the Decimator token is scaled by the Polynomial token from the Function Library to produce a 4-level amplitude of 0 V, 1 V, 2 V, and 3 V. The parameters of the Polynomial token are that the x^0 (offset) coefficient is 0, the x^1 (linear) coefficient is 0.24 ($0.24 \times 12.5 = 3$ V and with an offset of 0 V results in a 3 V output) and all other coefficients are zero.

The output of the Polynomial token is processed by the Symbol-to-Bit token, with parameters of two bits per symbol and the MSB inputted first. The three thresholds for the 4-level signal of the Symbol-to-Bit token are set to 0.5 V, 1.5 V, and 2.5 V, which are the optimum thresholds τ_{opt} of Equation 2.38 (Chapter 2) after scaling by the Polynomial token in Figures 3.14 and 3.15. The output of the Symbol-to-Bit token is inputted to the Gray Decode token from the Communications Library, with parameters of a threshold of 0.5 V, two bits per symbol and the MSB outputted first and one system simulation point per bit time $T_b = 1$ msec.

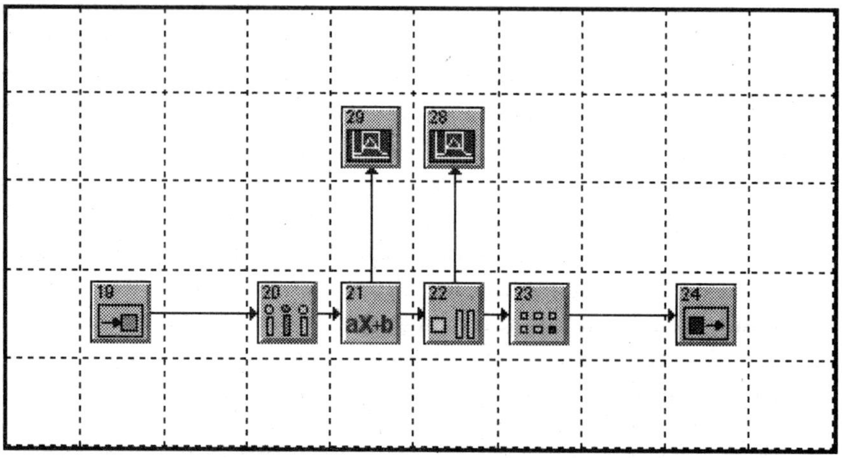

FIGURE 3.15 The 2-bit Gray decoder MetaSystem token of the ASK digital communication system in Figure 3.13.

The Decimator token output in Figure 3.13, within the MetaSystem, is delayed by 8 msec to correlate the source binary data with the output of the Gray decoder at the end of each bit time T_b. The value of this delay can be observed by the cross-correlation of the two signals in the BER analysis, as described in Chapter 1.

M-ary ASK Power Spectral Density

M-ary ASK transmits N bits per symbol ($N = \log_2 M$), but retains the same carrier and rectangular pulse shape $s_{PAM}(t)$ as binary ASK in Equation 3.15. The bandwidth of a ASK transmission is not affected by the pulse amplitude, but only by the pulse shape. The bi-sided power spectral density $PSD_{M\text{-}ary\,ASK}(f)$ for M-ary ASK then follows the analysis for binary ASK and results in the PSD given by Equation 3.39.

$$
\begin{aligned}
&PSD_{M\text{-}ary\,ASK}(f) = \\
&\frac{A_{avg}^2\,T_s}{16R_L}\,\mathrm{sinc}^2\big(2\pi(T_s/2)(f-f_c)\big) + \frac{A_{avg}^2}{4}\,\delta(f-f_c) + \\
&\frac{A_{avg}^2\,T_s}{16R_L}\,\mathrm{sinc}^2\big(2\pi(T_s/2)(f+f_c)\big) + \frac{A_{avg}^2}{4}\,\delta(f+f_c)
\end{aligned}
\tag{3.39}
$$

This is the same PSD as given in Equation 3.18 for a binary bandpass signal with the substitution here of the symbol time T_s, rather than the bit time T_b. The squared of the average amplitude $A_{avg}{}^2$ in Equation 3.39 is given by Equation 3.40, where P_j is the probability of occurrence of an M-ary ($M = 4$) symbol.

$$
A_{avg}^2 = \sum_{j=1}^{4} A^2\,m_j^2(t)\,P_j
\tag{3.40}
$$

Substituting the four equally likely ($P_1 = P_2 = P_3 = P_4$, $P_j = 0.25$) and equally spaced symbol amplitudes $A\,m_j(t) = 0$ V, 1.666 V, 3.333 V, and 5 V for each symbol time T_s here results in an average amplitude $A_{avg} = 3.118$ V. The symbol amplitudes can also be expressed as $A_j = 0$, A_v, $2A_v$ and $3A_v$, where $A_v = 1.666$ V and the amplitude difference between symbols is $A_v = 1.666$ V.

Setting the System Time: Number of Samples specification to 524 288 (2^{19}) points results in a spectral resolution of approximately 0.95 Hz. The normalized ($R_L = 1\ \Omega$) PSD of the M-ary ($M = 4$) ASK transmitter is shown scaled and as a single-sided spectrum in Figure 3.16 in dBm/Hz. Because the rectangular pulse width is the symbol time T_s, there are approximate nulls in the power spectral density that occur at the symbol rate $r_s = 1/T_s$ (500 symbols/sec or 500 Hz) and multiples of the symbol rate due to the sinc^2 term in Equation 3.43. Compare the 4-level ASK PSD to that of the binary ASK PSD, as shown in Figure 3.6.

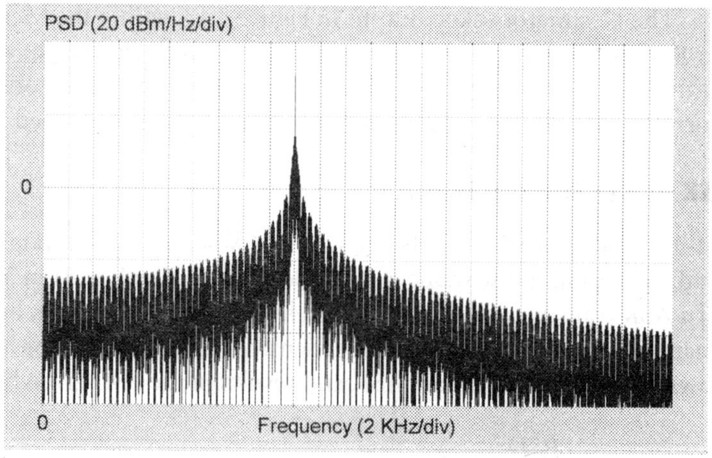

FIGURE 3.16 Power spectral density of a 1 kb/sec pseudonoise 4-ASK signal.

The bandwidth of the 4-ASK signal is again somewhat problematical. Defining the bandwidth as a percentage of the total power of the 4-ASK signal is again the usual course, because all of the available power would require an infinite bandwidth. The total power is easily calculated in the temporal domain. The power P in a 4-ASK signal for an arbitrary symbol per symbol time T_s, derived from Equation 3.38, is given by Equation 3.41.

$$P = \frac{A^2}{T_s R_L} \int_{(i-1)T_s}^{iT_s} m_j^2(t) \sin^2(2\pi f_c t)\, dt = \frac{A^2 m_j^2(t)}{2 R_L} \qquad (i-1)T_s \leq t \leq iT_s \quad j = 1, 2, 3, 4 \tag{3.41}$$

Equation 3.41 assumes that the ratio of the carrier frequency to the symbol rate f_c/r_s is an integer. The integration then over one symbol time T_s is for a whole number of cycles of the carrier sinusoid. The total power P_T then for the 4-ASK signal is given by Equation 3.42.

$$P_T = \frac{1}{2 R_L} \left[\sum_{j=1}^{4} A^2 m_j^2(t) P_j \right] = \frac{A_{avg}^2}{2 R_L} \tag{3.42}$$

The 4-ASK signal here has an average amplitude $A_{avg} = 3.118$ V and the normalized ($R_L = 1\ \Omega$) power is 4.86 W = 36.87 dBm ($10 \log_{10}[4860]$). Numerically integrating the power spectral density, Equation 3.39, from 0 to an arbitrary frequency and comparing the resulting power to the expected total power, Equation 3.42, provides the same data as that in Table 3.1 with the substitution of the symbol time T_s for the bit time T_b ($T_b = T_s/2$). The 4-ASK bandwidth here is one-half ($1/N$ with $M = 2^N$ and $M = 4$, $N = 2$) that of the binary ASK bandwidth as a percentage of the total power, but is the same as that for binary baseband rectangular PAM in Table 2.1.

The Correlation Receiver for M-ary ASK Signals

The optimum receiver for M-ary ASK bandpass signals is the matched receiver implemented as the correlator, as shown in Figure 3.13. The analysis here is similar to that of the correlation receiver for M-ary baseband signals in Chapter 2. The output of the correlator $z_j(t)$ ($j = 1, 2, \ldots, M$) is sampled at the end of each symbol period, $t = iT_s$, as given by Equation 3.43, which is the same as Equation 2.37.

$$z_j(iT_s) = \frac{1}{T_s} \int_{(i-1)T_s}^{iT_s} \gamma\, s_j(t) s_M(t)\, dt \qquad (i-1)T_s \le t \le iT_s \quad j = 1, 2, \ldots, M \tag{3.43}$$

The arbitrary correlator reference signal $s_M(t)$ is the M-ary bandpass signal that corresponds to the maximum amplitude symbol. Here it is assumed that the received signal is not attenuated by the channel ($\gamma = 1$) and there is no additive noise.

The received signal $s_j(t)$ ($j = 1, 2, \ldots, M$) is first multiplied by the reference signal $s_M(t)$ synchronized to the start of the symbol period and then integrated over that one symbol period. The output of the integrate-and-dump process is compared to multiple optimum thresholds τ_{opt} to decide which symbol was received.

There are three optimum thresholds for the 4-ASK bandpass digital communication system with equal apriori probabilities, as given by Equation 3.44, which is the same as Equation 2.38.

$$E_{d,symbol} = \int_{(i-1)T_s}^{iT_s} \left\{ \gamma \left[s_j(t) - s_k(t) \right] \right\}^2 dt \quad (i-1)T_s \le t \le iT_s \quad j, k = 1, 2, \ldots, M \tag{3.44}$$

Probability of Bit Error for M-ary ASK Signals

For AWGN, the noise processed by the optimum receiver again has a Gaussian probability distribution. The M-ary ASK received signal $s_j(t)$ is attenuated by the communication channel by an amount γ and can be considered here to have a normalized energy difference per symbol $E_{d,symbol}$ given by Equation 3.45.

$$E_{d,symbol} = \int_{(i-1)T_s}^{iT_s} \left\{ \gamma \left[s_j(t) - s_k(t) \right] \right\}^2 dt \quad (i-1)T_s \le t \le iT_s \quad j,k = 1, 2, \ldots, M \qquad (3.45)$$

If the energy difference per symbol $E_{d,symbol}$ is equal for all j, $k = 1, 2, \ldots, M$, then the probability of bit error for M-ary ASK signals corresponds to the analysis for binary asymmetrical signals, as given by Equations 3.10 and 3.11. The probability of symbol error P_s due to a symbol being misinterpreted as an adjacent symbol, with equally likely apriori probability of the M-ary bandpass symbols being transmitted ($P_j = 1/M, j = 1, 2, \ldots, M$) is given by Equation 3.46, which is the same as Equation 2.40 for M-ary baseband signals.

$$P_s = \frac{2(M-1)}{M} Q\left(\sqrt{\frac{E_{d,symbol}}{2N_o}} \right) \qquad (3.46)$$

The average energy difference per symbol $E_{avg,symbol}$ can be expressed in terms of the average amplitude A_{avg} as given by Equation 3.47.

$$E_{avg,\,symbol} = \int_{(i-1)T_s}^{iT_s} \left[\gamma A_{avg} \sin(2\pi f_c t) \right]^2 dt = \frac{\gamma^2 A_{avg}^2 T_s}{2} \qquad (3.47)$$

For the 4-ASK signal here, attenuation of the communication channel is assumed to be zero ($\gamma = 1$), $A_{avg} = 3.118$, $T_s = 2$ msec and assuming that the ratio of the carrier frequency to the symbol rate f_c/r_s is an integer, $E_{avg,symbol} = 9.72 \times 10^{-3}$ V^2-sec.
The energy difference per symbol $E_{d,symbol}$ is given by Equations 3.44 and 3.48.

$$E_{d,symbol} = \int_{(i-1)T_s}^{iT_s} \left\{ \gamma \left[A m_j(t) \sin(2\pi f_c t) - A m_k(t) \sin(2\pi f_c t) \right] \right\}^2 dt$$

$$j \ne k \quad j,k = 1, 2, 3, 4 \qquad (3.48)$$

$$E_{d,symbol} = \int_{(i-1)T_s}^{iT_s} \left[\gamma A_v \sin(2\pi f_c t) \right]^2 dt = \frac{\gamma^2 A_v^2 T_s}{2}$$

For the 4-ASK signal here, the attenuation of the communication channel is again assumed to be zero ($\gamma = 1$), $A = 5$ V, $m_j(t) = 0$ V, 1.666 V, 3.333 V, and 5 V, $T_s = 2$ msec, and the amplitude difference between symbols $A_v = 1.666$ V. From Equation 3.48 and assuming that the ratio of the carrier frequency to the symbol rate f_c/r_s is an integer, $E_{d,symbol} = 2.777 \times 10^{-3}$ V^2-sec. Noting that $E_{d,symbol} = 0.286$ $E_{avg,symbol}$ and $M = 4$ in Equation 3.46 gives Equation 3.49.

$$P_s = \frac{3}{2} Q\left(\sqrt{\frac{0.286\, E_{avg,symbol}}{2\, N_o}}\right) \tag{3.49}$$

Because each 4-level symbol represents two bits here, $E_{avg,symbol} = 2E_{b,4\text{-}level}$ and Equation 3.49 can be given as Equation 3.50.

$$P_s = \frac{3}{2} Q\left(\sqrt{\frac{0.286\, E_{b,4\text{-}level}}{N_o}}\right) \tag{3.50}$$

As for M-ary baseband signals in Chapter 2, the probability of bit error P_b is minimized for Gray-coded data and for 4-level bandpass signals results in Equation 3.51, which is the same as Equation 2.46 in Chapter 2.

$$P_{b,4\text{-}level,Gray\text{-}coded} = \frac{1}{2} P_s \tag{3.51}$$

The probability of bit error for the Gray-coded 4-ASK signal here is then given by Equation 3.52.

$$P_{b,4\text{-}level,Gray\text{-}coded} = \frac{3}{4} Q\left(\sqrt{\frac{0.286\, E_{b,4\text{-}level}}{N_o}}\right) \tag{3.52}$$

From the bit error analysis for 4-level PAM, as in Equation 2.44, the probability of bit error P_b without Gray-coding the source information for 4-ASK is given by Equation 3.53.

$$P_{b,4\text{-}level} = \frac{2}{3} P_s = Q\left(\sqrt{\frac{0.286\, E_{b,4\text{-}level}}{N_o}}\right) \tag{3.53}$$

Performance of M-ary ASK for the Optimum Receiver in AWGN

The M-ary ($M = 4$) ASK transmitter and optimum ASK receiver in Figure 3.13 can be evaluated for its performance in the presence of AWGN. A BER analysis is conducted with the same parameters for the binary bandpass ASK system as given in Table 3.2. Table 3.6 is a tabular list of the observed BER in a single trial of 10 000 information bits as a function of E_b/N_o to the theoretical P_b for Gray-coded source data, as given by Equation 3.52. The normalized energy per bit here is $E_{b,4\text{-}level} = E_{avg,symbol}/2 \approx 4.86 \times 10^{-3}$ V^2-sec.

A comparison of Tables 3.2 and 3.6 shows an approximate –2 dB E_b/N_o decrease in BER performance for Gray-coded 4-ASK over that for binary ASK (OOK), as predicted by Equations 3.11 and 3.52. The performance tradeoff is that 4-ASK has half the transmission bandwidth of binary ASK.

TABLE 3.6 Observed BER and Theoretical P_b as a Function of E_b/N_o in a Gray-Coded 4-ASK Digital Communication System with Optimum Receiver

E_b/N_o dB	N_o V²-sec	BER	P_b
∞	0	0	0
14	1.93×10^{-4}	2.8×10^{-3}	2.78×10^{-3}
12	3.07×10^{-4}	1.23×10^{-2}	1.27×10^{-2}
10	4.86×10^{-4}	3.25×10^{-2}	3.41×10^{-2}
8	7.70×10^{-4}	6.71×10^{-2}	6.68×10^{-2}
6	1.22×10^{-3}	1.091×10^{-1}	1.073×10^{-1}
4	1.93×10^{-3}	1.515×10^{-1}	1.493×10^{-1}
2	3.07×10^{-3}	1.998×10^{-1}	1.889×10^{-1}
0	4.86×10^{-3}	2.462×10^{-1}	2.224×10^{-1}

The observed BER in a single trial of 10 000 information bits as a function of E_b/N_o can also be readily compared to the theoretical P_b for straight-binary coded (non-Gray-coded) source data, as given by Equation 3.53, and listed in Table 3.7. The Gray Encoder and Gray Decoder tokens in the MetaSystems of the 4-ASK digital communication system in Figures 3.14 and 3.15 are deleted and the SystemVue simulation is executed.

TABLE 3.7 Observed BER and Theoretical P_b as a Function of E_b/N_o in a Straight-Binary Coded 4-ASK Digital Communication System with Optimum Receiver

E_b/N_o dB	N_o V²-sec	BER	P_b
∞	0	0	0
14	1.93×10^{-4}	4.4×10^{-3}	3.71×10^{-3}
12	3.07×10^{-4}	1.62×10^{-2}	1.69×10^{-2}
10	4.86×10^{-4}	4.79×10^{-2}	4.54×10^{-2}
8	7.70×10^{-4}	9.17×10^{-2}	8.90×10^{-2}
6	1.22×10^{-3}	1.425×10^{-1}	1.431×10^{-1} →

E_b/N_o dB	N_o V²-sec	BER	P_b
4	1.93×10^{-3}	1.994×10^{-1}	1.991×10^{-1}
2	3.07×10^{-3}	2.477×10^{-1}	2.519×10^{-1}
0	4.86×10^{-3}	2.822×10^{-1}	2.965×10^{-1}

A comparison of Tables 3.6 and 3.7 shows an approximate –2 dB E_b/N_o decrease in BER performance for straight-binary coded 4-ASK over that for Gray-coded 4-ASK, as predicted by Equations 3.52 and 3.53. The performance tradeoff is that Gray-coded 4-level ASK has an increase in complexity of the transmitter and the optimum receiver.

MULTILEVEL (M-ARY) FREQUENCY SHIFT KEYING

Multilevel (M-ary) frequency shift keying (FSK) transmits a symbol than represents $N = \log_2 M$ bits of information. An M-ary FSK signal *keys* the carrier frequency to one of M discrete frequencies during the symbol time T_s for the representation of N binary logic signals for the transmission of information. The modulated sinusoidal signal has an amplitude of A V, a carrier frequency of f_c Hz, and a 0° reference phase angle, as given by the analytical expression in Equation 3.54.

$$s_j(t) = A \sin(2\pi(f_c + k_f m_j(t))t) \quad (i-1)T_s \le t \le iT_s \quad j = 1, 2, \ldots, M \qquad (3.54)$$

SystemVue Simulation of M-ary FSK

ON THE CD

An M-ary FSK coherent digital communication system with AWGN and optimum receiver implemented as a correlator is shown in Figure 3.17 (see Fig3-17.svu on the CD-ROM). To facilitate the simulation in SystemVue, all of the Analysis tokens, except that for the BER token output as the Final Value token, are deleted in the simulation model Fig3-17DT. As described in Chapter 1 the SystemVue Textbook Edition does not permit tokens to be deleted.

The SystemVue simulation presented here is for an $M = 4$, 4-FSK, but can be easily extended to any 2^N-FSK ($N = \log_2 M$) digital communication system. This M-ary FSK system is similar to the Gray-coded M-ary ASK system shown in Figure 3.13. The 4-FSK signals are assumed to have equal apriori probabilities of occurrence ($P_1 = P_2 = P_3 = P_4$, $P_j = 0.25$).

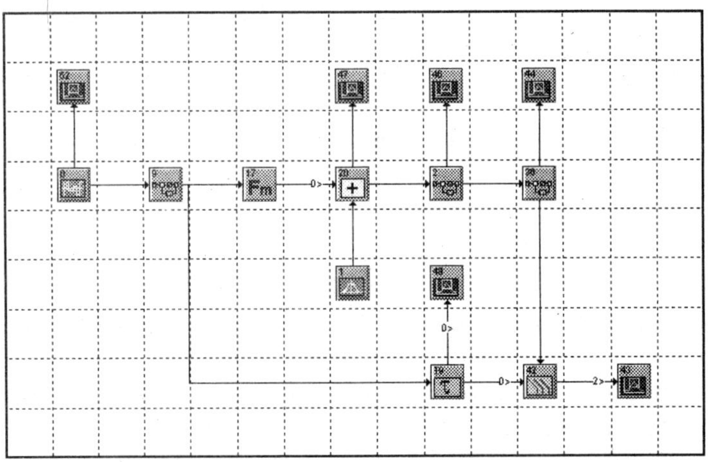

FIGURE 3.17 A 4-FSK coherent digital communication system with BER analysis.

The data source is the pseudonoise PN Sequence token from the Source Library, with parameters of a bit rate $r_b = 1$ kb/sec, amplitude ± 0.5 V and 0.5 V voltage and 0° phase offset. The PN Sequence token results in binary output amplitudes of 0 and 1 V for one bit time T_b. This data rate is the same as that used for the binary ASK, FSK, and PSK and M-ary ASK signals and facilitates the comparison of the power spectral densities.

The binary data is inputted to the same symbol encoding MetaSystem of M-ary ASK, as shown in Figure 3.14, but without the Gray Encode token. For M-ary FSK additive noise causes symbol errors to be equally likely among the $M - 1$ correlators corresponding to incorrect symbols and there is no advantage to Gray-coding the source information. There is a distinct advantage, however, to Gray code the information for either M-ary baseband PAM or M-ary bandpass ASK.

The symbol data 0 V, 1 V, 2 V, and 3 V is scaled by the Polynomial token from the Function Library to generate a 4-level amplitude of –3 V, –1 V, 1 V, and 3 V. The parameters of the Polynomial token though are different than that in Figure 3.17. The x^0 (offset) coefficient is –3, the x^1 (linear) coefficient is 2 (2×3 V = 6 V and with an offset of –3 V results in a 3 V output) and all other coefficients are zero. The output of the symbol encoding MetaSystem is the 4-level signal with amplitudes of –3 V, –1, 1 V, and 3 V at a data rate of 500 symbols/sec.

The non-linear FSK modulator is the Frequency Modulator token from the Function Library, with parameters of an amplitude of 5 V, a carrier frequency f_c = 20 kHz, 0° phase offset, and an initial modulation gain (or frequency deviation factor k_f) of 250 Hz/V. The factor k_f in Equation 3.54 is the *frequency deviation factor* (or the modulation gain) and the *frequency deviation* Δf is given by Equation 3.22.

The resulting four frequency deviations here are $\pm 3\text{ V} \times 250 = \pm 750$ Hz and ± 1 V $\times\ 250 = \pm\ 250$ Hz. Minimum frequency shift keying (MFSK) occurs when $\Delta f = n/2T_s = nr_b/4$ Hz $= n \times 250$ Hz, because $r_b = 1$ kb/sec here, where n is an integer.

The straight-binary coded 4-FSK signals $s_j(t)$ then are arbitrarily given by Equation 3.55.

$$\begin{aligned} s_1(t) &= 5\sin(2\pi(f_c - 3\Delta f)t) \quad b_{2i} = 0, b_{2i+1} = 0 \\ s_2(t) &= 5\sin(2\pi(f_c - \Delta f)t) \quad b_{2i} = 0, b_{2i+1} = 1 \\ s_3(t) &= 5\sin(2\pi(f_c + \Delta f)t) \quad b_{2i} = 1, b_{2i+1} = 0 \\ s_4(t) &= 5\sin(2\pi(f_c + 3\Delta f)t) \quad b_{2i} = 1, b_{2i+1} = 1 \\ & \qquad (i-1)T_s \le t \le iT_s \end{aligned} \tag{3.55}$$

The carrier frequency f_c here is 20 kHz, the frequency deviation Δf is 250 Hz, the source information di-bit is b_{2i}, b_{2i+1} and the symbol time is T_s.

The communication channel is represented by the Adder token and the Gaussian Noise token from the Source Library, with initial parameters of a standard deviation $\sigma = 0$ V and $\mu = 0$ V mean or a power spectral density $N_o = 0$ W/Hz.

The received signal is inputted to the 4-FSK correlation receiver MetaSystem, as shown in Figure 3.18. For each of the four correlators a Multiplier token multiplies the input signal with the coherent reference signal $s_j(t)$ ($j = 1, 2, 3, 4$), as given by Equation 3.54. Each of the outputs of the Multiplier tokens are processed by Integrate and Dump tokens from the Communications Library synchronized to the input signal. The parameters of the Integrate and Dump token are specified as an integration time equal to the symbol time T_s of 2 msec and a continuous output.

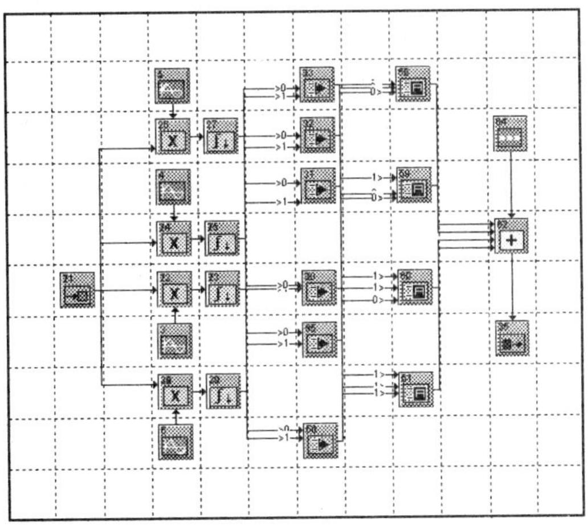

FIGURE 3.18 A 4-FSK correlation receiver MetaSystem of the FSK digital communication system in Figure 3.17.

The correlation receiver here uses the orthogonality definition in Equation 3.27. In the presence of noise, the received signal is not precisely orthogonal to the other $M-1$ coherent reference signals and their outputs at the end of a symbol time T_s is not zero. However, if the received signal is significantly stronger than the noise, its correlator output is the largest at the end of a symbol time T_s.

The outputs of the four correlators at the end of a symbol time T_s are processed to determine the largest value by the array of Analog Comparator tokens, multiple input AND gate tokens from the Logic Library, and an Adder token. The Analog Comparator provides both inverted logic and non-inverted logic outputs, which simplifies the logic.

The four AND gate tokens have logic false outputs of 0 V, but logic true outputs of 4 V, 3 V, 2 V, and 1 V for the four symbols here. Typically, for the 4-FSK receiver one correlator output is compared to the other three correlator outputs and, if the largest, its AND gate output is set to the intermediate voltage for that symbol (4 V, 3 V, 2 V, or 1 V). The other three AND gate outputs should be zero. The representation of the lowest amplitude symbol as 1 V (rather than 0 V) initially facilitates debugging during the simulation design process.

The output of the AND gates are summed with a constant parameter of –1 V from a Custom token by the Adder token to provide the symbol values of 3 V, 2 V, 1 V, and 0 V. The symbol decoding MetaSystem is the same as that for 4-ASK in Figure 3.15, except that the Gray Decode and Polynomial tokens are deleted, because Gray-coding is not required and scaling of the symbol amplitudes is accomplished in the correlation receiver MetaSystem of Figure 3.18.

The System Time: Sample Rate is set to 500 kHz. Two Sampler tokens are used to downsample the 500 kHz System Time rate to the 1 kb/sec bit rate or 1 kHz. The Sampler token parameters are set to a sample rate of 1 kHz, and an aperture time of 2 μsec (T_{system}, a single point sample).

The encoder Decimator output is delayed by 4 msec to correlate the data. The BER token parameters are set to output on every trial (number of trials = 1), with a threshold of 0.5 V for 4-FSK, and a 2 msec offset. The System Stop Time is set to 10 sec for convenience, which would process 10,000 information bits in possible error. The Gaussian Noise token is specified by the power spectral density (PSD) N_o W/Hz. Setting the Gaussian noise source PSD to 0 should produce 0 observed errors.

M-ary FSK Power Spectral Density

M-ary FSK transmits N bits per symbol ($N = \log_2 M$), but retains the same carrier and rectangular pulse shape $s_{PAM}(t)$, as binary FSK in Equations 3.24 and 3.25. The bi-sided power spectral density (PSD) for M-ary FSK then follows the analysis for binary FSK and results in the PSD similar to that given by Equation 3.36. The single-sided PSD, $PSD_{M\text{-}ary\ FSK\ ss}(f)$ (there are eight more components for the bi-sided PSD), for the 4-FSK signal replaces the bit time T_b with the symbol time T_s and has eight spectral components, as shown in Equation 3.56.

$$\begin{aligned}
&\frac{A^2T_b}{8R_L}\,\text{sinc}^2\left(2\pi\left(T_b/2\right)\left(f-f_c-\Delta f\right)\right)+\frac{A^2}{8R_L}\,\delta\left(f-f_c-\Delta f\right)+\\
&\frac{A^2T_b}{8R_L}\,\text{sinc}^2\left(2\pi\left(T_b/2\right)\left(f-f_c+\Delta f\right)\right)+\frac{A^2}{8R_L}\,\delta\left(f-f_c+\Delta f\right)+\\
&\frac{A^2T_b}{8R_L}\,\text{sinc}^2\left(2\pi\left(T_b/2\right)\left(f-f_c-3\Delta f\right)\right)+\frac{A^2}{8R_L}\,\delta\left(f-f_c-3\Delta f\right)+\\
&\frac{A^2T_b}{8R_L}\,\text{sinc}^2\left(2\pi\left(T_b/2\right)\left(f-f_c+3\Delta f\right)\right)+\frac{A^2}{8R_L}\,\delta\left(f-f_c+3\Delta f\right)
\end{aligned} \tag{3.56}$$

The minimum frequency deviation Δf is $1/2T_s = r_b/4$ Hz. With this value of Δf the peak of the PSD of each of the decomposed ASK signals in M-ary FSK would be at a null of the PSD of the other signal, as given by Equation 3.18. FSK using this value of Δf is known as *minimum frequency shift keying* (MFSK).

Setting the Gaussian noise PSD $N_o = 0$ W/Hz and increasing the System Time: Number of Samples specification to 524 288 (2^{19}) points results in a spectral resolution of approximately 0.95 Hz. The FSK (MFSK) minimum frequency deviation Δf is 250 Hz here, because $r_b = 1$ kb/sec. The normalized ($R_L = 1\ \Omega$) PSD of the Adder token channel output in Figure 3.17 without AWGN is shown scaled and as a single-sided spectrum in Figure 3.19 in dBm/Hz.

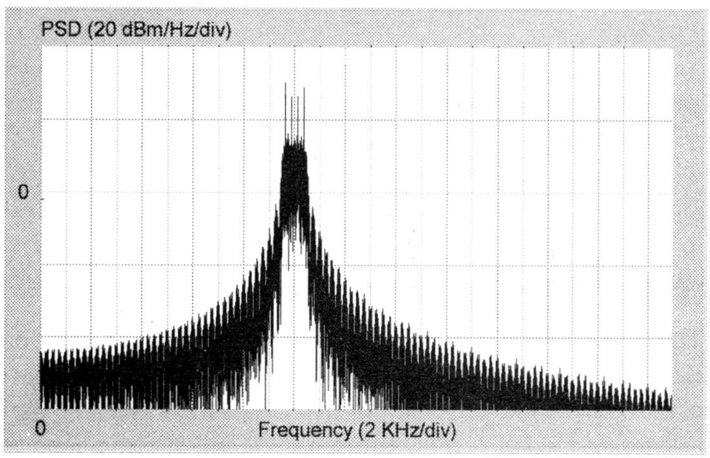

FIGURE 3.19 Power spectral density of a 1 kb/sec pseudonoise binary 4-FSK signal, $f_c = 20$ kHz, $\Delta f = \pm 250$ Hz, ± 750 Hz.

The carrier components ($\delta(f)$) are shown as spectral impulses at $f_c \pm \Delta f = 20 \pm 0.25$ kHz and $f_c \pm 3\Delta f = 20 \pm 0.75$ kHz, as given by Equation 3.56 and shown in Figure 3.19. This carrier frequency separation is the minimum possible, because each carrier spectral impulse is at the null of the PSD of another component. There are approximate nulls in the PSD that occur at the symbol rate $r_s = 1/T_s$ (500 b/sec or 500 Hz) and multiples of the symbol rate as upper and lower sidebands about the two of the four decomposed ASK carrier frequencies of $f_c = 20 \pm 750$ Hz due to the sinc^2 term in Equation 3.56.

The bandwidth of the M-ary FSK signal is defined as it was for binary FSK. The non-rigorous decomposition of an M-ary FSK signal into M binary ASK signals separated by $2\Delta f$, with the bandwidth of a binary ASK signal as a percentage of the total power given by Table 3.1, results in Table 3.8, where the symbol time T_s can be replaced by $(\log_2 M)T_b$. For the same percentage of the total power the bandwidth of the M-ary FSK signal increases linearly over that for the binary FSK signal.

For the 4-FSK (MFSK) signal here the 95% of the total power bandwidth is $6\Delta f + 2/T_b = 3.5$ kHz, while, in comparison, the binary FSK (MFSK) signal with the same bit rate of 1 kb/sec would have a bandwidth of 4.5 kHz, as in Table 3.3. However, an 8-FSK (MFSK) signal would have a 95% of the total power bandwidth of $14\Delta f + 4/3T_b = 4.833$ kHz and a 16-FSK (MFSK) signal would have a 95% of the total power bandwidth of $30\Delta f + 1/T_b = 8.5$ kHz for the same bit rate of 1 kb/sec.

TABLE 3.8 Bandwidth of an M-ary FSK Signal as a Percentage of the Total Power.

Bandwidth (Hz)	Percentage of Total Power'
$2(M \times 1)\Delta f + 4/T_s$	95%
$2(M \times 1)\Delta f + 6/T_s$	96.5%
$2(M \times 1)\Delta f + 8/T_s$	97.5%
$2(M \times 1)\Delta f + 10/T_s$	98%

The Correlation Receiver for M-ary FSK Signals

The optimum receiver for M-ary bandpass FSK signals is the matched receiver implemented as M correlators, as shown in Figure 3.17. The analysis here is similar to that of the correlation receiver for M-ary bandpass ASK signals. The output of the correlators $z_j(t)$ ($j = 1, 2, \ldots, M$) is sampled at the end of each symbol period, $t = iT_s$, as given by Equation 3.43. Each correlator reference signal $s_{ref,j}(t)$ is the bandpass signal for that particular symbol carrier frequency and deviation, as given by Equation 3.55.

The received signal $s_j(t)$ ($j = 1, 2, \ldots, M$) is first multiplied by the reference signal $s_{ref,j}(t)$ synchronized to the start of the symbol period and then integrated over that one symbol period. In the optimum receiver the M outputs of the correlators are compared and the largest output is selected as the symbol received.

Probability of Bit Error for M-ary FSK Signals

The probability of symbol error for coherent M-ary FSK is difficult to calculate, but is shown to be upper-bounded by Equation 3.57 [Simon95].

$$P_s = (M-1)\,Q\left(\sqrt{\log_2 M\left[\frac{E_b}{N_o}\right]}\right) \quad M \geq 4 \tag{3.57}$$

Because additive noise causes symbol errors to be equally likely among all the $M - 1$ correlators corresponding to incorrect symbols, and Gray-coding the source information provides no advantage here, the probability of bit error P_b for coherent M-ary FSK is related to symbol error P_s by Equation 3.58.

$$\begin{aligned} P_b &= \frac{M}{2\,(M-1)}P_s \\ P_b &= \frac{M}{2}Q\left(\sqrt{\log_2 M\left[\frac{E_b}{N_o}\right]}\right) \cdot M \geq 4 \end{aligned} \tag{3.58}$$

The M-ary bandpass FSK signal $s_j(t)$ is attenuated by the communication channel by an amount γ and has a normalized energy per symbol E_s^j given by Equation 3.59. The attenuation of the communication channel is assumed to be zero ($\gamma = 1$) and the 4-FSK signals are given by Equations 3.54 and 3.55.

$$E_s^{\,j} = \int_{(i-1)\,T_s}^{i\,T_s} s_j^{\,2}(t)\,dt \quad (i-1)T_s \leq t \leq iT_s \quad j = 1, 2, 3, 4 \tag{3.59}$$

From Equation 3.59 and assuming that the ratio of the carrier frequency and frequency deviation to the symbol rate f_c/r_s is an integer, the energy per symbol E_s^j for each 4-FSK signal $s_j(t)$ is equal, $E_s^j = E_s = 2.5 \times 10^{-2}$ V^2-sec here. For the 4-FSK signal the energy per bit $E_b = E_s/2 = 1.25 \times 10^{-2}$ V^2-sec.

Performance of M-ary FSK for the Optimum Receiver in AWGN

The M-ary ($M = 4$) FSK transmitter and optimum FSK receiver in Figure 3.17 can be evaluated for its performance in the presence of AWGN. A BER analysis is conducted with the same parameters for the binary bandpass FSK system as given in Table 3.4. Table 3.9 is a tabular list of the observed BER in a single trial of 10 000 information bits as a function of E_b/N_o to the theoretical upper-bound of P_b, as given by Equation 3.57. The energy per bit here is $E_b = 1.25 \times 10^{-2}$ V²-sec.

A comparison of Tables 3.4 and 3.9 shows an approximately +6 dB increase in BER performance for 4-FSK (MFSK) over that for binary FSK (MFSK). The performance tradeoff is that 4-FSK (MFSK) requires an additional transmission bandwidth of $6\Delta f = 1.5$ kHz here compared to the bandwidth of binary FSK (MFSK) with the same parameters. The SystemVue simulation performance for BER versus the upper-bound of P_b at an E_b/N_o ratio greater than 6 dB is attributed to the inherent statistical variation of the small (10^4) information bit sample here.

TABLE 3.9 Observed BER and Theoretical Upper-Bound of P_b as a Function of E_b/N_o in a 4-Level FSK (MFSK) Digital Communication System with Optimum Receiver

E_b/N_o dB	N_o V²-sec	BER	P_b (upper bound)
∞	0	0	0
14	4.98×10^{-4}	0	≈ 0
12	7.89×10^{-4}	0	≈ 10^{-8}
10	1.25×10^{-3}	2×10^{-4}	≈ 10^{-6}
8	1.98×10^{-3}	4×10^{-4}	≈ 10^{-4}
6	3.14×10^{-3}	3.9×10^{-3}	4.8×10^{-3}
4	4.98×10^{-3}	2.16×10^{-2}	2.52×10^{-2}
2	7.89×10^{-3}	5.79×10^{-2}	7.54×10^{-2}
0	1.25×10^{-2}	1.216×10^{-1}	1.586×10^{-1}

MULTILEVEL (M-ARY) PHASE SHIFT KEYING

A multilevel (M-ary) phase shift keying (PSK) signal shifts the phase angle of the carrier frequency to one of M discrete values during the symbol time T_s for the representation of $N = \log_2 M$ binary logic signals for the transmission of information. The modulated sinusoidal carrier signal has an amplitude of A V, a carrier frequency of f_c Hz, and a 0° reference phase angle, as given by the analytical expression in Equation 3.60.

$$s_j(t) = A\cos(2\pi f_c t + k_p m_j(t)) \quad (i-1)T_s \le t \le iT_s \quad j = 1, 2, \ldots, M \qquad (3.60)$$

A convenient set of equally spaced phase angles $k_p\, m_j(t)$ for 4-PSK, also known as quaternary (Q) PSK, is ± 45° and ± 135° for one symbol time T_s. QPSK is the terminology often used for 4-PSK and the signal then is given by Equation 3.61.

$$\begin{aligned}
s_1(t) &= A\cos(2\pi f_c t + 45°) && \text{di-bit } 10 \quad b_{2i}b_{2i+1} = 11 \\
s_2(t) &= A\cos(2\pi f_c t + 135°) && \text{di-bit } 00 \quad b_{2i}b_{2i+1} = 00 \\
s_3(t) &= A\cos(2\pi f_c t + 225°) && \text{di-bit } 01 \quad b_{2i}b_{2i+1} = 01 \\
s_4(t) &= A\cos(2\pi f_c t + 315°) && \text{di-bit } 11 \quad b_{2i}b_{2i+1} = 10 \\
& && (i-1)T_s \le t \le iT_s
\end{aligned} \tag{3.61}$$

The input information is straight-binary coded data (b_{2i}, b_{2i+1}), but is Gray-coded (di-bit) so that if a symbol error is interpreted as an adjacent symbol it would now only produce one bit in error. The Gray code improves the BER performance of the M-ary digital communication system, as described in Chapter 2.

Using the trigonometric identity $A\cos(u + v) = A\cos(u)\cos(v) - A\sin(u)\sin(v)$, Equation 3.61 for QPSK can be described in terms of an *in-phase* (I, cosine) and *quadrature* (Q, sine) components, as given by Equation 3.62.

$$\begin{aligned}
s_1(t) &= A_v\cos(2\pi f_c t) - A_v\sin(2\pi f_c t) && \text{di-bit } 10 \quad b_{2i}b_{2i+1} = 11 \\
s_2(t) &= -A_v\cos(2\pi f_c t) - A_v\sin(2\pi f_c t) && \text{di-bit } 00 \quad b_{2i}b_{2i+1} = 00 \\
s_3(t) &= -A_v\cos(2\pi f_c t) + A_v\sin(2\pi f_c t) && \text{di-bit } 01 \quad b_{2i}b_{2i+1} = 01 \\
s_4(t) &= A_v\cos(2\pi f_c t) + A_v\sin(2\pi f_c t) && \text{di-bit } 11 \quad b_{2i}b_{2i+1} = 10 \\
& && (i-1)T_s \le t \le iT_s
\end{aligned} \tag{3.62}$$

From Equation 3.61, the carrier amplitude $A_v = 0.707A$ in Equation 3.62. The QPSK signal can be decomposed into two binary PSK signals with same carrier frequency f_c, with one corresponding to a cosine carrier (I component) and the other corresponding to the sine carrier (Q component), as given by Equation 3.62. Because of the I-Q decomposition of the *quaternary* PSK signal, this bandpass modulation method is also often referred to as a *quadrature* PSK signal or, redundantly, as QPSK.

SystemVue Simulation of M-ary PSK

ON THE CD

An M-ary PSK coherent digital communication system with AWGN and optimum receiver implemented as a correlator is shown in Figure 3.20 (see Fig3-20.svu on the CD-ROM). To facilitate the simulation in SystemVue, all of the Analysis tokens, except the BER token output as the Final Value token, are deleted in the simulation model Fig3-20DT. As described in Chapter 1 the SystemVue Textbook Edition does not permit tokens to be deleted.

M-ary PSK transmits a symbol that represents $N = \log_2 M$ bits of information. The SystemVue simulation presented here is for an $M = 4$, 4-PSK signal, but can be easily extended to any 2^N-PSK ($N = \log_2 M$) digital communication system. This M-ary PSK system is similar to the Gray-coded M-ary ASK system, as shown in Figure 3.13. The 4-PSK signals are assumed to have equal apriori probabilities of occurrence ($P_1 = P_2 = P_3 = P_4$, $P_j = 0.25$).

The data source is the pseudonoise PN Sequence token from the Source Library, with parameters of a bit rate $r_b = 1$ kb/sec, amplitude $\pm$ 0.5 V and 0.5 V voltage and 0° phase offset. The PN Sequence token results in binary output amplitudes of 0 and 1 V for one bit time T_b. This data rate is the same as that used for the binary ASK, FSK, and PSK and the M-ary ASK and FSK signals and facilitates the comparison of the power spectral densities.

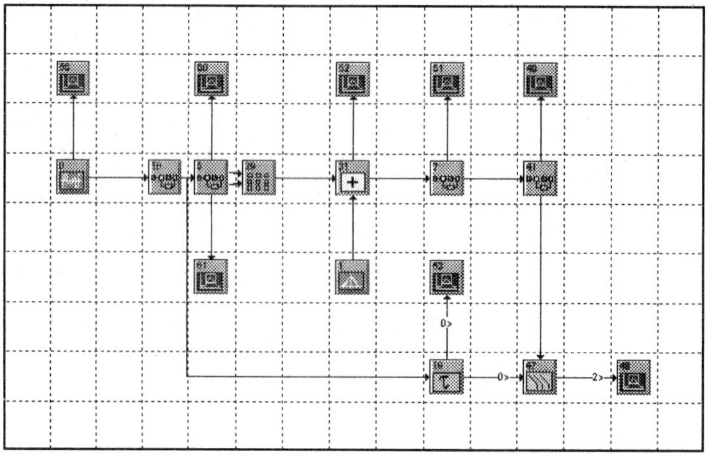

FIGURE 3.20 A 4-PSK (QPSK) coherent digital communication system with BER analysis.

The System Time: Sample Rate is set to 500 kHz. The output of the PN Sequence token is inputted to a Gray encoder MetaSystem, as shown in Figure 3.14, but without the Polynomial token. The output of the Bit-to-Symbol token of the Gray encoder MetaSystem is the 4-level signal with amplitudes of 0 V, 1 V, 2 V, and 3 V at a rate of 500 symbols/sec and no scaling by the Polynomial token is required here. The output of the Bit-to-Symbol token is upsampled by the Resampler token from the Operator Library to restore the System Time: Sample Rate to 500 kHz.

The non-linear M-ary PSK modulator is the I-Q component generator MetaSystem, as shown in Figure 3.21, and the Quadrature Modulator token from the Communications Library, as shown in Figure 3.20. SystemVue also provides the PSK Carrier token from the Source library, which can be used here directly, but the discrete token I-Q component generator and the Quadrature Modulator token illustrate the M-ary PSK modulation principle in more detail.

The input to the I-Q component generator MetaSystem is a 4-level signal with amplitudes of 0 V, 1 V, 2 V, and 3 V at a data rate of 500 symbols/sec, which represent the di-bits 00, 01, 10, and 11. The most significant bit (MSB) represents the cosine I-component and the least significant bit (LSB) represents the sine Q-component in Equation 3.61.

A symbol voltage greater than a threshold of 1.5 V is an MSB of 1 and an I-component of 1 V. A symbol voltage less than a threshold of 1.5 V is an MSB of 0 and an I-component of –1 V.

For the LSB of the di-bit, the thresholds are more complicated. A symbol voltage greater than a threshold of 2.5 V or greater than 0.5 V and less than 1.5 V is an LSB of 1 and a Q-component of 1 V. A symbol voltage less than a threshold of 0.5 V or greater than 1.5 V and less than 2.5 V is an LSB of 0 and a Q-component of –1 V.

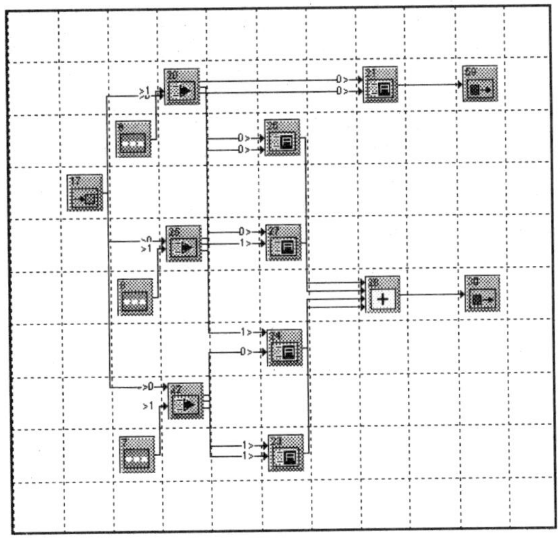

FIGURE 3.21 The I and Q component generator MetaSystem token of the 4-PSK (QPSK) digital communication system.

The three Analog Comparators from the Logic Library provide the three thresholds of 0.5V, 1.5 V, and 2.5 V for the process to determine the I-Q components. The Analog Comparator provides both inverted logic and non-inverted logic outputs, which simplifies the logic.

One of the five AND gate tokens has a logic false output of –1 V, but a logic true output of 1 V with its inputs tied together to provide a voltage level buffer for the I-component. The four other AND gate tokens have logic false outputs of 0 V, but logic true outputs of either −1 V or 1 V and are summed by the Adder token to provide the Q-component.

The I-Q components are inputted to the Quadrature Modulator token from the Communications Library, the I-Q modulator as shown in Figure 3.20. The Quadrature Modulator token, with parameters that are an amplitude $A_v = 0.707\ A = 0.707 \times 5 = 3.536$ V, a carrier frequency $f_c = 20$ kHz, and an initial phase of 0°, generates the in-phase and quadrature cosine and sine carrier components, as given by Equation 3.62.

The I-Q modulator in Figure 3.21 is complicated, but illustrates the concept of multilevel baseband symbol to PSK signal translation. The Analog Comparator, AND and Adder tokens provide the mapping of the Gray-coded baseband symbols to the 4-PSK signal output from the Quadrature Modulator token.

The I-Q component generator in the MetaSystem of Figure 3.21 can be simplified by the QAM Mapper token from the Communications Library. The QAM Mapper token is also used in the SystemVue simulation of a quadrature amplitude modulation (QAM) coherent digital communication system.

ON THE CD

An external file (see QPSK.txt on the CD-ROM) for the I-Q modulator used with the QAM Mapper token for 4-PSK is given in Listing 3.1. The QAM Mapper token greatly simplifies the I-Q component generator of the 4-PSK (QPSK) digital communication system by deleting the Analog Comparator, AND, and Adder tokens.

LISTING 3.1 External File for the QAM Mapper Token to Be Used in the I-Q Component Generator of the 4-PSK (QPSK) Digital Communication System

```
n=4
0   -1.   -1.
1   -1.    1.
2    1.    1.
3    1.   -1.
```

The communications channel is represented by an Adder token and a Gaussian Noise token from the Source Library, with initial parameters of a standard deviation $\sigma = 0$ V and $\mu = 0$ V mean or a power spectral density $N_o = 0$ W/Hz.

The optimum receiver is an in-phase and quadrature (I-Q) correlator MetaSystem, as shown in Figure 3.22. The QPSK signal can be decomposed into two binary PSK signals with quadrature (I, Q) components. Two Multiplier tokens multiply the input signal with the coherent quadrature reference signals $s_{ref\,I}(t) = A_v \cos(2\pi f_c t)$ and $s_{ref\,Q}(t) = A_v \sin(2\pi f_c t)$. The reference signals are provided by two Sinusoid tokens from the Source Library, with parameters of an amplitude of $A_v = 3.536$ V peak, a frequency of $f_c = 20$ kHz, and 0° phase offset. The outputs of the Multiplier tokens are processed by two Integrate and Dump tokens from the Communications Library synchronized to the input signal. The parameters of the

Integrate and Dump tokens are specified as an integration time equal to the symbol time $T_s = 2$ msec and a continuous output.

The two Analog Comparator token comparison inputs are set to the optimum threshold τ_{opt} given by Equation 3.3, because the decomposed QPSK signals are symmetrical. The optimum thresholds τ_{opt} are 0 V here, because $z_0(iT_b) = -6.25$ V and $z_1(iT_b) = 6.25$ V, as given by Equations 3.43 and 3.62. The Custom token from the Source Library provides the constant comparator input, with parameters of one output with an algebraic simulation equation that is $p(0) = 0$. The comparator input could also have been left unconnected.

The cosine I-component Analog Comparator provides an output of 2 V or 0 V representing the MSB of the di-bit and the sine Q-component Analog Comparator provides an output of 1 V or 0 V representing the LSB. An Adder token sums the two I and Q component Analog Comparator outputs to provide a symbol voltage of 0 V, 1 V, 2 V, or 3 V.

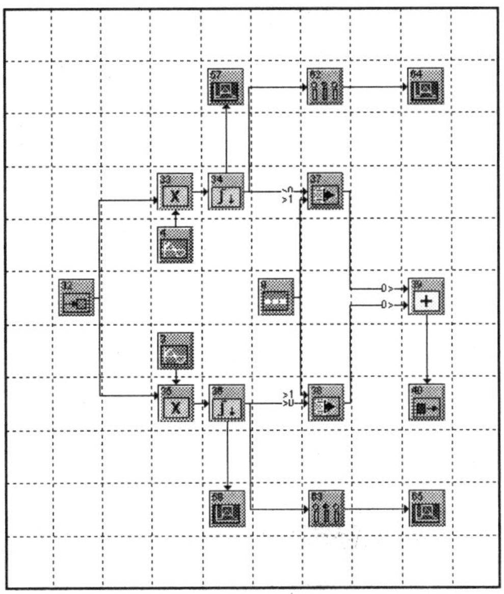

FIGURE 3.22 The I-Q correlator MetaSystem token of the 4-PSK (QPSK) digital communication system.

The output of the I-Q correlator MetaSystem is inputted to the 2-bit Gray decoder MetaSystem, as shown in Figure 3.15. Here the signal is downsampled by the Decimator token by a factor of 1000, producing one sample point per symbol time $T_s = 2$ msec. The output of the Decimator token is processed by the Symbol-to-Bit token, with parameters of two bits per symbol and the MSB inputted first. The three thresholds for the 4-level signal of the Symbol-to-Bit token are set to 0.5 V, 1.5 V, and 2.5 V. The output of the Symbol-to-Bit token is inputted to the Gray Decode token from the Communications Library, with parameters of a threshold of 0.5 V, two bits per symbol and the MSB outputted first and one system simulation point per bit time $T_b = 1$ msec.

The Decimator token output in Figure 3.20, within the MetaSystem, is delayed by 8 msec to correlate the source binary data with the output of the Gray decoder at the end of each bit time T_b. The value of this delay can be observed by the cross-correlation of the two signals in the BER analysis, as described in Chapter 1.

The System Time: Sample Rate is set to 500 kHz. The BER token parameters are set to output on every trial (number of trials = 1), with a threshold of 0.5 V for QPSK, and a 2 msec offset. The System Stop Time is set to 10 sec for convenience, which would process 10 000 information bits in possible error. The Gaussian Noise token is specified by the power spectral density (PSD) N_o W/Hz. Setting the Gaussian noise source PSD to 0 should produce 0 observed errors.

M-ary PSK Power Spectral Density

The M-ary ($M = 4$) PSK or QPSK signal can be decomposed into an in-phase (I) and quadrature (Q) binary PSK signal. Each decomposed binary PSK signal has a PSD, $PSD_{QPSK\,I,\,Q}(f)$, then given by Equation 3.63, which is derived from Equation 3.36 for the binary PSK signal.

$$P_{QPSK\;I,Q}(f) = \frac{1}{4R_L}\left[\frac{A^2 T_s}{4}\,\text{sinc}^2\big(2\pi(T_s/2)(f-f_c)\big) + \frac{A^2 T_s}{4}\,\text{sinc}^2\big(2\pi(T_s/2)(f+f_c)\big)\right] \quad (3.63)$$

Because $T_s = 2T_b$, the bandwidth of the QPSK signal requires only half the bandwidth of a binary PSK signal with the same bit rate r_b bits/sec. The power P in a QPSK signal per symbol time T_s, derived from Equation 3.59, is given by Equation 3.64. Unlike M-ary ASK, the power in a QPSK signal does not depend upon the apriori probability of symbol occurrence P_j.

$$P = \frac{A^2}{T_s R_L}\int_{(i-1)T_S}^{iT_S} \cos^2(2\pi f_c t + k_p \text{m}_j(t))\,\text{d}t = \frac{A^2}{2R_L} \quad (3.64)$$

$$(i-1)T_s \le t \le iT_s \quad j = 1, 2, 3, 4$$

Equation 3.63 assumes that the ratio of the carrier frequency to the symbol rate f_c/r_s is an integer. The integration then over one symbol time T_s is for a whole number of cycles of the carrier sinusoid. The QPSK signal has an amplitude $A = 5$ V and the normalized ($R_L = 1\ \Omega$) power is 12.5 W = 40.97 dBm (10 $\log_{10}$[12500]). The bandwidth of a QPSK signal as a percentage of the total power in the signal is given by Table 3.10.

TABLE 3.10 Bandwidth of a QPSK Signal as a Percentage of the Total Power

Bandwidth (Hz)	Percentage of Total Power
$1/T_b$	90%
$1.5/T_b$	93%
$2/T_b$	95%
$3/T_b$	96.5%
$4/T_b$	97.5%
$5/T_b$	98%

Setting the Gaussian noise PSD $N_o = 0$ W/Hz and increasing the System Time: Number of Samples specification to 524 288 (2^{19}) points results in a spectral resolution of approximately 0.95 Hz. The normalized ($R_L = 1\ \Omega$) PSD of the Adder token channel output in Figure 3.20 without AWGN is shown scaled and as a single-sided spectrum in Figure 3.23 in dBm/Hz.

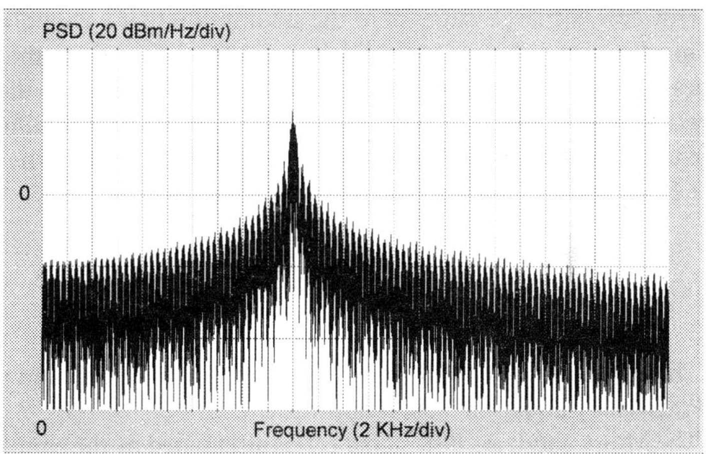

FIGURE 3.23 Power spectral density of a 1 kb/sec pseudonoise QPSK signal.

The PSD of the QPSK signal is similar to that of the binary PSK signal in Figure 3.12. There are approximate nulls in the PSD that occur here at the symbol rate $r_s = 1/T_s$ (500 b/sec or 500 Hz) and multiples of the symbol rate centered about the carrier frequency $f_c = 20$ kHz due to the sinc^2 term in Equation 3.62.

Probability of Bit Error for M-ary PSK Signals

The probability of symbol error P_s for coherent M-ary PSK is given by Equation 3.65 [Carlson02].

$$P_s = 2Q\left(\sqrt{2\log_2 M\left[\frac{E_b}{N_o}\right]\sin^2\frac{\pi}{M}}\right) \tag{3.65}$$

Because the source information is Gray-coded here, symbol errors due to additive noise occur only for adjacent symbols and the probability of bit error P_b is given by Equation 3.66.

$$P_b = \frac{1}{\log_2 M}P_s$$

$$P_b = \frac{2}{\log_2 M}Q\left(\sqrt{2\log_2 M\left[\frac{E_b}{N_o}\right]\sin^2\frac{\pi}{M}}\right) \tag{3.66}$$

However, for large values of M the assumption that a symbol error only occurs due to adjacent symbols is less valid, because the difference in the phase shift φ becomes increasingly smaller as M increases. For this worst case there are $M - 1$ incorrect symbols and in $M/2$ of these symbols a given bit will differ from the same bit in the correct symbol. The upper and lower bound then for the probability of bit error P_b for coherent M-ary PSK is given by Equation 3.67.

$$\frac{1}{\log_2 M}P_s \le P_b \le \frac{M}{2}\frac{1}{M-1}P_s \quad M \ge 4 \tag{3.67}$$

The coherent M-ary PSK signal here has $M = 4$ (QPSK) and the upper-bound for P_b, given by Equation 3.66, is appropriate.

The M-ary bandpass PSK signal $s_j(t)$ is attenuated by the communication channel by an amount γ and has a normalized energy per symbol E_s^j given by Equation 3.59. The attenuation of the communication channel is assumed to be zero ($\gamma = 1$) and the 4-PSK signals are given by Equations 3.60 and 3.61.

From Equation 3.59 with $T_s = 2$ msec and assuming that the ratio of the carrier frequency to the symbol rate f_c/r_s is an integer, the energy per symbol E_s^j for each 4-PSK signal $s_j(t)$ is equal, $E_s^j = E_s = 2.5 \times 10^{-2}$ V^2-sec here. For the 4-PSK signal the energy per bit $E_b = E_s/2 = 1.25 \times 10^{-2}$ V^2-sec, which is the same as the 4-FSK signal here.

Performance of M-ary PSK for the Optimum Receiver in AWGN

The M-ary PSK transmitter and optimum bandpass receiver in Figure 3.20 can be evaluated for its performance in the presence of AWGN. A BER analysis is conducted with the same parameters as that for M-ary ASK and M-ary FSK. The System Stop Time is set to 10 sec for convenience, which would process 10 000 information bits in possible error. To facilitate the simulation in SystemVue, all of the Analysis tokens, except that for the BER token are deleted from the QPSK digital communication system with BER analysis in Figure 3.20. The Final Value token from the Sink Library records the total number of errors in 10 000 trials, as shown in Figure 3.7 for binary ASK.

Table 3.11 is a tabular list of the observed BER in a single trial of 10 000 information bits as a function of E_b/N_o to the theoretical upper bound P_b for Gray-coded QPSK, as given by Equation 3.66. The normalized energy per bit here is $E_b = 1.25 \times 10^{-2}$ V^2-sec.

A comparison of Table 3.11 with Table 3.6 shows a + 6 dB increase in BER performance for M-ary PSK (QPSK) over that for M-ary ASK (4-ASK) using the optimum receiver with the same 95% of the total power bandwidth. The SystemVue simulation performance for BER versus the upper-bound of P_b at an E_b/N_o ratio greater than 4 dB is attributed to the inherent statistical variation of the small (10^4) information bit sample here.

A comparison of Table 3.11 with Table 3.9 shows approximately the same BER performance for M-ary PSK (QPSK) and that for M-ary FSK (MFSK, 4-FSK). However, at 2 kHz (regardless of M) M-ary PSK has a 95% of the total power bandwidth that is less than that of M-ary FSK (4-FSK), which has 95% of the total power bandwidth of 3.5 kHz for the same bit rate of 1 kb/sec.

TABLE 3.11 Observed BER and Theoretical Upper-Bound P_b as a Function of E_b/N_o in a Gray-Coded QPSK Digital Communication System with Optimum Receiver

E_b/N_o dB	N_o V²-sec	BER	P_b (upper-bound)
∞	0	0	0
14	4.98×10^{-4}	0	0
12	7.89×10^{-4}	0	10^{-8}
10	1.25×10^{-3}	0	10^{-6}
8	1.98×10^{-3}	3×10^{-4}	10^{-4}
6	3.14×10^{-3}	4.2×10^{-3}	2.4×10^{-3}
4	4.98×10^{-3}	1.46×10^{-2}	1.25×10^{-3}
2	7.89×10^{-3}	3.69×10^{-2}	3.75×10^{-3}
0	1.25×10^{-2}	7.73×10^{-1}	7.85×10^{-3}

QUADRATURE AMPLITUDE MODULATION

The multilevel (M-ary) modulation methods utilized thus far encoded information as the parameters of the amplitude, frequency or phase of a sinusoidal carrier. The M-ary ASK, FSK and PSK have been compared to the binary ASK, FSK, and PSK for their bit error rate (BER) performance and bandwidth efficiency. Quadrature amplitude modulation (QAM) encodes information as the amplitude and phase of a sinusoidal carrier and, for the first time, two parameters are used to produce M discernable symbols that are bandwidth efficient and have sufficient BER performance.

A QAM signal modifies the amplitude and shifts the phase angle of the carrier frequency to one of M discrete sets of values during the symbol time T_s for the representation of $N = \log_2 M$ binary logic signals for the transmission of information. The QAM in-phase (I) cosine and quadrature (Q) sine carrier signals have a carrier frequency of f_c Hz, and a 0° reference phase angle, as given by the analytical expression in Equation 3.68.

$$s_j(t) = A_I^j \cos(2\pi f_c t) + A_Q^j \sin(2\pi f_c t) \quad (i-1)T_s \le t \le iT_s \quad j = 1, 2, \ldots, M \qquad (3.68)$$

The in-phase $A_I{}^j$ and quadrature $A_Q{}^j$ carrier amplitudes determine the phase shift and the resulting complex phasor *constellation plot* for the QAM signal. The constellation plot is usually symmetrical in the four quadrants. The QAM signal, as did the QPSK signal, displays an inherent I-Q decomposition in Equation 3.68.

The input information is straight-binary coded data ($b_i, b_{i+1}, \ldots, b_{i+M-1}$), but is Gray-coded ($N$-bit) so that if a symbol error is interpreted as an adjacent symbol it would now only produce one bit in error. Gray-coding is easier to implement within a given quadrant of the constellation plot and improves the BER performance of the M-ary digital communication system, as described in Chapter 2.

SystemVue Simulation of QAM

ON THE CD

A quadrature amplitude modulation (QAM) coherent digital communication system with AWGN is shown in Figure 3.24 (see Fig3-24.svu on the CD-ROM). To facilitate the simulation in SystemVue, all of the Analysis tokens, except that for the BER token output as the Final Value token, are deleted in the simulation model Fig3-24DT. As described in Chapter 1 the SystemVue Textbook Edition does not permit tokens to be deleted.

QAM transmits a symbol that represents $N = \log_2 M$ bits of information. The SystemVue simulation presented here is for a 16-QAM signal, but can be easily extended to any 2^N-QAM digital communication system.

The data source is the pseudonoise PN Sequence token from the Source Library, with parameters of a bit rate $r_b = 1$ kb/sec, amplitude $\pm$ 0.5 V and 0.5 V voltage and 0° phase offset. The PN Sequence token results in binary output amplitudes of 0 and 1 V for one bit time T_b. This data rate is the same as that used for the binary ASK, FSK, and PSK and the M-ary ASK, FSK, and PSK signals and facilitates the comparison of the power spectral densities.

The System Time: Sample Rate is set to 500 kHz. The output of the PN Sequence token is inputted to the MetaSystem, as shown in Figure 3.25. The input binary data is downsampled by the Decimator token from the Operator Library by a factor of 500, producing one sample point per bit time $T_b = 1$ msec. The System Time: Sample Rate of 500 kHz divided by the decimation factor of 500 is 1 kHz or a period of 1 msec. The output of the Decimator token is inputted to the Bit-to-Symbol token from the Communications Library, with parameters of a threshold of 0.5 V, four bits per symbol and the MSB inputted first.

The output of the Bit-to-Symbol token is the 16-level signal with amplitudes of 0 to 15 V in steps of 1 V at a data rate of 500 symbols/sec. The output of the Bit-to-Symbol token is upsampled by the Resampler token from the Operator Library to restore the System Time: Sample Rate to 500 kHz.

The output of the bit-to-symbol MetaSystem in Figure 3.25 is input to the QAM Mapper token from the Communications Library in Figure 3.24, with parameters of a constellation size of 16 and an external file that provide the in-phase $A_I{}^j$ and quadrature $A_Q{}^j$ carrier amplitudes, as given by Equation 3.68. The external file (see 16QAMV_32.txt on the CD-ROM) is given in Listing 3.2.

ON THE CD

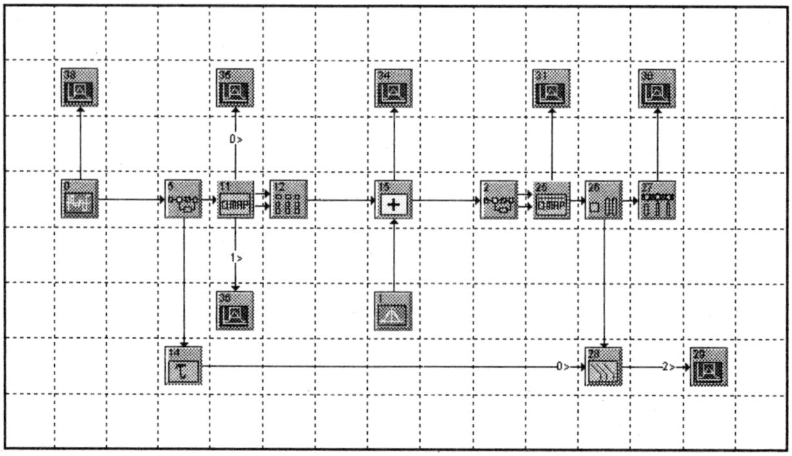

FIGURE 3.24 A 16-QAM coherent digital communication system with BER analysis.

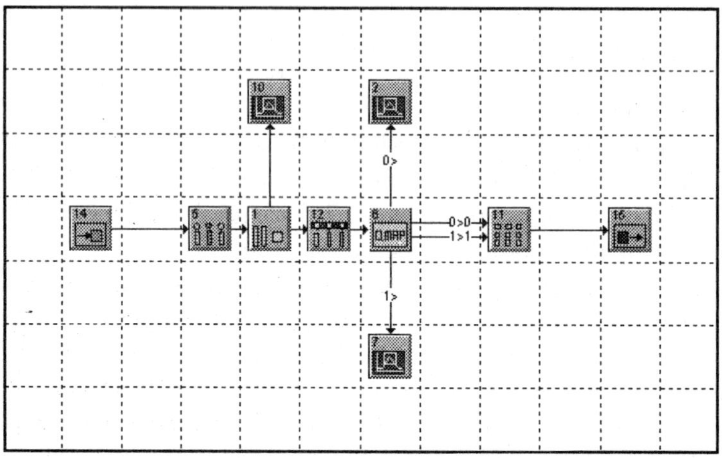

FIGURE 3.25 The bit-to-symbol MetaSystem of the 16-QAM digital communication system.

LISTING 3.2 External File for the QAM Mapper Token of the 16-QAM Digital Communication System

```
n=16
0    -1.    1.
1    -3.    1.
2    -1.    3.
3    -3.    3.
```

```
4    1.    1.
5    1.    3.
6    3.    1.
7    3.    3.
8   –1.   –1.
9   v1.   –3.
10  –3.   –1.
11  –3.   –3.
12   1.   –1.
13   3.   –1.
14   1.   –3.
15   3.   –3.
```

The QAM Mapper token external file defines the number of points in the constellation plot ($n = 16$ here) and each entry contains the symbol voltage inputted from the Bit-to-Symbol token and the mapped in-phase and quadrature voltage output. If the input symbol voltage is 5 V, then the output in-phase (I) voltage is 1 V and the output quadrature (Q) voltage is 3 V. The complexity of the mapping of the multilevel baseband symbol to the 16-QAM signal is simplified by the use of the QAM Mapper token.

The QAM Mapper token external file in Listing 3.2 defines the constellation plot for the CCITT (*Comité Consultatif International Téléphonique et Télégraphique*, an international standards organization) V.32 9.6 kb/sec nonredundant coding modem [Haykin01]. Nonredundant coding divides the input data stream into four consecutive data bits (*quad-bits*).

The I and Q component modulator is the Quadrature Modulator from the Communications Library, with parameters that are an amplitude $A = 1$ V, a carrier frequency $f_c = 20$ kHz, and an initial phase of 0°. The maximum amplitude of the I-Q voltage from the QAM Mapper, as given by Listing 3.2, is $A_{IQ} = \sqrt{(3^2 + 3^2)} = 4.247$ V. The peak output voltage from the Quadrature Modulator then is $A_{QAM} = AA_{IQ} = 4.247$ V and the two other output voltages are approximately $\sqrt{(3^2 + 1^2)} = 3.162$ V and $\sqrt{(1^2 + 1^2)} = 1.414$ V.

The communication channel is represented by an Adder token and a Gaussian Noise token from the Source Library, with initial parameters of a standard deviation $\sigma = 0$ V and $\mu = 0$ V mean or a power spectral density $N_o = 0$ W/Hz.

The optimum receiver is an in-phase and quadrature (I-Q) correlator Meta-System, as shown in Figure 3.26. This I-Q correlator is similar to that for 4-PSK (QPSK) digital communication system, as shown in Figure 3.22. Two Multiplier tokens multiply the input signal with the coherent quadrature reference signals $s_{refI}(t) = A_v \cos(2\pi f_c t)$ and $s_{refQ}(t) = A_v \sin(2\pi f_c t)$. The reference signals are provided by two Sinusoid tokens from the Source Library, with parameters of an amplitude $A_v = 2$ V peak, a carrier frequency $f_c = 20$ kHz, and 0° phase offset.

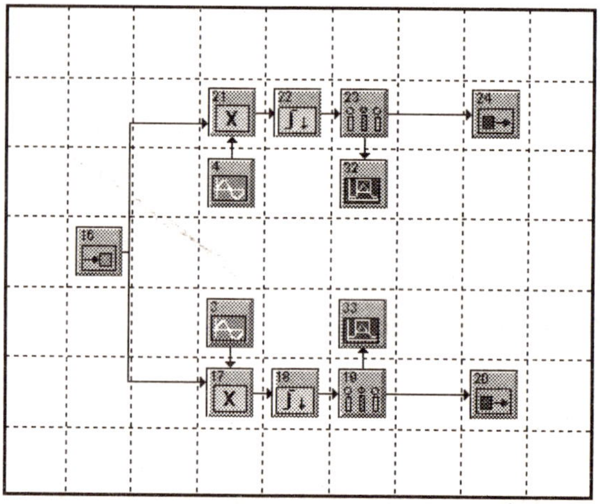

FIGURE 3.26 The I-Q correlator MetaSystem of the 16-QAM digital communication system.

The outputs of the Multiplier tokens are processed by two Integrate and Dump tokens from the Communications Library synchronized to the input signal. The output of the I-Q integrate-and-dump operation $z_j(iT_s)$ is given by Equation 3.43 and, with no attenuation by the communication channel ($\gamma = 1$), are equal to $A_I^j A_v/2 = A_I^j$ and $A_Q^j A_v/2 = A_Q^j$, because $A_v = 2$ V here.

The parameters of the Integrate and Dump tokens are specified as an integration time equal to the symbol time T_s of 4 msec and a continuous output, which is downsampled by two Decimator tokens by a factor of 2000, producing one sample point per symbol time $T_s = 4$ msec. The System Time: Sample Rate of 500 kHz divided by the decimation factor of 2000 is 250 Hz or a period of 4 msec. The sampled data from the Decimator tokens then are the in-phase A_I^j and quadrature A_I^j carrier amplitudes.

The output of the I-Q correlator MetaSystem is inputted to the QAM Demapper token from the Communications Library in Figure 3.24, with parameters of a constellation size of 16 and an external file that provide the in-phase A_I^j and quadrature A_Q^j carrier amplitudes. The external file is the same as that for the QAM Mapper token (16QAMV_32.txt), as given in Listing 3.2.

The output of the QAM Demapper token is inputted to a Symbol-to-Bit token, with parameters of four bits per symbol and the MSB inputted first. The three thresholds for the 4-level signal of the Symbol-to-Bit token are set to 0.5 V, 1.5 V, and 2.5 V. The output of the Symbol-to-Bit is inputted to the BER token for performance analysis and also to a Resampler token from the Operator Library to restore the System Time: Sample Rate to 500 kHz for data display.

The data from the Decimator token in the bit-to-symbol MetaSystem of Figure 3.25 input is delayed by 8 msec to correlate the data. The BER token parameters are set to output on every trial (number of trials = 1), with a threshold of 0.5 V for QAM, and a 0 sec offset. The System Stop Time is set to 10 sec for convenience, which would process 10 000 information bits in possible error. The Gaussian Noise token is specified by the power spectral density (PSD) N_o W/Hz. Setting the Gaussian noise source PSD to 0 should produce 0 observed errors.

QAM Power Spectral Density

The QAM signal can be decomposed into in-phase and quadrature signals and displays the characteristics of an M-ary ASK and M-ary PSK signal. The QAM signal has a PSD, $PSD_{QAM}(f)$, given by Equation 3.69, which is derived from Equation 3.39 for the M-ary ASK signal. The discrete carrier component ($\delta(f)$) does not appear in the PSD of a QAM signal, as it did in Equation 3.39 for M-ary ASK.

$$PSD_{QAM}(f) = \frac{1}{4R_L}\left[\frac{A_{avg}^2 T_s}{4}\operatorname{sinc}^2\left(2\pi\left(T_s/2\right)\left(f-f_c\right)\right)+\frac{A_{avg}^2 T_S}{4}\operatorname{sinc}^2\left(2\pi\left(T_s/2\right)\left(f+f_c\right)\right)\right] \quad (3.69)$$

Because $T_s = 4T_b$ here, the bandwidth of the 16-QAM signal requires only one-quarter the bandwidth of a binary PSK signal with the same bit rate r_b bits/sec. The square of the average amplitude $A_{avg}{}^2$ in Equation 3.69 is given by Equation 3.70, where P_j is the probability of occurrence of a QAM (16-QAM here) symbol.

$$A_{avg}^2 = \sum_{j=1}^{16}\left[(A_I^j)^2 + (A_Q^j)^2\right]P_j \quad (3.70)$$

Substituting the 16 equally likely ($P_1 = P_2 = \ldots = P_{16}$, $P_j = 0.0625$) and the in-phase $A_I{}^j$ and quadrature $A_Q{}^j$ carrier amplitudes, as given in Listing 3.2, for each symbol time T_s here results in an average amplitude $A_{avg} = 3.162$ V.

The total power in a QAM signal is given by Equation 3.42. Unlike M-ary PSK, the power in a QAM signal does depend upon the apriori probability of symbol occurrence P_j. The 16-QAM signal here has an amplitude $A_{avg} = 3.162$ V and the normalized ($R_L = 1\ \Omega$) power is 12.5 W = 40.97 dBm (10 $\log_{10}$[12500]), as given by Equation 3.42. The bandwidth of a QAM signal as a percentage of the total power in the signal is given by Table 3.12, where $T_s = (\log_2 M)\ T_b$.

Setting the Gaussian noise PSD $N_o = 0$ W/Hz and increasing the System Time: Number of Samples specification to 524 288 (2^{19}) points results in a spectral resolution of approximately 0.95 Hz. The normalized ($R_L = 1\ \Omega$) PSD of the Adder token channel output in Figure 3.24 without AWGN is shown scaled and as a single-sided spectrum in Figure 3.23 in dBm/Hz.

TABLE 3.12 Bandwidth of a QAM Signal as a Percentage of the Total Power

Bandwidth (Hz)	Percentage of Total Power
$2/T_s$	90%
$3/T_s$	93%
$4/T_s$	95%
$6/T_s$	96.5%
$8/T_s$	97.5%
$10/T_s$	98%

The PSD of the 16-QAM signal is similar to that of the QPSK signal in Figure 3.23, as shown in Figure 3.27. There are approximate nulls in the PSD that occur here at the symbol rate $r_s = 1/T_s$ (250 b/sec or 250 Hz) and multiples of the symbol rate centered about the carrier frequency $f_c = 20$ kHz due to the sinc^2 term in Equation 3.69.

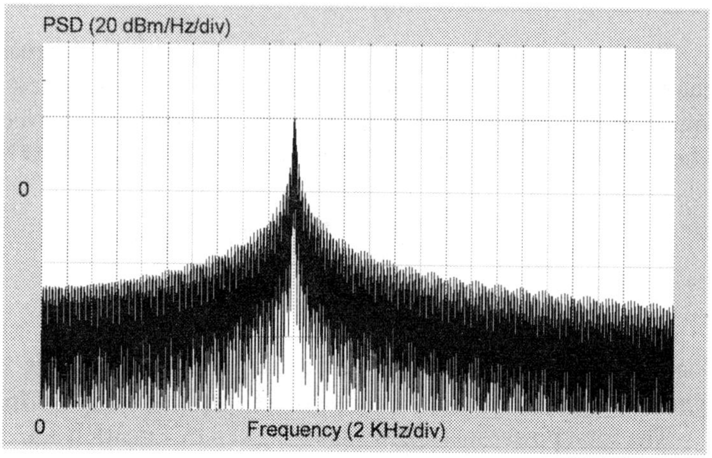

FIGURE 3.27 Power spectral density of a 1 kb/sec pseudonoise 16-QAM signal.

Probability of Bit Error for QAM Signals

The probability of symbol error for coherent QAM with a rectangular shape complex phasor constellation plot, as given by Listing 3.2 for 16-QAM, is tightly upper-bounded by Equation 3.71 [Proakis01].

$$P_s \leq 4\mathrm{Q}\left(\sqrt{\frac{3\,E_{avg,symbol}}{(M-1)\,N_o}}\right) \tag{3.71}$$

The average energy per bit $E_{avg,b} = E_{avg,symbol} / \log_2 M$ and $T_s = \log_2 M\ T_b$, as given by Equation 3.72.

$$\begin{aligned} E_{avg,symbol} &= \frac{\gamma^2 A_{avg}^2\, T_s}{2} \\ E_{avg,b} &= \frac{\gamma^2 A_{avg}^2\, T_s}{2\,\log_2 M} = \frac{\gamma^2 A_{avg}^2\, T_b}{2} \end{aligned} \tag{3.72}$$

For the QAM signal here, $M = 16$, the attenuation of the communication channel is assumed to be zero ($\gamma = 1$), $A_{avg} = 3.162$ V, the bit time $T_b = 1$ msec and the average energy per bit $E_{avg,b} = 5 \times 10^{-3}$ V^2-sec.

The quad-bits of the CCITT V.32 9.6 kb/sec nonredundant coding modem are Gray-coded in each of the four quadrants of the complex phasor constellation plot and the probability of bit error P_b is then given by Equation 3.73.

$$P_b \leq \frac{4}{\log_2 M}\mathrm{Q}\left(\sqrt{\frac{3\,(\log_2 M)\,E_{avg,b}}{(M-1)\,N_o}}\right) \tag{3.73}$$

Performance of QAM for the Optimum Receiver in AWGN

The QAM transmitter and optimum QAM receiver in Figure 3.24 can be evaluated for its performance in the presence of AWGN. A BER analysis is conducted with the same parameters for the Gray-coded QPSK signal as given in Table 3.11. Table 3.13 is a tabular list of the observed BER in a single trial of 10 000 information bits as a function of E_b/N_o to the theoretical Gray-coded source data, as given by Equation 3.73. The normalized average energy per bit here is $E_{avg,b} = E_{avg,symbol}/4 = 5 \times 10^{-3}$ V^2-sec.

A comparison of Tables 3.11 and 3.13 shows an approximate –4 dB E_b/N_o decrease in BER performance for 16-QAM over that for QPSK. The SystemVue simulation performance for BER versus the upper-bound of P_b at an E_b/N_o ratio greater than 6 dB is attributed to the inherent statistical variation of the small (10^4) information bit sample here. The performance tradeoff is that 16-QAM has half the transmission bandwidth of QPSK. A comparison of Tables 3.7 and 3.13 shows an approximate +4 dB E_b/N_o increase in BER performance for 16-QAM over that for 4-ASK and in half the transmission bandwidth.

TABLE 3.13 Observed BER and Theoretical Upper-Bound P_b as a Function of E_b/N_o in a 16-QAM (CCITT V.32) Digital Communication System with Optimum Receiver.

E_b/N_o dB	N_o V²-sec	BER	P_b (upper-bound)
∞	0	0	0
14	1.99×10^{-4}	0	$\approx 5 \times 10^{-6}$
12	3.16×10^{-4}	1×10^{-4}	$\approx 10^{-4}$
10	5.00×10^{-4}	2.9×10^{-3}	2.4×10^{-3}
8	7.92×10^{-4}	1.53×10^{-2}	1.25×10^{-2}
6	1.25×10^{-3}	4.03×10^{-2}	3.67×10^{-2}
4	1.99×10^{-3}	6.87×10^{-2}	7.78×10^{-2}
2	3.16×10^{-3}	1.236×10^{-1}	1.297×10^{-1}
0	5.00×10^{-3}	1.756×10^{-1}	1.867×10^{-1}

DIFFERENTIAL PHASE SHIFT KEYING

The binary and multilevel (M-ary) digital bandpass modulation systems described thus far have utilized the optimum receiver implemented as a correlator. However, the correlation receiver requires a reference signal $s_{ref}(t)$ that is precisely synchronized in phase and frequency with the input signal to be demodulated. Noncoherent receivers do not require a reference signal and are simpler, but are less accurate than coherent receivers.

Differential phase shift keying (DPSK) is a noncoherent binary PSK digital communication system. For a coherent binary PSK digital communication system the individual information bit b_i directly affects the phase of the transmitted carrier, as given by Equation 3.30. DPSK transmits information as the difference between the carrier phase φ_i transmitted for the current bit b_i and the carrier phase φ_{i-1} transmitted for the previous bit b_{i-1}, as given by Equation 3.74.

$$\begin{aligned} \varphi_i &= \varphi_{i-1} && \text{if } b_i = 1 \\ \varphi_i &= \varphi_{i-1} + \pi && \text{if } b_i = 0 \end{aligned} \tag{3.74}$$

DPSK requires a one bit initialization time at the beginning of each data transmission. The DPSK modulated sinusoidal carrier has the same analytical expression as that for the binary PSK signal in Equation 3.30. If the differentially encoded data sequence is d_i, then Table 3.14 illustrates the generation of the DPSK signal. The differential encoded data sequence d_i is the *complement* (NOT) of the exclusive OR

(XOR) logic operation between the input binary data b_i and the delayed differential encoded data sequence d_{i-1}.

TABLE 3.14 Input Binary Data b_i, Differentially Encoded Binary Data d_i, and Transmitted Phase φ_i (Radians) for a DPSK Signal

b_i	d_{i-1}	d_i	φ_i
		1	0
1	1	1	0
0	1	0	π
0	0	1	0
1	1	1	0
0	1	0	π
0	0	1	0
1	1	1	0
1	1	1	0

The binary data sequence b_i is recovered from the phase difference $\varphi_i - \varphi_{i-1}$ from Table 3.14 and Equation 3.74. Although DPSK is a noncoherent digital communication system and does not require a reference signal $s_{ref}(t)$, the optimum receiver is the correlator and precise timing of the bit time T_b is still required. Methods for the synchronization of the bit time in the receiver are presented in Chapter 4, "Synchronization and Equalization."

SystemVue Simulation of DPSK

ON THE CD

A DPSK noncoherent digital communication system with AWGN and optimum receiver implemented as a correlator is shown in Figure 3.28 (see Fig3-28.svu on the CD-ROM). To facilitate the simulation in SystemVue, all of the Analysis tokens, except that for the BER token output as the Final Value token, are deleted in the simulation model Fig3-28DT. As described in Chapter 1 the SystemVue Textbook Edition does not permit tokens to be deleted.

The SystemVue simulation is similar to that of the binary PSK communication system in Figure 3.11. The data source is the polar pseudonoise PN Sequence token from the Source Library, with parameters of a binary amplitude of ± 0.5 V, a bit rate $r_b = 1$ kb/sec, and 0.5 V voltage and 0° phase offset. This PN Sequence token will transmit binary unipolar rectangular pulses for the input data b_i with an amplitude of 0 V or 1 V.

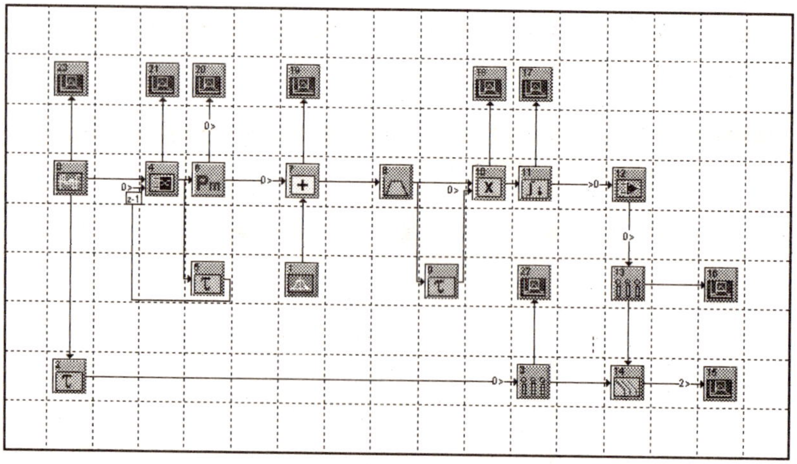

FIGURE 3.28 A DPSK noncoherent digital communication system with BER analysis.

The differential encoded data sequence d_i is generated by the XOR token from the Logic Library with a threshold parameter of 0.5 V. The complement of the XOR logic operation is formed by setting the parameter for a logic true output to 0 V and a logic false output to 1 V. The output of the XOR token is delayed by approximately one bit time T_b by the Delay token from the Operator Library and fed back to the input of the XOR token.

The output of the XOR token shows logic instability or glitches because of the feedback of the delayed differential encoded data sequence d_{i-1}. These glitches impact the performance of the PSK modulator and are obviated by setting the delay to $T_b - T_{system} = 0.998$ msec to provide a clean logic signal.

The non-linear binary PSK modulator is a Phase Modulator token from the Function Library, with parameters of an amplitude $A = 5$ V, a carrier frequency $f_c = 20$ kHz, a modulation gain (or phase deviation factor k_p) of 0.5 2π (radians)/V, and 180° phase offset. This phase offset provides the initial phase $\theta = 180° = \pi$ radians, as given by Equations 3.30 and 3.32 for binary PSK.

Here the requisite PSK phase shift φ for binary 0 when $m_j(t) = 0$ V is $(0 \text{ V} \times 0.5\ 2\pi/\text{V}) + \pi = \pi$ radians. The phase shift φ for binary 1 when $m_j(t) = 1$ V is $\Delta\varphi + \theta = (1 \text{ V} \times 0.5\ 2\pi/\text{V}) + \pi = 0$ radians.

The communication channel is represented by the Adder token and the Gaussian Noise token from the Source Library, with initial parameters of a standard deviation $\sigma = 0$ V and $\mu = 0$ V mean or a power spectral density $N_o = 0$ W/Hz.

The optimum receiver for DPSK is the correlation receiver, as described in Chapter 2 and shown in Figure 2.25, but modified for passband noncoherent demodulation, as shown in Figure 3.28. The Linear System Filter token from the Operator Library provides a bandpass filter (BPF), which is selected to be a 9-pole Butterworth filter, with a reasonable bandwidth of $2/T_b = 2$ kHz, centered at the carrier frequency $f_c = 20$ kHz. This bandwidth would pass approximately 90% of the total power in the DPSK signal.

A Multiplier token multiplies the input signal with the noncoherent reference signal $s_{ref}(t)$. The reference signal $s_{ref}(t)$ is derived from the received signal delayed by one bit time T_b by the Delay token from the Operator Library. The output of the Multiplier token is processed by the Integrate and Dump token from the Communications Library synchronized to the input signal. The parameters of the Integrate and Dump token are specified as an integration time equal to the bit time T_b of 1 msec and a continuous output.

The Analog Comparator token comparison input is set to the optimum threshold τ_{opt} given by Equation 3.3, because the DPSK signals are symmetrical. The optimum threshold τ_{opt} is 0 V here, because $z_0(iT_b) = -12.5$ V and $z_1(iT_b) = 12.5$ V, as given by Equation 3.2. The comparator input has been left unconnected, but a Custom token from the Source Library could provide the input, with parameters of one output with an algebraic simulation equation that is p(0) = 0.

The System Time: Sample Rate is set to 500 kHz. Two Sampler tokens are used to downsample the 500 kHz System Time rate to the 1 kb/sec bit rate or 1 kHz. The Sampler token parameters are set to a sample rate of 1 kHz, and an aperture time of 2 μsec (T_{system}, a single point sample).

The PN Sequence token data Sampler input is delayed by 2 msec to correlate the data. The BER token parameters are set to output on every trial (number of trials = 1), with a threshold of 0.5 V for DPSK, and a 2 msec offset. The System Stop Time is set to 10 sec for convenience, which would process 10 000 information bits in possible error. The Gaussian Noise token is specified by the power spectral density (PSD) N_o W/Hz. Setting the Gaussian noise source PSD to 0 should produce 0 observed errors.

DPSK Power Spectral Density

The DPSK signal is source coded, but transmitted as a binary PSK signal. The bisided power spectral density (PSD) of the DPSK signal then is given by Equation 3.36 and shown in Figure 3.12. The bandwidth of a bandpass DPSK signal is also the same as that for the binary PSK and binary ASK signal as a percentage of total power, as given by Table 3.1. The power per bit time T_b in a DPSK signal is the same as that for a binary PSK signal, as given by Equation 3.37.

Probability of Bit Error for DPSK Signals

DPSK signals are binary and orthogonal over an interval of two bit times $2T_b$, as given by Equation 3.75 for the definition of an orthogonal signal $o(t)$ over a time interval.

$$o(t) = \int_{(i-1)T_b}^{(i+1)T_b} s_0(t)\, s_1(t)\, dt = 0 \tag{3.75}$$

For DPSK, the transmission of a binary 0 using $s_0(t)$ advances the phase of the carrier by 180° and for a binary 1 using $s_1(t)$ the phase of the carrier is unchanged over two bit times $2T_b$, as given by Equation 3.76.

$$\begin{aligned}
s_0(t) &= A_v \sin(2\pi f_c t) && (i-1)T_b \le t \le T_b \\
s_0(t) &= A_v \sin(2\pi f_c t + \pi) && T_b \le t \le (i+1)T_b \\
\\
s_1(t) &= A_v \sin(2\pi f_c t) && (i-1)T_b \le t \le T_b \\
s_1(t) &= A_v \sin(2\pi f_c t) && T_b \le t \le (i+1)T_b
\end{aligned} \tag{3.76}$$

The DPSK signals satisfy Equation 3.75 and the probability of error P_e for coherent correlation demodulation of orthogonal signals is given by Equation 3.77 [Haykin01].

$$P_e = \frac{1}{2}\exp\left(-\frac{E}{2N_0}\right) \tag{3.77}$$

In Equation 3.77 E is the energy per bit. DPSK signals have an equivalent bit interval $T_{DPSK} = 2T_b$ and the effective energy per bit then is twice that for an orthogonal signal over a single bit duration ($E_b = 2E$). The probability of bit error P_b for the noncoherent demodulation of DPSK then is upper-bounded by Equation 3.78, where P_b is not dependent upon the Q function, but rather uses the exponential function [Sklar01].

$$P_b = \frac{1}{2}\exp\left(-\frac{E_b}{N_0}\right) \tag{3.78}$$

Performance of DPSK for the Optimum Receiver in AWGN

The DPSK transmitter and optimum bandpass receiver in Figure 3.28 can be evaluated for its performance in the presence of AWGN. A BER analysis is conducted with the same parameters as that for binary PSK. The System Stop Time is set to 10 sec for convenience, which would process 10 000 information bits in possible error. To facilitate the simulation in SystemVue, all of the Analysis tokens, except that for the BER token are deleted from the DPSK digital communication system with BER analysis in Figure 3.28. The Final Value token from the Sink Library records the total number of errors in 10 000 trials, as shown in Figure 3.7 for binary ASK.

Table 3.15 is a tabular list of the observed BER in a single trial of 10 000 information bits as a function of E_b/N_o to the theoretical upper-bound P_b for DPSK, as given by Equation 3.78. The amplitude $A_v = 5$ V and the bit time $T_b = 1$ msec here and the normalized energy per bit is $E_b = 1.25 \times 10^{-2}$ V^2-sec, as given by Equations 3.4 and 3.76. The attenuation of the communication channel is taken to be zero ($\gamma = 1$).

A comparison of Table 3.5 with Table 3.15 shows a –4 dB decrease in BER performance for the noncoherent demodulation of DPSK over that for the coherent demodulation of binary PSK using the optimum receiver. The SystemVue simulation performance for BER versus the upper-bound of P_b at an E_b/N_o ratio greater than 4 dB is attributed to the inherent statistical variation of the small (10^4) information bit sample here. The tradeoff is that the noncoherent DPSK correlator does not require a reference signal $s_{ref}(t)$.

TABLE 3.15 Observed BER and Theoretical Upper-Bound P_b as a Function of E_b/N_o in a DPSK Digital Communication System with Noncoherent Correlation Receiver

E_b/N_o dB	N_o V^2-sec	BER	P_b (upper-bound)
∞	0	0	0
12	7.89×10^{-4}	0	6.6×10^{-8}
10	1.25×10^{-3}	2×10^{-4}	2.3×10^{-5}
8	1.98×10^{-3}	4.9×10^{-3}	1.8×10^{-3}
6	3.14×10^{-3}	1.71×10^{-2}	9.3×10^{-3}
4	4.98×10^{-3}	4.48×10^{-2}	4.06×10^{-2}
2	7.89×10^{-3}	9.96×10^{-2}	1.025×10^{-1}
0	1.25×10^{-2}	1.813×10^{-1}	1.839×10^{-1}

DIFFERENTIAL QUATERNARY PHASE SHIFT KEYING

Differential quaternary phase shift keying (DQPSK) is a noncoherent 4-level PSK digital communication system that is similar to DPSK. A coherent quaternary PSK (QPSK) digital communication system can be decomposed into an in-phase (I) and quadrature (Q) binary PSK signal, as given by Equation 3.62. The DQPSK signal can be analyzed in a manner analogous to both DPSK and QPSK. DQPSK transmits information as the difference between the carrier phase φ_i transmitted for the current *di-bit* $b_{2i}b_{2i+1}$ and the carrier phase φ_{i-1} transmitted for the previous *di-bit* $b_{2i-2}b_{2i-1}$, as given by Equation 3.79.

$$\begin{aligned} \varphi_i &= \varphi_{i-1} && \text{if } b_{2i}b_{2i+1} = 00 \\ \varphi_i &= \varphi_{i-1} + \pi/2 && \text{if } b_{2i}b_{2i+1} = 01 \\ \varphi_i &= \varphi_{i-1} + \pi && \text{if } b_{2i}b_{2i+1} = 10 \\ \varphi_i &= \varphi_{i-1} + 3\pi/2 && \text{if } b_{2i}b_{2i+1} = 11 \end{aligned} \tag{3.79}$$

DQPSK requires a two bit initialization time at the beginning of each data transmission. The DQPSK signal shifts the phase angle φ_i of the carrier during the symbol time T_s, as given by Equation 3.79. The modulated sinusoidal carrier has an amplitude of A V, a carrier frequency of f_c Hz and a 0° reference phase angle, as given by the analytical expression in Equation 3.80.

$$s_i(t) = A\cos(2\pi f_c t + \varphi_i) \quad (i-1)T_s \le t \le i\,T_s \tag{3.80}$$

From Equation 3.79, the current phase angle φ_i is given by Equation 3.81.

$$\begin{aligned} &\varphi_i = \varphi_{i-1} + \Delta\varphi(d_i) \qquad d_i = 0, 1, 2, 3 \\ &\text{where} \quad \Delta\varphi(0) = 0 \quad \Delta\varphi(1) = \pi/2 \\ &\qquad\qquad \Delta\varphi(2) = \pi \quad \Delta\varphi(3) = 3\pi/2 \end{aligned} \tag{3.81}$$

The di-bit $b_{2i}b_{2i+1}$ is the symbol d_i in Equation 3.81. Using the trigonometric identity $A\cos(u + v) = A\cos(u)\cos(v) - A\sin(u)\sin(v)$, Equation 3.80 can be expressed as Equation 3.82.

$$\begin{aligned} s_i(t) &= A\left[A_I(i)\cos(2\pi f_c t) - A_Q(i)\sin(2\pi f_c t)\right] \quad (i-1)T_s \le t \le iT_s \\ \text{where} \quad A_I(i) &= A_I(i-1)\cos(\Delta\varphi(d_i)) - A_Q(i-1)\sin(\Delta\varphi(d_i)) \\ A_Q(i) &= A_I(i-1)\sin(\Delta\varphi(d_i)) + A_Q(i-1)\cos(\Delta\varphi(d_i)) \end{aligned} \tag{3.82}$$

$A\,A_I(i)$ and $A\,A_Q(i)$ are the amplitudes of the in-phase and quadrature components of the differentially encoded DQPSK signal and both are functions of their previous values and the current input di-bit $b_{2i}b_{2i+1}$ or the symbol d_i.

Although DQPSK is a noncoherent digital communication system and does not require a reference signal $s_{ref}(t)$, the optimum receiver is the correlator and precise timing of the symbol time T_s is still required. Methods for the synchronization of the symbol time in the receiver are presented in Chapter 4.

The binary data sequence b_i is recovered from the phase difference $\varphi_i - \varphi_{i-1}$. The phase angle φ_i of the received DQPSK signal is estimated from the arctangent ($\tan^{-1} x$) of the ratio of the quadrature Q_R and in-phase I_R components of the correlation receiver during that symbol time T_s. From Equation 3.79 the phase difference $\varphi_i - \varphi_{i-1}$ is also analogous to the cosine and sine of the phase difference. Computation of the arctangent can then be obviated by using Cartesian coordinates and the trigonometric identities for the cosine or sine of the arctangent of an angle, as given by Equation 3.83.

$$\begin{aligned} \cos\left[\tan^{-1}\left(\frac{Q_R(i)}{I_R(i)}\right)\right] &= \frac{I_R(i)}{\sqrt{I_R^2(i)+Q_R^2(i)}} \\ \sin\left[\tan^{-1}\left(\frac{Q_R(i)}{I_R(i)}\right)\right] &= \frac{Q_R(i)}{\sqrt{I_R^2(i)+Q_R^2(i)}} \end{aligned} \tag{3.83}$$

It is assumed that the received in-phase I_R and quadrature Q_R components at any symbol time have the same magnitude. From the trigonometric identities $\cos(u-v) = \cos(u)\cos(v) + \sin(u)\sin(v)$ and $\sin(u-v) = \sin(u)\cos(v) - \cos(u)\sin(v)$ and Equation 3.83, the in-phase R_I and quadrature R_Q differentially decoded symbol components are given by Equation 3.84.

$$\begin{aligned} R_I &= I_R(i)\,I_R(i) + Q_R(i)\,Q_R(i-1) \\ R_Q &= Q_R(i)\,I_R(i) - I_R(i)\,Q_R(i-1) \end{aligned} \tag{3.84}$$

The binary data sequence b_i is then recovered by mapping the R_I and R_Q components to the estimated di-bit and binary data $b_{2i}b_{2i+1}$.

SystemVue Simulation of DQPSK

ON THE CD

A DQPSK noncoherent digital communication system with AWGN and optimum receiver implemented as a correlator is shown in Figure 3.29 (see Fig3-29.svu on the CD-ROM). To facilitate the simulation in SystemVue, all of the Analysis tokens, except that for the BER token output as the Final Value token, are deleted in the simulation model Fig3-29DT. As described in Chapter 1, the SystemVue Textbook Edition does not permit tokens to be deleted.

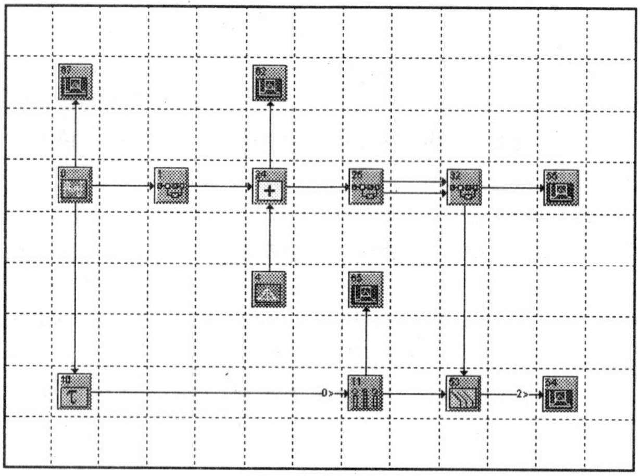

FIGURE 3.29 A DQPSK noncoherent digital communication system with BER analysis.

The SystemVue simulation is similar to that of the QPSK communication system in Figure 3.20. The data source is the polar pseudonoise PN Sequence token from the Source Library, with parameters of a binary amplitude of ± 0.5 V, a bit rate r_b = 1 kb/sec, and 0.5 V voltage and 0° phase offset. This PN Sequence token will transmit binary unipolar rectangular pulses for the input data b_i with an amplitude of 0 V or 1 V.

The differential encoded DQPSK signal is generated by the MetaSystem shown in Figure 3.30. The input binary data b_i signal is downsampled by the Decimator Token from the Operator Library from the System Time: Sample Rate of 2 MHz to the data rate of 1 kHz. The Bit-to-Symbol token from the Communications Library provides the 4-level signal with amplitudes of 0 V, 1 V, 2 V, and 3 V at a rate of 500 symbols/sec. The output of the Bit-to-Symbol token is upsampled by the Resampler token from the Operator Library to restore the System Time: Sample Rate of 2 MHz.

The output of the Resampler token is inputted to the QAM Mapper token from the Communications Library. The QAM Mapper token external file (see ON THE CD DQPSK.txt on the CD-ROM), as given by Listing 3.3, defines the number of points in the constellation plot (n = 4 here) and each entry contains the symbol voltage inputted from the Bit-to-Symbol token and the mapped in-phase and quadrature voltage output. If the input symbol voltage is 2 V, then the output in-phase (I) voltage is –1 V and the output quadrature (Q) voltage is 0 V.

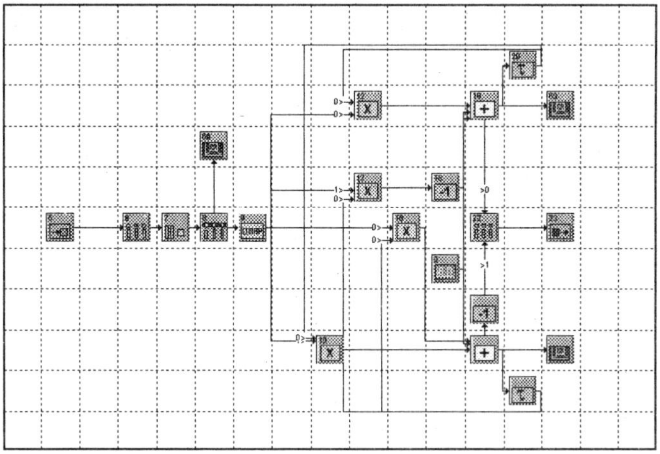

FIGURE 3.30 The differential encoder and I-Q modulator MetaSystem token of the DQPSK digital communication system.

LISTING 3.3 External File for the QAM Mapper Token of the DQPSK Digital Communication System

```
n=4
0    1.    0.
1    0.    1.
2   -1.    0.
3    0.   -1.
```

The outputs of the QAM Mapper token are input to a complex processor implemented with four Multiplier tokens, two Adder tokens, two Negate tokens, and two Delay tokens from the Main and Operator Libraries, as shown in the MetaSystem of Figure 3.30. This encode processor outputs the in-phase $A_I(i)$ and quadrature $A_Q(i)$ voltage amplitudes of the differentially encoded DQPSK signal, as given by Equation 3.82.

The differential process can be initialized with a Pulse Train token from the Source Library, providing the sequence of $A_I(-1) = 1$ and $A_Q(-1) = 1$, with the parameters of an amplitude of 1 V, a pulse width of 2 msec (the symbol time T_s here), and a frequency of 50 msec (a period of 20 sec, longer than the SystemVue simulation Stop Time of 10 sec). The output of the processor is inputted to the Quadrature Modulator token from the Communications Library, with parameters that are an amplitude $A = 5$ V, a carrier frequency $f_c = 20$ kHz, and an initial phase of 0°.

The communication channel is represented by an Adder token and a Gaussian Noise token from the Source Library, with initial parameters of a standard deviation $\sigma = 0$ V and $\mu = 0$ V mean or a power spectral density $N_o = 0$ W/Hz.

The optimum receiver for DQPSK is the correlation receiver modified for passband I-Q noncoherent demodulation, as shown in the MetaSystem of Figure 3.31. The DQPSK signal optimum receiver is similar to the DPSK receiver shown in Figure 3.28.

The Linear System Filter token from the Operator Library provides a bandpass filter (BPF), which is selected to be a 9-pole Butterworth filter, with a reasonable bandwidth of $2/T_s = 1$ kHz ($T_s = 2$ msec here), centered at the carrier frequency $f_c = 20$ kHz. This bandwidth would pass approximately 90% of the total power in the DQPSK signal.

The quadrature reference signal $s_{ref\,Q}(t)$ is derived from the received signal by a 90° phase shifter implemented as a Delay token with a delay of 12.5 μsec, which is a period of $1/4f_c$ for a carrier frequency $f_c = 20$ kHz. Because the requisite delay is 12.5 μsec, the System Time: Sample Rate is raised to 2 MHz, from the usual 500 kHz, for a simulation sampling time T_{system} resolution of 0.5 μsec here.

A Multiplier token multiplies the input signal with the I and Q noncoherent reference signals $s_{ref\,I}(t)$ and $s_{ref\,Q}(t)$. The reference signals are derived from the direct and phase shifted received signal delayed by one symbol time T_s by the Delay token from the Operator Library.

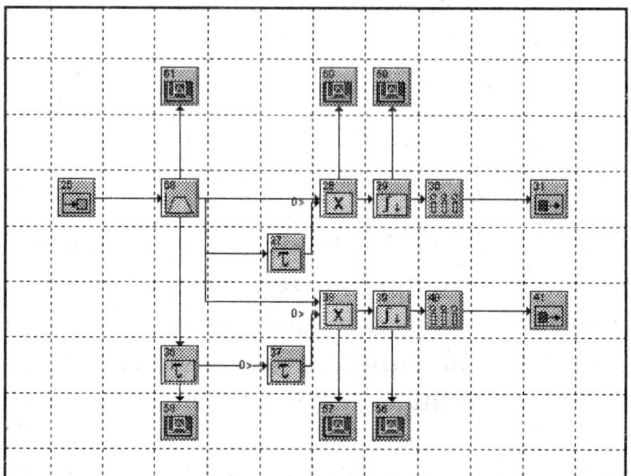

FIGURE 3.31 The noncoherent correlation receiver MetaSystem token of the DQPSK digital communication system.

The output of the Multiplier tokens are processed by Integrate and Dump tokens from the Communications Library synchronized to the input signal. The parameters of the Integrate and Dump tokens are specified as an integration time equal to the symbol time T_s of 2 msec and a continuous output. The output of the Integrate and Dump tokens are downsampled to the symbol rate $r_s = 500$ Hz by Sampler tokens from the Operator Library to provide the in-phase I_R and quadrature Q_R components of the correlation receiver during a symbol time T_s.

The quadrature Q_R and in-phase I_R components of the correlation receiver are input to a complex processor implemented with four Multiplier tokens, two Adder tokens, one Negate tokens, and two Delay tokens from the Main and Operator Libraries, as shown in the MetaSystem of Figure 3.32. This decode processor outputs the in-phase R_I and quadrature R_Q differentially decoded symbol components as given by Equation 3.84.

The output of the differential decoder is inputted to the QAM Demapper token from the Communications Library, with parameters of a constellation size of 4 and an external file, which provide the in-phase A_I and quadrature A_Q carrier amplitudes. The external file is the same as that for the QAM Mapper token (DQPSK.txt), as given in Listing 3.3. The output of the QAM Demapper token is inputted to a Symbol-to-Bit token, with parameters of two bits per symbol and the MSB inputted first. The three thresholds for the 4-level signal of the Symbol-to-Bit token are set to 0.5 V, 1.5 V, and 2.5 V.

The output of the Symbol-to-Bit in the differential decoder MetaSystem is inputted to the BER token for performance analysis. The BER token parameters are set to output on every trial (number of trials = 1), with a threshold of 0.5 V for QAM, and a 0 sec offset. The System Stop Time is set to 10 sec for convenience, which would process 10 000 information bits in possible error. The Gaussian Noise token is specified by the power spectral density (PSD) N_o W/Hz. Setting the Gaussian noise source PSD to 0 should produce 0 observed errors.

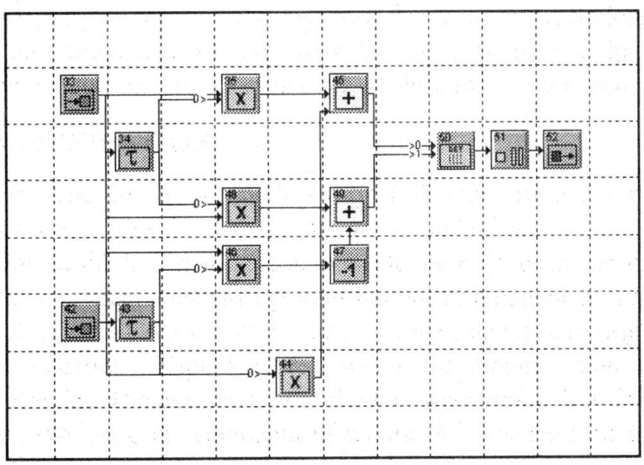

FIGURE 3.32 The differential decoder MetaSystem token of the DQPSK digital communication system.

DQPSK Power Spectral Density

The DQPSK signal is differentially encoded, but transmitted as a quaternary (4-level) PSK (QPSK) signal and can also be decomposed into an in-phase and quadrature binary PSK signal. The bi-sided power spectral density (PSD) of each component of the DQPSK signal is given by Equation 3.63 and shown in Figure 3.23 for QPSK. The bandwidth of a bandpass DQPSK signal is then also the same as that for the QPSK signal as a percentage of total power, as given by Table 3.10. The power per symbol time T_s in a DQPSK signal is the same as that for a QPSK signal, as given by Equation 3.64.

Probability of Bit Error for DQPSK Signals

The probability of symbol error for differential quaternary PSK is approximated by Equation 3.85 [Lindsey91].

$$P_s \approx 2Q\left(\sqrt{\frac{1.172\,E_b}{N_o}}\right) \tag{3.85}$$

From Equations 3.59 and 3.80, the energy per symbol E_s is given by Equation 3.86.

$$E_s = \frac{\gamma^2 A^2 T_s}{2} \tag{3.86}$$

The attenuation of the communication channel is assumed to be zero ($\gamma = 1$), $A = 5$ V, the symbol time $T_s = 2$ msec and the energy per symbol $E_s = 2.5 \times 10^{-2}$ V^2-sec. For DQPSK the energy per bit E_b is $E_s/2 = 1.25 \times 10^{-2}$ V^2-sec.

Assuming that the probability of occurrence for each of the four error regions is equally likely and from Equation 2.44, the probability of bit error for DQPSK then is given by Equation 3.87.

$$P_b \approx \frac{4}{3}Q\left(\sqrt{\frac{1.172\,E_b}{N_o}}\right) \tag{3.87}$$

Performance of DQPSK for the Optimum Receiver in AWGN

The DQPSK transmitter and optimum noncoherent bandpass receiver in Figure 3.29 can be evaluated for its performance in the presence of AWGN. A BER analysis is conducted with the same parameters as that for QPSK. The System Stop Time is set to 10 sec for convenience, which would process 10 000 information bits in possible error. To facilitate the simulation in SystemVue, all of the Analysis windows, except the BER token are deleted from the DQPSK digital communication system with BER analysis in Figure 3.29. The Final Value token from the Sink Library records the total number of errors in 10 000 trials, as shown in Figure 3.7 for binary ASK.

Table 3.16 is a tabular list of the observed BER in a single trial of 10 000 information bits as a function of E_b/N_o to the theoretical P_b for DQPSK, as given by Equation 3.87. The amplitude $A = 5$ V and the symbol time $T_s = 2$ msec here and the normalized energy per bit is $E_b = 1.25 \times 10^{-2}$ V^2-sec, as given by Equations 3.59 and 3.86. The attenuation of the communication channel is taken to be zero ($\gamma = 1$).

A comparison of Table 3.15 with Table 3.16 shows an approximate –2 dB decrease in BER performance for the noncoherent demodulation of DQPSK over that for the noncoherent demodulation of DPSK using the optimum receiver. The SystemVue simulation performance for BER versus P_b at an E_b/N_o ratio greater than 6 dB is attributed to the inherent statistical variation of the small (10^4) information bit sample here. The tradeoff is that although DQPSK requires one half the bandwidth of DPSK, it is substantially more complex.

TABLE 3.16 Observed BER and Theoretical P_b as a Function of E_b/N_o in a DQPSK Digital Communication System with Noncoherent Correlation Receiver

E_b/N_o dB	No V2-sec	BER	P_b
∞	0	0	0
12	7.89×10^{-4}	0	6.4×10^{-5}
10	1.25×10^{-3}	2.1×10^{-3}	4.0×10^{-4}
8	1.98×10^{-3}	8.9×10^{-3}	4.4×10^{-3}
6	3.13×10^{-3}	3.82×10^{-2}	2.05×10^{-2}
4	4.97×10^{-3}	5.55×10^{-2}	5.69×10^{-2}
2	7.89×10^{-2}	$1.205 \times 10_{-1}$	1.159×10^{-1}
0	1.25×10^{-2}	2.622×10^{-1}	2.519×10^{-1}

NONCOHERENT DEMODULATION OF BINARY FREQUENCY SHIFT KEYING

Differential binary and quaternary phase shift keying (DPSK and DQPSK) utilize noncoherent demodulation, where the reference signal $s_{ref}(t)$ for the correlator is derived from the received signal itself. Binary frequency shift keying (FSK) can also be demodulated noncoherently to simplify the receiver and provide a degree of a design advantage. The binary FSK signal utilizes a noncoherent demodulator, which does not require either a derived reference signal or bit time synchronization.

SystemVue Simulation of Noncoherent Binary FSK

ON THE CD

The noncoherent demodulator of a BFSK signal consists of two bandpass filters and envelope detectors, followed by a binary threshold comparator, as shown in Figure 3.33 (see Fig3-33.svu on the CD-ROM). To facilitate the simulation in SystemVue, all of the Analysis tokens, except that for the BER token output as the Final Value token, are deleted in the simulation model Fig3-33DT. As described in Chapter 1 the SystemVue Textbook Edition does not permit tokens to be deleted.

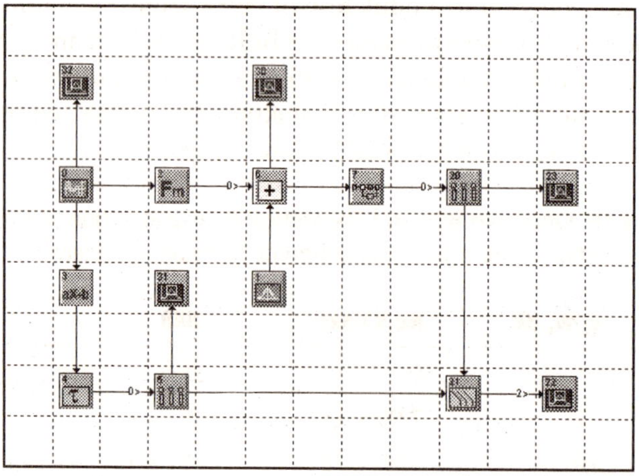

FIGURE 3.33 A binary FSK noncoherent digital communication system with BER analysis.

The SystemVue simulation is similar to that of the coherent BFSK communication system in Figure 3.8. The data source is the polar pseudonoise PN Sequence token from the Source Library, with parameters of a binary amplitude of ± 1 V, a bit rate r_b = 1 kb/sec, and 0 V voltage and 0° phase offset. This PN Sequence token will transmit binary polar rectangular pulses for the input data b_i with an amplitude of ± 1 V.

The non-linear FSK modulator is the Frequency Modulator token from the Function Library, with parameters of an amplitude A = 5 V, a carrier frequency f_c = 20 kHz, 0° phase offset, and an initial modulation gain (or frequency deviation factor k_f) of 2000 Hz/V. The resulting frequency deviation Δf here is ± 1 V × 2000 = ± 2 kHz, as given by Equation 3.22. This frequency deviation Δf = 2 kHz is not the minimum, which is $r_b/2$ = 500 Hz here, but is appropriate for the noncoherent demodulation of binary FSK. The bandpass filtering of the deviated carrier frequencies $f_c \pm \Delta f$ is facilitated if they are well separated in frequency.

The communication channel is represented by the Adder token and the Gaussian Noise token from the Source Library, with initial parameters of a power spectral density $N_o = 0$ W/Hz. The System Time: Sample Rate is set to 500 kHz.

The noncoherent binary FSK demodulator is the MetaSystem shown in Figure 3.34. The Linear System Filter tokens from the Operator Library provide two bandpass filters (BPF), which are selected to be a 9-pole Butterworth filter, with a reasonable bandwidth of $2/T_b = 2$ kHz, centered at the carrier frequency and the frequency deviation, or $f_c \pm \Delta f = 20 \pm 2$ kHz. This bandwidth would pass at least 90% of the total power in the binary FSK signal.

The output of each BPF is inputted to the envelope detector, which consists of the Half Wave Rectifier token from the Function Library and a lowpass filter (LPF). The Half Wave Rectifier token is an ideal diode with a conduction threshold of 0 V. The LPF is the Linear System Filter token selected to be a 9-pole Butterworth filter, again with a bandwidth of $2/T_b = 2$ kHz.

The positive voltage output of the BPF and envelope detector, which processes the signal centered at $f_c - \Delta f = 20 - 2$ kHz $= 18$ kHz, is negated by the Negate token from the Operator Library because this signal represents a binary 0. This demodulated signal and the positive voltage output of the BPF and envelope detector, which processes the signal centered at $f_c + \Delta f = 20 + 2$ kHz $= 22$ kHz, are summed by an Adder token.

The output of the Adder token is inputted to an Analog Comparator token with the comparison input set to 0 V (unconnected). The binary FSK signal can be decomposed as the sum of two amplitude shift keyed (ASK) signals, as given by Equation 3.24. The demodulated noncoherent binary FSK signal, from the BPF and envelope detector, then is a polar signal derived from the two summed on-off keyed (OOK) ASK signals. This noncoherent binary FSK demodulator is not optimal because the envelope detector and comparator are not equivalent to correlation.

Two Sampler tokens are used to downsample the 500 kHz System Time rate to the 1 kb/sec bit rate or 1 kHz. The Sampler token parameters are set to a sample rate of 1 kHz, and an aperture time of 2 μsec (T_{system}, a single point sample).

The PN Sequence token data Sampler input is delayed by 2 msec to correlate the data. The BER token parameters are set to output on every trial (number of trials = 1), with a threshold of 0.5 V for binary FSK, and a 1 msec offset. The System Stop Time is set to 10 sec for convenience, which would process 10 000 information bits in possible error. The Gaussian Noise token is specified by the power spectral density (PSD) N_o W/Hz. Setting the Gaussian noise source PSD to 0 should produce 0 observed errors.

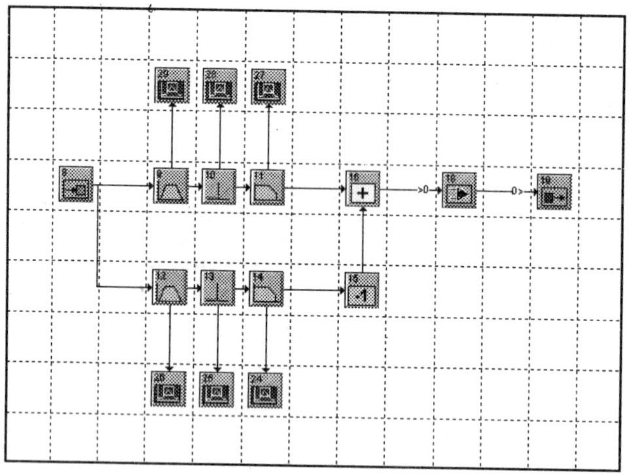

FIGURE 3.34 The noncoherent demodulator MetaSystem token of the binary FSK digital communication system.

Probability of Bit Error for Noncoherent Binary FSK Signals

The probability density function (*pdf*) of a binary FSK signal with additive white Gaussian noise (AWGN) is found to be a Rician-distributed random variable when the normalized power of the received signal $(\gamma A)^2$ is significantly higher than the normalized noise power σ_o^2, as given by Equation 3.88 [Papoulis02].

$$\text{pdf}_{\text{Rician}}(r) = 0 \qquad r < 0$$

$$\text{pdf}_{\text{Rician}}(r) = \sqrt{\frac{r}{2\pi\,\gamma A\,\sigma_o^2}}\,\exp\left[\frac{-(r-\gamma A)^2}{2\,\sigma_o^2}\right] \qquad r \geq 0 \tag{3.88}$$

However, the pdf of only AWGN is found to be a Rayleigh-distributed random variable, as given by Equation 3.89 [Papoulis02].

$$\text{pdf}_{\text{Rayleigh}}(r) = 0 \qquad r < 0$$

$$\text{pdf}_{\text{Rayleigh}}(r) = \frac{r}{\sigma_o^2}\exp\left[\frac{-r^2}{2\,\sigma_o^2}\right] \qquad r \geq 0 \tag{3.89}$$

The probability of bit error P_b^j due to either a binary 1 being transmitted and a binary 0 being received (P_b^1) or a binary 0 being transmitted and a binary 1 being received (P_b^0) is given by Equation 3.90 [Stern04].

$$P_b^j = \int_0^\infty \mathrm{pdf}_{\mathrm{Rician}}(m) \int_m^\infty \mathrm{pdf}_{\mathrm{Rayleigh}}(r)\, dr\, dm \tag{3.90}$$

Assuming that the apriori probabilities of transmission of a binary 1 and binary 0 are equally likely ($P_1 = P_0 = 0.5$) and the noncoherent demodulators are symmetrical, the probability of bit error P_b for noncoherent binary FSK, from Equations 2.27 and 3.90, is given by Equation 3.91 [Shanmugan83].

$$P_b^j = \frac{1}{2}\exp\left[\frac{-(\gamma A)^2}{4\sigma_o^2}\right] \tag{3.91}$$

The normalized energy per bit Eb for a binary FSK signal is $(\gamma A)^2\, T_b/2$. The average noise power σ_o^2 at the output of a bandpass filter with a bandwidth B Hz is given by Equation 3.92.

$$\sigma_o^2 = 2B\frac{N_o}{2} = 2r_b N_o = \frac{2N_o}{T_b} \tag{3.92}$$

The bandwidth B of the bandpass filter is taken to be $2r_b$ Hz. The probability of bit error P_b then is given by Equation 3.93.

$$P_b = \frac{1}{2}\exp\left[\frac{-E_b}{4N_o}\right] \tag{3.93}$$

Performance of Noncoherent Binary FSK in AWGN

The binary FSK transmitter and noncoherent bandpass receiver in Figure 3.33 can be evaluated for its performance in the presence of AWGN. A BER analysis is conducted with the same parameters as that for coherently demodulated binary FSK. The System Stop Time is set to 10 sec for convenience, which would process 10 000 information bits in possible error. To facilitate the simulation in SystemVue, all of the Analysis windows, except the BER token are deleted from the noncoherent binary FSK digital communication system with BER analysis in Figure 3.33. The Final Value token from the Sink Library records the total number of errors in 10 000 trials, as shown in Figure 3.7 for binary ASK.

Table 3.17 is a tabular list of the observed BER in a single trial of 10 000 information bits as a function of E_b/N_o to the theoretical P_b for noncoherent binary FSK, as given by Equation 3.93. The amplitude $A = 5$ V, the bit time $T_b = 1$ msec and the

TABLE 3.17 Observed BER and Theoretical P_b as a Function of E_b/N_o in a Binary FSK Digital Communication System with Noncoherent Receiver

E_b/N_o dB	N_o V²-sec	BER	P_b
∞	0	0	0
16	3.13×10^{-4}	0	2.3×10^{-5}
14	4.97×10^{-4}	3×10^{-4}	9.2×10^{-4}
12	7.89×10^{-4}	7.8×10^{-3}	9.5×10^{-3}
10	1.25×10^{-3}	3.03×10^{-2}	4.10×10^{-2}
8	1.98×10^{-3}	8.51×10^{-2}	1.031×10^{-1}
6	3.13×10^{-3}	1.604×10^{-1}	1.842×10^{-1}
4	4.97×10^{-3}	2.491×10^{-1}	2.666×10^{-1}
2	7.89×10^{-3}	3.163×10^{-1}	3.365×10^{-1}
0	1.25×10^{-2}	3.776×10^{-1}	3.894×10^{-1}

normalized energy per bit $E_b = 1.25 \times 10^{-2}$ V²-sec, as given by Equations 3.4 and 3.21. The attenuation of the communication channel is taken to be zero ($\gamma = 1$).

A comparison of Table 3.17 with Table 3.15 shows an approximate –4 dB decrease in BER performance for the noncoherent demodulation of binary FSK over that for the noncoherent demodulation of binary DPSK using the optimum receiver. Here the comparison is made between two binary, but noncoherently demodulated digital communication systems. The tradeoff is that although binary FSK requires more bandwidth by an amount equal to $2\Delta f$ Hz for any percentage of the total power compared to binary DPSK and does not perform as well, it is substantially less complex.

NONCOHERENT DEMODULATION OF BINARY AMPLITUDE SHIFT KEYING

Binary amplitude shift keying (ASK) can also be demodulated noncoherently to simplify the receiver and provide a degree of a design advantage. The bandpass filter, envelope detector and analog comparator, as in the noncoherent demodulator of FSK shown in Figure 3.34, are used here. The ASK signal utilizes on-off keying (OOK) and the noncoherent demodulator does not require either a derived reference signal or bit time synchronization.

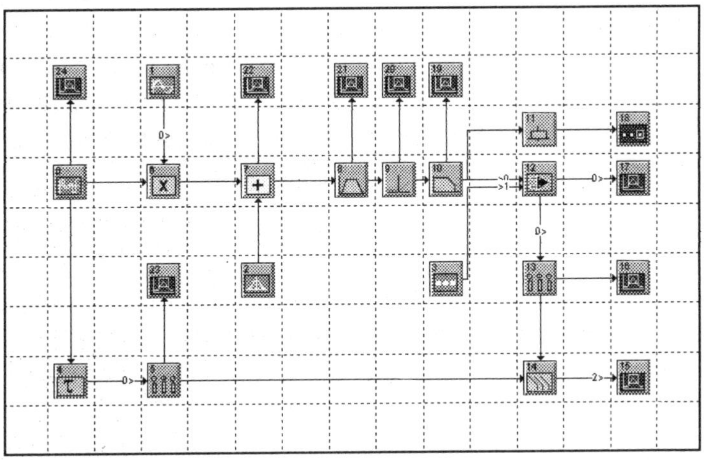

FIGURE 3.35 A binary ASK (OOK) noncoherent digital communication system with BER analysis.

SystemVue Simulation of Noncoherent Binary ASK

ON THE CD

The noncoherent demodulator of a binary ASK (OOK) signal consists of a bandpass filter and an envelope detector, followed by a binary threshold comparator, as shown in Figure 3.35 (see Fig3-35.svu on the CD-ROM). To facilitate the simulation in SystemVue, all of the Analysis tokens, except that for the BER token output as the Final Value token, are deleted in the simulation model Fig3-35DT. As described in Chapter 1 the SystemVue Textbook Edition does not permit tokens to be deleted.

The SystemVue simulation is similar to that of the coherent ASK communication system in Figure 3.4. The data source is the polar pseudonoise PN Sequence token from the Source Library, with parameters of a binary amplitude of $\pm$ 0.5 V, a bit rate r_b = 1 kb/sec, and 0.5 V voltage and 0° phase offset. This PN Sequence token will transmit binary polar rectangular pulses for the input data b_i with an amplitude of either 1 V or 0 V.

The carrier generator is the Sinusoid token from the Source Library, with parameters of an amplitude A = 5 V peak, a carrier frequency f_c = 20 kHz, and 0° phase offset. The non-linear binary ASK modulator is the Multiplier token.

The communication channel is represented by the Adder token and the Gaussian Noise token from the Source Library, with initial parameters of a standard deviation σ = 0 V and μ = 0 V mean or a power spectral density N_o = 0 W/Hz.

The noncoherent binary ASK demodulator is a bandpass filter, envelope detector and analog comparator. The Linear System Filter token from the Operator

Library provides a bandpass filter (BPF), which is selected to be a 9-pole Butterworth filter, with a reasonable bandwidth of $2/T_b = 2$ kHz, centered at the carrier frequency f_c. This bandwidth would pass at least 90% of the total power in the binary ASK signal.

The output of the BPF is inputted to the envelope detector, which consists of the Half Wave Rectifier token from the Function Library and a lowpass filter (LPF). The Half Wave Rectifier token is an ideal diode with a conduction threshold of 0 V. The LPF is the Linear System Filter token selected to be a 9-pole Butterworth filter, again with a bandwidth of $2/T_b = 2$ kHz.

The positive voltage output of the BPF and envelope detector, which processes the signal centered at the carrier frequency $f_c = 20$ kHz, is inputted to an Analog Comparator token with the comparison input set to a threshold voltage. The threshold τ is derived from the Custom token from the Source Library, which provides the Analog Comparator token comparison input from one output with an algebraic simulation equation that is $p(0) = \tau$ V. This noncoherent binary ASK demodulator is not optimal because the envelope detector and comparator are not equivalent to correlation.

Setting the threshold τ analytically requires the analysis of the transient response of the BPF and envelope detector to the binary ASK signal, which is somewhat involved. However, an Average token from the Operator Library can be used to compute the result in the SystemVue simulation, as shown in Figure 3.35.

The BPF output is inputted to the Average token, with a time window parameter of 10 sec, the System Stop Time. A Final Value token from the Sink Library displays the result. To facilitate the simulation in SystemVue, all of the Analysis windows, except the Final Value token are deleted from the noncoherent BFSK digital communication system with BER analysis in Figure 3.35. With no additive noise, the threshold $\tau = 0.83$ V.

Two Sampler tokens are used to downsample the 500 kHz System Time rate to the 1 kb/sec bit rate or 1 kHz. The Sampler token parameters are set to a sample rate of 1 kHz, and an aperture time of 2 μsec (T_{system}, a single point sample).

The PN Sequence token data Sampler input is delayed by 2 msec to correlate the data. The BER token parameters are set to output on every trial (number of trials = 1), with a threshold of 0.5 V for binary FSK, and a 1 msec offset. The System Stop Time is set to 10 sec for convenience, which would process 10 000 information bits in possible error. The Gaussian Noise token is specified by the power spectral density (PSD) N_o W/Hz. Setting the Gaussian noise source PSD to 0 should produce 0 observed errors.

Probability of Bit Error for Noncoherent Binary ASK Signals

The probability density function (*pdf*) of a binary ASK (OOK) signal with additive white Gaussian noise (AWGN) is also found to be a Rician-distributed random variable when the normalized power of the received signal $(\gamma A)^2$ is significantly higher than the normalized noise power σ_o^2, as given by Equation 3.88 [Papoulis02]. However, the pdf of only AWGN is a Rayleigh-distributed random variable, as given by Equation 3.89 [Papoulis02].

The probability of bit error P_b due to a binary 1 being transmitted and a binary 0 being received and a binary 0 being transmitted and a binary 1 being received is given by Equation 3.94 [Stern04].

$$P_b^j = \int_0^\infty \mathrm{pdf}_{\mathrm{Rician}}(m) + \int_m^\infty \mathrm{pdf}_{\mathrm{Rayleigh}}(r)\, dr\, dm \tag{3.94}$$

Assuming that the apriori probabilities of transmission of a binary 1 and binary 0 are equally likely ($P_1 = P_0 = 0.5$) and that the normalized received signal power $(\gamma A)^2$ is significantly higher than the normalized noise power σ_o^2, the probability of bit error Pb for noncoherent binary ASK (OOK), from Equations 2.27 and 3.94, is given by Equation 3.95 [Shanmugan83].

$$P_b \approx \frac{1}{2} \mathrm{Q}\left(\frac{\gamma A}{2\sigma_o}\right) + \frac{1}{2} \exp\left[\frac{-(\gamma A)^2}{8\sigma_o^2}\right] \tag{3.95}$$

The normalized energy per bit E_b for a binary ASK (OOK) signal is $(\gamma A)^2\, T_b/4$ and the normalized noise power at the output of a bandpass filter with a bandwidth of B Hz is given by Equation 3.92. The probability of bit error P_b then is given by Equation 3.96.

$$P_b \approx \frac{1}{2} \mathrm{Q}\left(\sqrt{\frac{E_b}{2N_o}}\right) + \frac{1}{2} \exp\left[\frac{-E_b}{4N_o}\right] \tag{3.96}$$

The theoretical optimum threshold τ_{opt} for the binary ASK (OOK) signal is given by Equation 3.97 [Shanmugan83].

$$\tau_{opt} \approx \frac{\gamma A}{2} \sqrt{1 + \frac{8\sigma_o^2}{(\gamma A)^2}} \tag{3.97}$$

Equation 3.97 assumes that the noncoherent receiver utilizes an ideal bandpass and lowpass filter in the envelope detector. Because the normalized power of the received signal $(\gamma A)^2$ is significantly higher than the normalized noise power σ_o^2, the theoretical optimum threshold $\tau_{opt} \approx \gamma A/2$ V. However, the theoretical optimum threshold is not practical, because the bandpass and lowpass filters are, in fact, not ideal.

Performance of Noncoherent Binary ASK in AWGN

The binary ASK transmitter and noncoherent bandpass receiver in Figure 3.35 can be evaluated for its performance in the presence of AWGN. A BER analysis is conducted with the same parameters as that for coherently demodulated binary ASK. The System Stop Time is set to 10 sec for convenience, which would process 10 000 information bits in possible error. To facilitate the simulation in SystemVue, all of the Analysis windows, except the BER token and the Final Value token for the estimated threshold are deleted from the noncoherent binary ASK digital communication system with BER analysis in Figure 3.35. The Final Value token from the Sink Library records the total number of errors in 10 000 trials, as shown in Figure 3.7 for binary ASK.

Table 3.18 is a tabular list of the observed BER in a single trial of 10 000 information bits as a function of E_b/N_o to the theoretical P_b for noncoherent binary ASK, as given by Equation 3.96. The amplitude $A = 5$ V and the bit time $T_b = 1$ msec here and the normalized energy per bit is $E_b = 6.25 \times 10^{-3}$ V^2-sec, as given by Equations 3.4 and 3.14. The attenuation of the communication channel is taken to be zero ($\gamma = 1$).

A comparison of Table 3.18 with Table 3.17 shows an approximate –2 dB decrease in BER performance for the noncoherent demodulation of binary ASK (OOK) with a fixed threshold over that for the noncoherent demodulation of binary FSK. Here the comparison is again made between two binary, but noncoherently demodulated digital communication systems. The tradeoff is that, although binary ASK does not perform as well, it requires less bandwidth by an amount equal to $2\Delta f$ Hz for any percentage of the total power compared to binary FSK and is substantially less complex.

TABLE 3.18 Observed BER and Theoretical P_b as a Function of E_b/N_o in a Binary ASK Digital Communication System with Noncoherent Receiver and a Fixed Threshold $\tau = 0.83$ V

E_b/N_o dB	N_o V^2-sec	BER	P_b
∞	0	0	0
16	1.56×10^{-4}	3×10^{-4}	2.23×10^{-5}
14	2.48×10^{-4}	3×10^{-3}	1.01×10^{-3}
12	3.94×10^{-4}	1.72×10^{-2}	2.14×10^{-2}
10	6.25×10^{-4}	4.69×10^{-2}	4.75×10^{-2}
8	9.90×10^{-4}	1.131×10^{-1}	1.222×10^{-1}
6	1.56×10^{-3}	1.983×10^{-1}	2.225×10^{-1}

→

E_b/N_o dB	N_o V²-sec	BER	P_b
4	2.48×10^{-3}	2.846×10^{-1}	3.317×10^{-1}
2	3.94×10^{-3}	3.645×10^{-1}	4.293×10^{-1}
0	6.25×10^{-3}	4.271×10^{-1}	5.084×10^{-1}

Threshold for Demodulation of Noncoherent Binary ASK

The theoretical optimum threshold τ_{opt} for the binary ASK (OOK) signal is given by Equation 3.97 and is $\gamma A/2 = 2.5$ V here. The carrier amplitude $A = 5$ V and attenuation of the communication channel is taken to be zero ($\gamma = 1$). The outputs of the Half Wave Rectifier token and the Linear System Filter token lowpass filter (LPF) of the envelope detector in Figure 3.35 with no additive noise ($N_o = 0$) and a System Stop Time of 100 msec is shown in Figure 3.36.

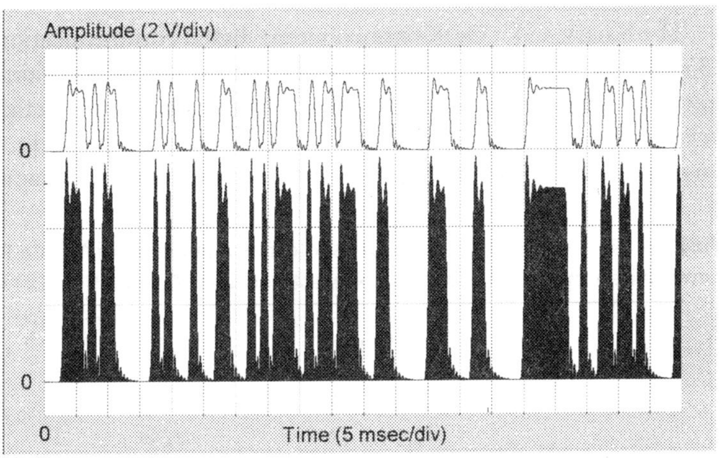

FIGURE 3.36 Half Wave Rectifier token (bottom) and the Linear System Filter token LPF (top) outputs of the binary ASK (OOK) noncoherent digital communication system.

The Half Wave Rectifier token (bottom of Figure 3.36) seems to have a peak amplitude of at least 5 V, which would assure that the theoretical optimum threshold τ_{opt} is 2.5 V as expected. However, the output of the Linear System Filter token LPF (top of Figure 3.36) shows only a peak value of approximately 1.6 V. Figure 3.37 shows the output of the Half Wave Rectifier token with an expanded time scale. The Linear System Filter token bandpass filter (BPF) is also non-ideal and the transient rise-time of the response is clearly demonstrated in Figure 3.37.

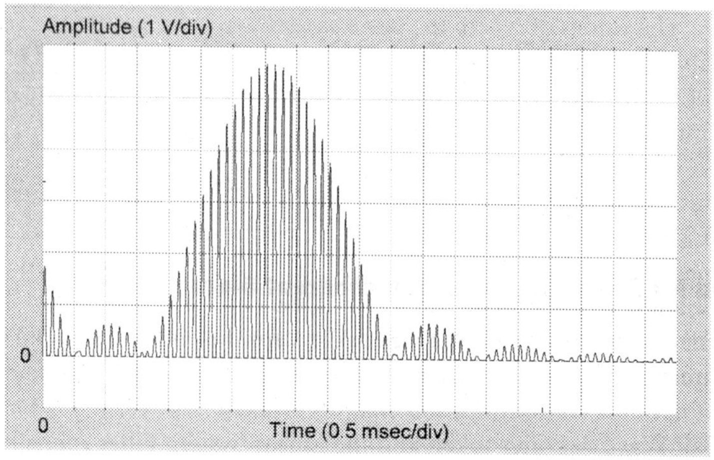

FIGURE 3.37 Half Wave Rectifier token outputs of the binary ASK (OOK) noncoherent digital communication system.

The binary ASK (OOK) noncoherent digital communication system in Figure 3.35 uses the Average token from the Operator Library to obtain an estimate of the practical threshold τ here. Table 3.19 is a tabular list of the practical thresholds τ observed in the SystemVue simulation as a function of the additive white Gaussian noise (AWGN) power spectral density N_o, as described in Chapter 1.

TABLE 3.19 Observed Practical Threshold τ as a Function of N_o in a Binary ASK Digital Communication System with Noncoherent Receiver ($E_b = 6.25 \times 10^{-3}$ V²-sec).

N_o V²-sec	τ V
0	0.83
1.56×10^{-4}	0.91
2.48×10^{-4}	0.93
3.94×10^{-4}	0.97
6.25×10^{-4}	1.03
9.90×10^{-4}	1.08
1.56×10^{-3}	1.18
2.48×10^{-3}	1.31
3.94×10^{-3}	1.47
6.25×10^{-3}	1.70

The values of N_o are the same as that used to measure the BER performance in Table 3.18 and the System Stop Time is set to 10 sec here. In a practical binary ASK (OOK) noncoherent digital communication system, the threshold can be observed continuously and adjusted to improve the BER performance.

The optimum threshold τ_{opt} for the binary ASK (OOK) coherent digital communication system, as shown in Figure 3.4, can also be adjusted to improve the BER performance. Binary ASK is an asymmetrical signal and $\tau_{opt} \neq 0$. Binary frequency and phase shift keyed (FSK and PSK) signals are symmetrical and $\tau_{opt} = 0$ if the additive noise has a zero mean value. For the binary FSK and PSK coherent digital communications systems in Figures 3.8 and 3.11, adjusting the optimum threshold topt is not required with AWGN.

CONSTELLATION PLOTS

Complex modulation schemes in the presence of distortion and noise are often displayed as a phasor constellation plot. Figure 3.38 shows a scatter plot for the decimated I-Q correlator output in the MetaSystem of Figure 3.26 for the 16 level quadrature amplitude modulated (16-QAM) system with no noise. The SystemVue formulation of a constellation plot is as an Analysis Window overlay. The constellation points (small circles) lie in the complex plane and are described by the in-phase $A_I{}^j$ and quadrature $A_Q{}^j$ carrier amplitudes, as given in Listing 3.2. The solid lines indicate all the possible transitions from one constellation point to another.

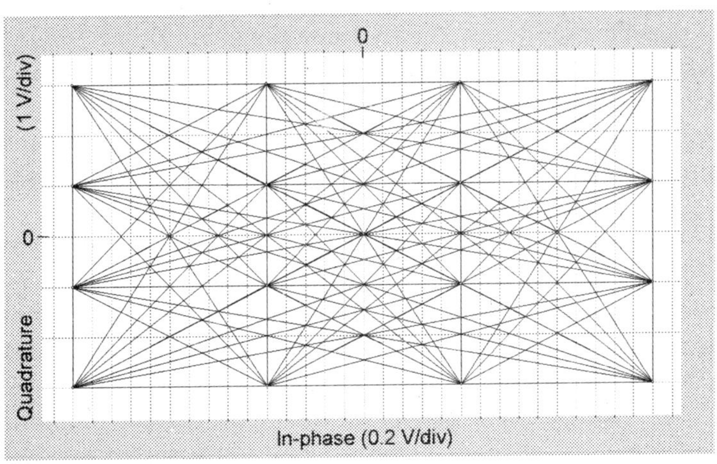

FIGURE 3.38 Scatter plot of the 16-QAM digital communication system with no noise ($E_b/N_o = \infty$).

Figure 3.39 shows a scatter plot for the decimated I-Q correlator output in the MetaSystem of Figure 3.26 with additive white Gaussian Noise (AWGN). The normalized noise power spectral density $N_o = 1.99 \times 10^{-4}$ and the ratio of energy per bit to the noise power $E_b/N_o = 14$ dB here. The locations of the constellation points in the scatter plot are difficult to discern.

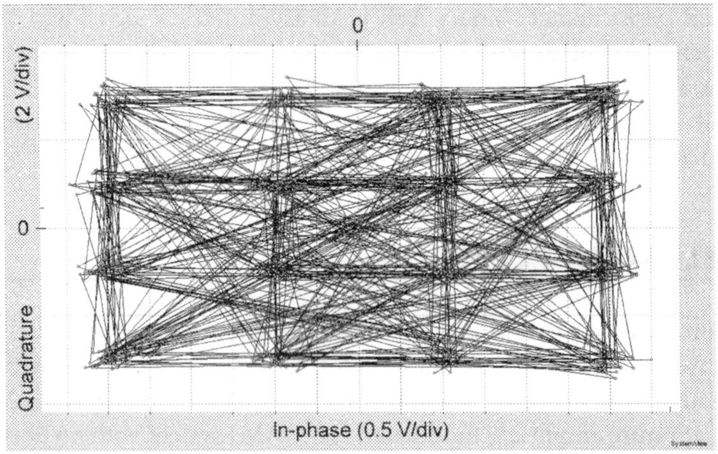

FIGURE 3.39 Scatter plot of the 16-QAM digital communication system with $E_b/N_o = 14$ dB.

Figure 3.40 shows the constellation plot without the solid lines that indicate all the possible transitions. The 16 decision regions for the 16-QAM digital communication system are also shown as dark lines on the in-phase and quadrature complex phasor constellation plot. For this value of $E_b/N_o = 14$ dB there are no constellation points close to the boundary of the decision regions. There are no discernable errors in the SystemVue simulation with the transmission information bits at $E_b/N_o = 14$ dB, as shown in Table 3.13.

Figure 3.41 is the constellation plot for the decimated I-Q correlator output in the MetaSystem of Figure 3.22 for the four level phase shift key (4-PSK or QPSK) system with additive white Gaussian Noise (AWGN). The normalized noise power spectral density $N_o = 4.98 \times 10^{-4}$ and the ratio of energy per bit to the noise power $E_b/N_o = 14$ dB here.

The four decision regions for the QPSK digital communication system are again shown as dark lines on the in-phase and quadrature complex phasor constellation plot. For this value of $E_b/N_o = 14$ dB there are no constellation points even close to the boundary. There are no discernable errors in the SystemVue simulation with the transmission information bits at $E_b/N_o = 14$ dB, as shown in Table 3.11.

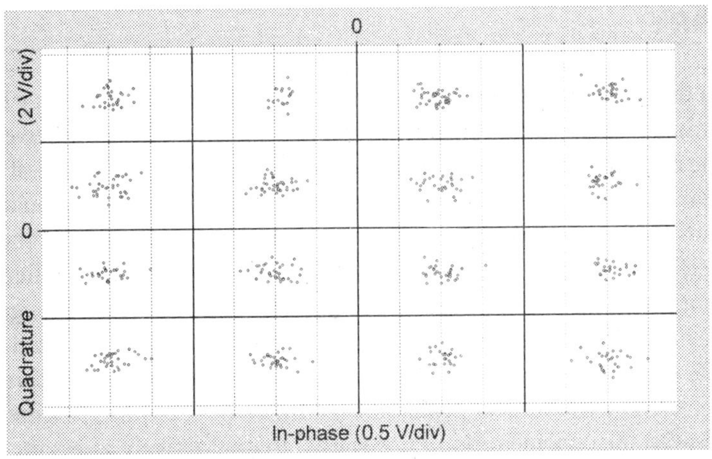

FIGURE 3.40 Constellation points and decision regions of the 16-QAM digital communication system $E_b/N_o = 14$ dB.

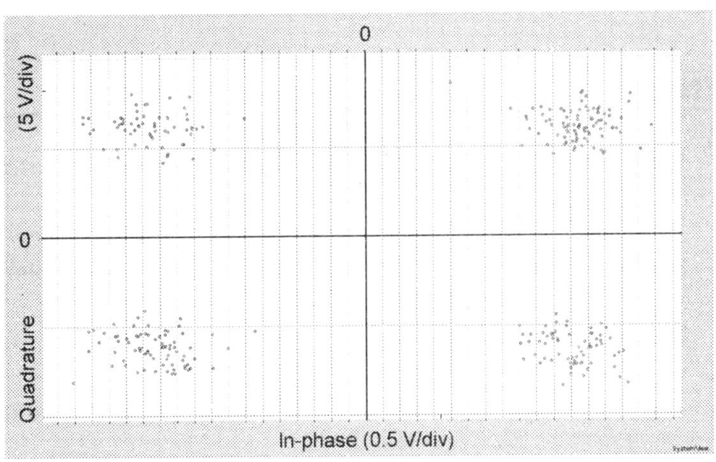

FIGURE 3.41 Constellation points and decision regions of the QPSK digital communication system $E_b/N_o = 14$ dB.

A comparison of Figure 3.40 for 16-QAM and Figure 3.41 for QPSK shows that the clustering of data points is approximately within 0.75 V of the decision boundary for 16-QAM, but within 4 V for QPSK at the same E_b/N_o ratio. The constellation plot then confirms the performance indicated by the bit error rate (BER) for QPSK in Table 3.11 and 16-QAM in Table 3.13.

SUMMARY

In this chapter, SystemVue has been used to design and analyze bandpass modulation and demodulation systems. The optimum bandpass receiver was shown to be the correlation receiver. The amplitude, frequency, and phase of a carrier are modulated with binary and multilevel sources of information. Quadrature amplitude modulation (QAM) is shown to be a bandwidth efficient method of modulation. Differential phase shift keyed (DPSK) modulation and noncoherent frequency and amplitude shift keyed modulation do not require a reference signal in the receiver. For all of these systems the bit error performance was analyzed in the presence of additive white Gaussian noise (AWGN) for the simple and optimum correlation receiver. The theoretical probability of bit error (P_b) was presented for each system. Finally, the constellation plot was used to qualitatively assess the observed bit error rate (BER) performance of the bandpass digital communication system with AWGN. In the next chapter carrier, phase and symbol synchronization, channel equalization and channel models are presented as integral components of a digital communication system.

REFERENCES

[Carlson02] Carlson, A. Bruce, et al., *Communication Systems.* McGraw-Hill, 2002.

[Haykin01] Haykin, Simon, *Communication Systems.* Wiley, 2001.

[Lathi98] Lathi, B.P., *Modern Digital and Analog Communication Systems.* Oxford University Press, 1998.

[Lindsey91] Lindsey, William, et al., *Telecommunciation Systems Engineering.* Dover, 1991.

[Papoulis02] Papoulis, Athanasios, *Probability, Random Variables and Stochastic Processes.* McGraw-Hill, 2002.

[Proakis01] Proakis, John, *Digital Communications.* McGraw-Hill, 2001.

[Shanmugan83] Shanmugan, K. Sam, *Digital and Analog Communication Systems.* Wiley, 1983.

[Simon95] Simon, Marvin, et al., *Digital Communications Techniques: Signal Design and Detection.* Prentice Hall, 1995.

[Sklar01] Sklar, Bernard, *Digital Communications.* Prentice Hall, 2001.

[Stern04] Stern, Harold, et al., *Communication Systems Analysis and Design.* Prentice Hall, 2004.

4 Synchronization and Equalization

In This Chapter

- Acquisition and Tracking of Synchronization
- Carrier Frequency and Phase Synchronization
- Symbol Synchronization
- Equalization of Bandlimited Channels
- Channel Models

The transmitter and receiver of a digital communication system are not initially synchronized in any manner, and the channel that is used to communicate may be less than ideal. The simulation of a digital communication system is often simplified by first assuming an ideal linear channel and the perfect synchronization of the carrier frequency and phase (for bandpass systems) and the symbol or bit timing (for baseband and bandpass systems). The SystemVue simulations of baseband and bandpass digital communication systems presented in Chapter 2, "Baseband Modulation and Demodulation," and Chapter 3, "Bandpass Modulation and Demodulation," have tacitly made this assumption so that only their degraded performance due to additive white Gaussian noise (AWGN) could be assessed.

Practical digital communication systems must cope with the initial lack of synchronization and must track and acquire the needed synchronization information.

The communication channel can also be nonlinear and cause additional performance degradations. Synchronization information must be continuously derived from the received signal, even in the presence of both AWGN and a nonlinear communication channel. However, as presented in this chapter, investigating the effect of synchronization and a nonlinear communication channel separately in the performance of a digital communication system facilitates the analysis.

The source files for these SystemVue simulation studies in synchronization, equalization, and channel modeling are located in the Chapter 4 folder on the CD-ROM, and are identified by the figure number (such as Fig4-1.svu). Appendix A includes a complete description of the contents of the CD-ROM.

ON THE CD

ACQUISITION AND TRACKING OF SYNCHRONIZATION

The continuous acquisition and tracking of the carrier frequency, carr[illegible] symbol timing of a digital communication signal is a requisite for the o[illegible] herent reception of a digitally modulated signal. *Synchronization* is th[illegible] insuring that events (frequency, phase, and timing) in the transmitter [illegible] spond exactly—including a compensated time delay in the transmiss[illegible] same events in the received signal.

The estimation of carrier frequency and phase is known as *car*[illegible] *carrier synchronization*. The estimation of the start and stop times o[illegible] tion signal is known as *clock recovery* or *symbol synchronization* [Proakis[illegible] two modes of synchronization can either be coincident or occur sequentially. Noncoherent reception of a bandpass digital communication signal or a baseband signal only requires symbol synchronization.

Synchronization can be implemented either as a data-aided or a nondata-aided process. In a data-aided process, synchronization information is transmitted as a time division multiplexed (TDM) *preamble* to the source information transmission. This technique is used in digital satellite and wireless communication, where the minimization of the time required to synchronize a data transmission is a requisite. However, the TDM preamble reduces the data throughput efficiency [Haykin01]. The data-aided synchronization approach is not considered here.

The nondata-aided synchronization process avoids the use of the TDM preamble, but the receiver has the task of acquiring and tracking the synchronization information from the modulated signal. Although throughput is improved, more time is required to synchronize a data transmission, and there is an increase in the complexity of the receiver.

Synchronization is essentially a parameter estimation problem and the algorithmic approach for solving it is the maximum likelihood estimation (MLE). The solution to the MLE of carrier and symbol synchronization utilizes discrete time signal processing and iteration. However, a traditional approach to the acquisition and tracking of carrier and symbol synchronization is the use of the nonlinear phase locked loop (PLL) and the Costas loop.

CARRIER FREQUENCY AND PHASE SYNCHRONIZATION

Bandpass digital communication systems with a discernable carrier component in the received spectrum can readily exact the carrier frequency and phase synchronization event. The power spectral density (PSD) of a 1 kb/sec binary amplitude shift keyed (ASK), on-off keying (OOK) signal is shown in Figure 3.5. The prominent carrier component can be extracted even though the OOK carrier is only intermittently present. Finally, the PSD of a 1 kb/sec binary phase shift keyed (PSK) signal is shown in Figure 3.12. The carrier can also be extracted even though there is no prominent spectral carrier component.

The phase locked loop (PLL) is used to track the frequency and phase of a bandpass digital communication signal. The PLL consists of three components, as shown in Figure 4.1 (see Fig4-1.svu on the CD-ROM). The Multiplier token simulates a *phase comparator*. The *loop filter* is the Linear Systems Filter token from the Operator Library as a transfer function or Laplace filter in the *s-domain*. Finally, the Frequency Modulator token simulates a *voltage controlled oscillator* (VCO).

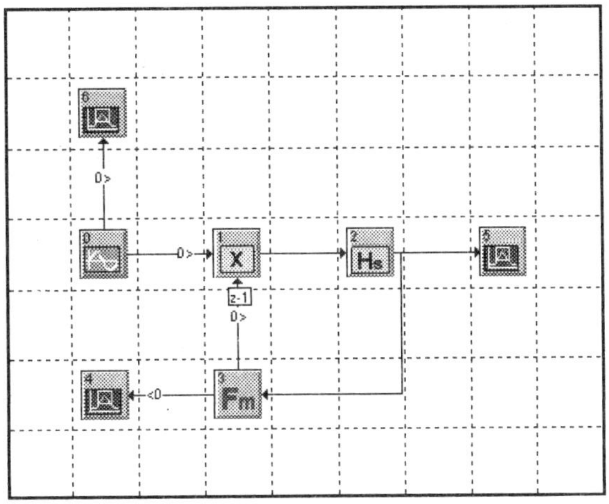

FIGURE 4.1 Phase locked loop test system.

The theoretical Laplace filter is appropriate for the loop filter of this initial PLL description, with parameters of a transfer function given by $1/(7.96\times10^{-5}s+1)$, a 1-pole Butterworth filter with a –3 dB bandwidth of 2 kHz. The Frequency Modulator token from the Function Library simulates the VCO, with parameters of an amplitude A_o = 1 V, a frequency f_o = 19.8 kHz, a modulation gain or frequency deviation factor k_f = 1000 Hz/V, and 0° phase offset. The initial VCO frequency f_o = 19.8 kHz is deliberately chosen to be 200 Hz lower in frequency, and with a phase offset from that of the sinusoidal source.

The Sinusoid token from the Source Library provides the test signal input with an amplitude A_i = 1 V, a frequency f_i = 20 kHz, and 30° phase offset. The System Time: Sample Rate is set to 500 kHz, well above the 20 kHz rate of the sinusoidal input, and the System Stop Time is arbitrarily set to 20 msec.

Figure 4.2 shows the SystemVue simulation acquisition and tracking carrier frequency error signal outputted from the loop filter of the PLL and inputted to the VCO. The error signal stabilizes at approximately 3 msec with a mean amplitude of 0.2 V, which increases the mean frequency of the VCO to 19.8 kHz + 0.2 V × 1000 Hz/V = 20 kHz, the nominal frequency of the sinusoidal test signal input. However, this simple PLL with a limited loop filter shows a ripple on the carrier frequency error signal of approximately ± 25 mV. This is equivalent to a frequency deviation $\Delta f = \pm 0.025$ V × 1000 Hz/V = ± 25 Hz or approximately 1% of the carrier frequency.

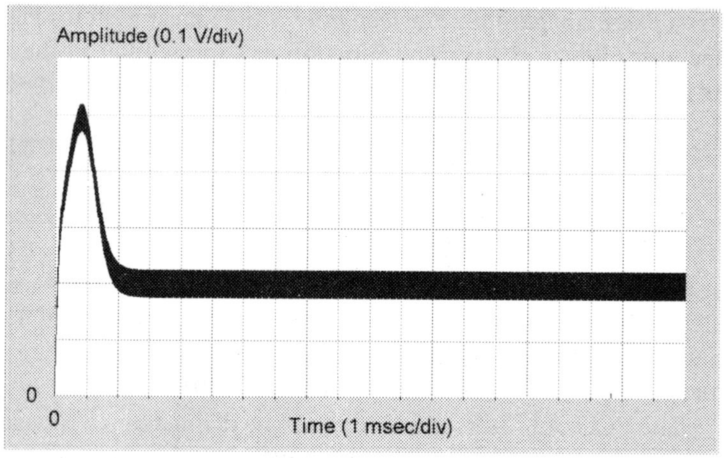

FIGURE 4.2 Acquisition and tracking carrier frequency error signal of the phase locked loop test system.

ON THE CD

SystemVue provides the PLL token from the Communications Library, as shown in Figure 4.3 (see Fig4-3.svu on the CD-ROM). The PLL token parameters are a VCO frequency f_{VCO} = 19.8 kHz, 0° VCO phase, and a modulation gain of 1000 Hz/V, which is the same as the simple PLL shown in Figure 4.1. The initial VCO frequency of the PLL token f_{VCO} = 19.8 kHz is again deliberately chosen to be 200 Hz lower in frequency, and with a phase offset from that of the sinusoidal source.

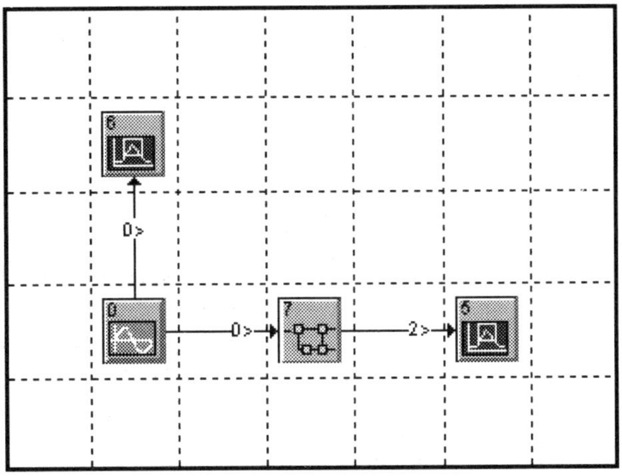

FIGURE 4.3 PLL token test system.

The Sinusoid token from the Source Library provides the test signal input with an amplitude A_i = 1 V, a frequency f_i = 20 kHz, and 30° phase offset. The System Time: Sample Rate is set to 500 kHz, well above the 20 kHz rate of the sinusoidal input, and the System Stop Time is arbitrarily set to 20 msec.

However, the loop filter configuration differs from that shown in Figure 4.1. A 9-pole low-pass Bessel filter specified by its –3 dB bandwidth precedes the actual loop filter, which is a Laplace transfer function given by $1 + a/s + b/s^2 = (s^2 + as + b)/s^2$. The parameter of the Bessel filter is a –3 dB bandwidth of 2 kHz, which is the same as that for the 1-pole Butterworth loop filter of the simple PLL shown in Figure 4.1.

The loop filter parameters $a = 0$ and $b = 0$ aren't used here since the Bessel filter is available in the PLL token. The design of the coefficients of the Laplace transfer function is complicated, and utilizes advanced concepts from control theory [Sklar01]. Often, the coefficients are merely selected to provide additional low frequency response for transient signals [Lindsey91].

The VCO output frequency can also be divided by an integer factor N, which is utilized when the PLL is a frequency synthesizer, but here $N = 1$ [Stern04]. The PLL phase comparator input and output can be *hard-limited* to remove amplitude variations. The PLL token provides a carrier frequency error signal that can demodulate a frequency modulated (FM) signal, a carrier phase error signal that can demodulate a phase modulated (PM) signal, and the VCO output signal.

Figure 4.4 shows the SystemVue simulation acquisition and tracking carrier frequency error signal outputted from the PLL token. The error signal stabilizes at approximately 5 msec with an amplitude of 0.2 V with no appreciable ripple. The mean frequency of the PLL token output is 19.8 kHz + 0.2 V × 1000 Hz/V = 20 kHz, which is the nominal frequency of the sinusoidal test signal input, as expected.

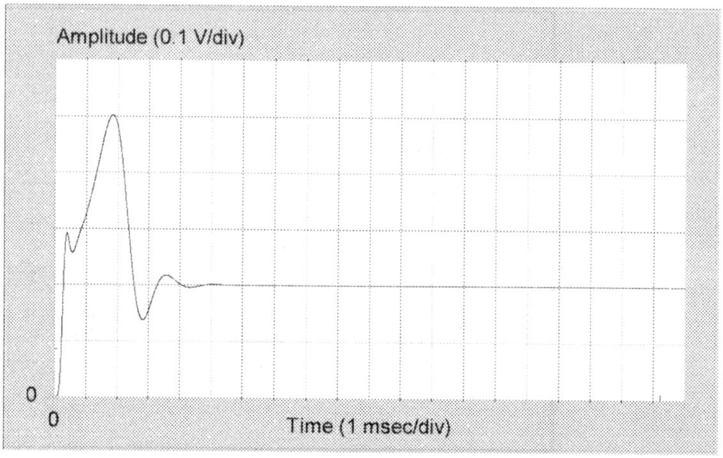

FIGURE 4.4 Acquisition and tracking carrier frequency error signal of the PLL token.

Although the PLL can acquire and track a continuous sinusoid, digital communication systems can feature a carrier frequency that is only intermittently available, as in ASK (OOK), or in *suppressed carrier* analog and digital communication systems [Proakis01]. The Costas loop is essentially two PLLs operating parallel with a common VCO, which performs better than the single PLL in these applications [Carlson02].

ON THE CD

Figure 4.5 shows a Costas loop test system (see Fig4-5.svu on the CD-ROM). The Sinusoid token from the Source Library provides the test signal input with an amplitude $A_i = 1$ V, a frequency $f_i = 20$ kHz, and 0° phase offset. The output of the Sinusoid token is inputted to the Multiplier token. The Pulse Train token from the Source Library provides the other input to the Multiplier token, with parameters of an amplitude $A = 1$ V, a frequency $f_o = 200$ Hz, a pulse width $\tau = 2.5$ msec (50% duty cycle), and 0 V voltage and 0° phase offset. The output of the Multiplier token, an ASK (OOK) sinusoidal signal, is inputted to the Costas Loop token.

The System Time: Sample Rate is set to 5 MHz, well above the 20 kHz rate of the sinusoidal input, which facilitates the accurate difference between the negated test signal input signal and the Costas Loop token VCO quadrature output signal in the Analysis token, as shown in Figure 4.5. The System Stop Time is arbitrarily set to 20 msec.

The Costas Loop token parameters are a VCO frequency $f_{VCO} = 19.8$ kHz, 0° VCO phase, and a modulation gain of 1000 Hz/V, which is the same as the PLL token test system shown in Figure 4.3. The Costas Loop token provides baseband (demodulated) in-phase (I) and quadrature (Q) output signals, and a VCO (carrier frequency) I and Q output signals.

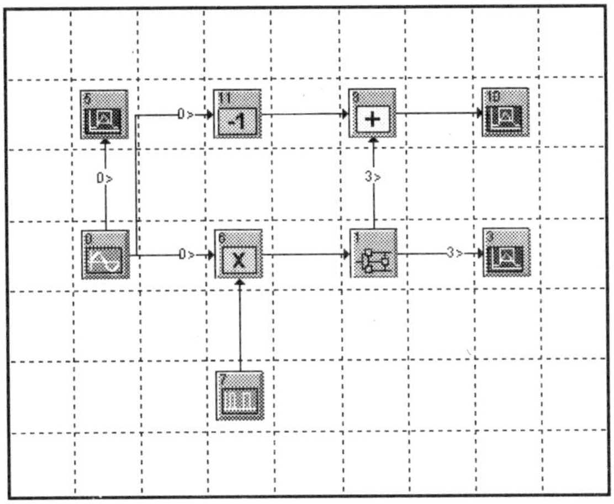

FIGURE 4.5 Costas Loop token test system.

The loop filter configuration also differs from that in Figure 4.3, and the loop filter is only a Laplace transfer function given by $1 + a/s + b/s^2 = (s^2 + as + b)/s^2$. The loop filter parameters are $a = 1000$ and $b = 0$ here. The resulting transfer function is $1 + 1000/s = (s + 1000)/s$, which has a *break frequency* in the Bode magnitude plot at $f = 1000/2\,\pi \approx 159$ Hz. Beyond $f = 159$ Hz, the magnitude of the transfer function asymptotically approaches 1, which is appropriate for the ASK (OOK) signal with a 200 Hz on-off keying (binary 1, 0) rate here.

The Sinusoid token output is negated by the Negate token from the Operator Library and inputted to the Adder token. The other input to the Adder token is the Costas Loop token VCO quadrature output signal. The difference between the test input signal and the quadrature output signal of the Costas Loop token test system stabilizes at approximately 10 msec with an amplitude of ± 15 mV with some ripple. The mean frequency of the Costas Loop token output is 20 kHz, the nominal frequency of the sinusoidal test signal input, as expected.

ON THE CD

The SystemVue simulation of carrier synchronization for binary ASK (OOK) is shown in Figure 4.6 (see Fig4-6.svu on the CD-ROM). The binary ASK (OOK) coherent digital communication system was shown in Figure 3.4. A comparison of Figures 3.4 and 4.6 shows that the original Sinusoid token, with parameters of a carrier frequency $f_c = 20$ kHz and an amplitude $A_c = 5$ V, as the coherent reference signal in the receiver is replaced by the VCO quadrature signal output of the Costas Loop token.

The Linear System Filter token from the Operator Library provides a bandpass filter (BPF), which is a 9-pole Butterworth filter centered at the carrier frequency f_c = 20 kHz with an initial bandwidth of 400 Hz. The received signal from the communication channel is inputted to the BPF, and the output of the BPF is then inputted to the Costas Loop token. The Gain token from the Operator Library, with a parameter of a linear gain of 5, increases the fixed 1 V peak output of the Costas Loop token VCO quadrature signal to 5 V peak, which is the same as the amplitude of the coherent reference signal in Figure 3.4.

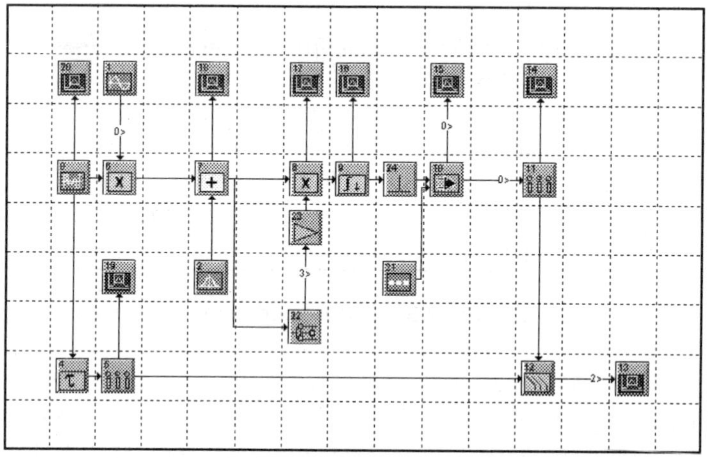

FIGURE 4.6 Carrier synchronization for a binary ASK (OOK) digital communication system.

The demodulation of the binary ASK (OOK) signal may result in phase ambiguity or the ± 180° phase rotation of the VCO output, because of the period of time when the signal is keyed off. The output of the correlation receiver is inputted to the Rectifier token from the Function Library to remove the resulting ± *A* V ambiguity, due to the phase rotation, before processing by the threshold detector of the Analog Comparator.

The SystemVue simulation of the binary ASK (OOK) digital communication system is described in Chapter 3. All other SystemVue tokens and token parameters are as presented there. The System Time: Sample Rate is set to 500 kHz, well above the nominal 1 kb/sec bit rate and 20 kHz carrier frequency. The System Stop Time is set to 10 sec for convenience, which would process 10 000 information bits in possible error.

To facilitate the simulation in SystemVue, all of the Analysis tokens, except that for the BER (bit error rate) token output as the Final Value token, are deleted in the simulation model Fig4-6DT. As described in Chapter 1, "Communication Simulation Techniques," the SystemVue Textbook Edition does not permit tokens to be deleted. The Gaussian Noise token is specified by the power spectral density (PSD) N_o W/Hz. Setting the Gaussian noise source PSD to 0 should produce 0 observed errors.

The BER analysis for the binary ASK (OOK) digital communication system is described in Chapter 3. The normalized energy difference per bit here remains $E_d \approx 1.25 \times 10^{-2}$ V^2-sec, as given by Equations 3.10 and 3.15. Table 4.1 is a tabular list of the observed BER with the optimum receiver BER (opt) of the binary ASK (OOK) signal, and the theoretical probability of bit error (P_b) for comparison with the observed BER here, with carrier synchronization BER (synch).

The original BER analysis is listed in Table 3.2. A comparison of Table 4.1 with Table 3.2 indicates that there is an approximate –4 dB decrease in performance attributed to the *narrowband* noise after the BPF, with a 20 kHz center frequency and 400 Hz bandwidth, at the input to the Costas Loop token. Decreasing the bandwidth of the BPF to 100 Hz improves the BER to only an approximate –2 dB decrease in performance, at an increase in the specification for the carrier frequency stability (100 Hz /20 kHz = 0.5% rather than 400 Hz /20 kHz = 2%).

TABLE 4.1 Observed BERs and Theoretical P_b as a Function of E_d/N_o in a Binary ASK (OOK) Digital Communication System with Optimum Receiver and Carrier Synchronization

E_d/N_o dB	N_o V^2-sec	BER (opt)	P_b	BER (synch)
∞	0	0	0	0
20	1.25×10^{-4}			2.9×10^{-3}
18	1.98×10^{-4}			5.8×10^{-3}
16	3.14×10^{-4}			9.6×10^{-3}
14	4.98×10^{-4}			1.96×10^{-2}
12	7.89×10^{-4}	1.9×10^{-3}	2.5×10^{-3}	3.08×10^{-2}
10	1.25×10^{-3}	1.23×10^{-2}	1.25×10^{-2}	5.45×10^{-2}
8	1.98×10^{-3}	3.61×10^{-2}	3.75×10^{-2}	1.045×10^{-1}
6	3.14×10^{-3}	8.15×10^{-2}	7.93×10^{-2}	1.796×10^{-1}
4	4.98×10^{-3}	1.325×10^{-1}	1.314×10^{-1}	2.577×10^{-1}
2	7.89×10^{-3}	1.884×10^{-1}	1.872×10^{-1}	
0	1.25×10^{-2}	2.387×10^{-1}	2.393×10^{-1}	

ON THE CD

The SystemVue simulation of carrier synchronization for binary PSK is shown in Figure 4.7 (see Fig4-7.svu on the CD-ROM). The binary PSK coherent digital communication system is shown in Figure 3.11. A comparison of Figures 3.11 and 4.7 again shows that the original Sinusoid token, with parameters of a carrier frequency f_c = 20 kHz and an amplitude A_c = 5 V, as the coherent reference signal in

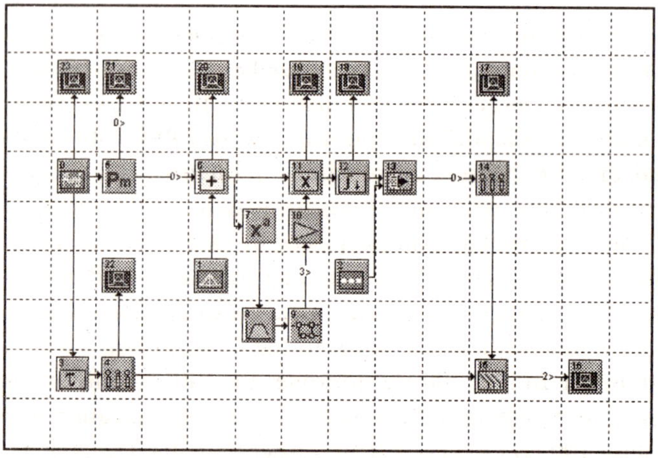

FIGURE 4.7 Carrier synchronization for a binary PSK digital communication system.

the receiver is replaced by the VCO frequency output of the PLL token. However, the parameters of the PLL token are not the same as that in Figure 4.3.

The carrier synchronization for binary PSK processes the square of the received signal, since there is no discernable carrier component (a suppressed carrier) [Proakis05]. The square of the received signal contains a spectral component at twice the carrier frequency, because of the square of the received binary PSK bandpass signal [Sklar01].

The received signal from the communication channel is first inputted to a *square law* device implemented as the Power token from the Function Library, with an exponent of 2. The output of the Power token is inputted to a bandpass filter (BPF). The Linear System Filter token from the Operator Library provides the BPF, which is a 9-pole Butterworth filter centered at twice the carrier frequency $2f_c = 40$ kHz, and with a bandwidth of 400 Hz.

The PLL token parameters are a VCO frequency $f_{VCO} = 39.8$ kHz, 0° VCO phase, and a modulation gain of 10 Hz/V. The initial VCO frequency is approximately twice that of the carrier frequency $f_c = 20$ kHz, and is deliberately chosen to be 200 Hz lower in frequency, and with a phase offset from twice the carrier frequency.

The parameter of the 9-pole Bessel filter is a –3 dB bandwidth of 5 kHz. The loop filter parameters are $a = 0.1$ and $b = 0$, for improved performance of the PLL at the 1 kb/sec data rate here [Lindsey91]. Finally, the VCO frequency output signal is divided by the integer factor $N = 2$ to provide a 20 kHz reference signal from the nominal 40 kHz VCO frequency output signal.

The Gain token from the Operator Library, with a parameter of a linear gain of 5, increases the fixed 1 V peak output of the PLL token VCO divider signal to 5 V peak, which is the same as the amplitude of the coherent reference signal in Figure 3.11. The demodulation of the binary PSK signal does not result in a phase ambiguity with the additive white Gaussian noise (AWGN) communication channel in use here.

ON THE CD

The SystemVue simulation of the binary PSK digital communication system is described in Chapter 3. All other SystemVue tokens, token parameters, and SystemVue specifications are as presented there. To facilitate the simulation in SystemVue, all of the Analysis tokens, except that for the BER token output as the Final Value token, are deleted in the simulation model Fig4-7DT. As described in Chapter 1, the SystemVue Textbook Edition does not permit tokens to be deleted. The Gaussian Noise token is specified by the power spectral density (PSD) N_o W/Hz. Setting the Gaussian noise source PSD to 0 should produce 0 observed errors.

The BER analysis for the binary PSK digital communication system is described in Chapter 3. The normalized energy per bit here remains $E_b \approx 1.25 \times 10^{-2}$ V^2-sec, as given by Equations 3.4 and 3.33. Table 4.2 is a tabular list of the observed BER with the optimum receiver BER (opt) of the binary PSK signal, and the theoretical probability of bit error (P_b) for comparison with the observed BER here with carrier synchronization BER (synch). The original BER analysis is listed in Table 3.5.

A comparison of Table 4.2 with Table 3.5 indicates that there is only approximately a –1 dB decrease in performance attributed to the narrowband noise after the BPF, with a 40 kHz center frequency and 400 Hz bandwidth, at the input to the PLL token.

TABLE 4.2 Observed BERs and Theoretical P_b as a Function of E_b/N_o in a Binary PSK Digital Communication System with Optimum Receiver and Carrier Synchronization

E_b/N_o dB	N_o V^2-sec	BER (opt)	P_b	BER (synch)
∞	0	0	0	0
10	1.25×10^{-3}	0	4.05×10^{-6}	0
8	1.98×10^{-3}	2×10^{-4}	2.06×10^{-4}	1.1×10^{-3}
6	3.14×10^{-3}	2.9×10^{-3}	2.41×10^{-3}	6.7×10^{-3}
4	4.98×10^{-3}	1.24×10^{-2}	1.25×10^{-2}	2.14×10^{-2}
2	7.89×10^{-3}	3.77×10^{-2}	3.75×10^{-2}	5.89×10^{-2}
0	1.25×10^{-2}	8.02×10^{-2}	7.93×10^{-2}	1.121×10^{-2}

SYMBOL SYNCHRONIZATION

The decoding of either a baseband or bandpass digital communication system requires the synchronization of the symbol or bit time. Open-loop symbol synchronizers generate a signal at the symbol rate by processing the baseband data signal or the demodulated, decoded bandpass signal, which is a baseband data signal, with filtering and nonlinear devices [Sklar01]. However, open-loop symbol synchronization methods also have a mean tracking error that may be small for a high signal-to-noise ratio (SNR), but never zero.

Closed-loop symbol synchronizers use comparators that process the incoming data signal and a variable symbol clock to reduce the ultimate *asynchrony*. The early/late gate synchronizer is a configuration that utilizes the concept of the correlation receiver and the feedback comparison loop to provide symbol synchronization. Figure 4.8 shows the configuration of the early/late gate synchronizer.

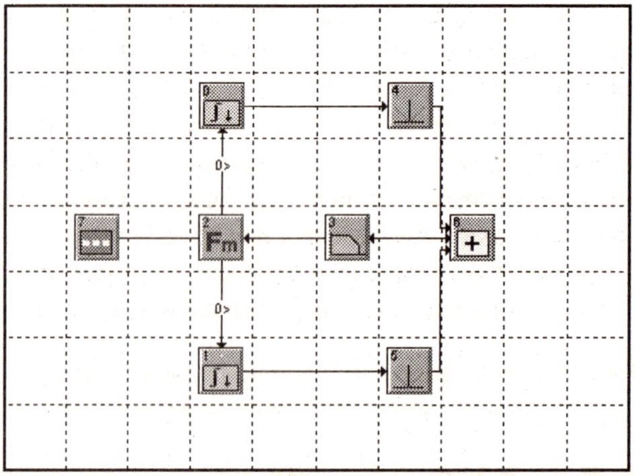

FIGURE 4.8 Early/late gate synchronizer (not a SystemVue simulation model).

The baseband data signal is inputted to two integrators that process the signal over two different intervals. The late gate, the upper integrator in Figure 4.8, integrates from d to the symbol time T_s (or the bit time T_b). The early gate integrates from 0 to $T_s - d$. The times 0 and T_s are the current best estimate of the beginning and end of the symbol period. The difference of the absolute values of the output of the integrators is a measure of the symbol timing error.

The error is fed back through a low-pass filter to a voltage controlled oscillator (VCO or FM modulator) that determines the new estimate of the beginning (0) and end (T_s) of the symbol period. When perfect synchronization occurs, the error is zero. The range for the value of the parameter d is $T_s/2 < d < T_s$.

Figure 4.8 is not a SystemVue simulation model because the available integrate token does not have a control input to begin and end the integration. However, the

Communications Library provides the Bit Synchronizer token, which is a complete early/late gate synchronizer.

ON THE CD

The SystemVue simulation of symbol synchronization for a binary baseband input signal with input jitter is shown in Figure 4.9 (see Fig4-9.svu on the CD-ROM). The binary data source with jitter is configured as the Sawtooth token from the Source Library, with parameters of an amplitude $A = 0.5$ V, a frequency $f_o = 100$ Hz, and a –0.25 V voltage and 0° phase offset. The output of the Sawtooth token is inputted to the Frequency Modulator token from the Function Library, with parameters of an amplitude $A_c = 1$ V, a frequency $f_c = 500$ Hz, 0° phase offset, and an initial modulation gain (or frequency deviation factor) $k_f = 1000$ Hz/V.

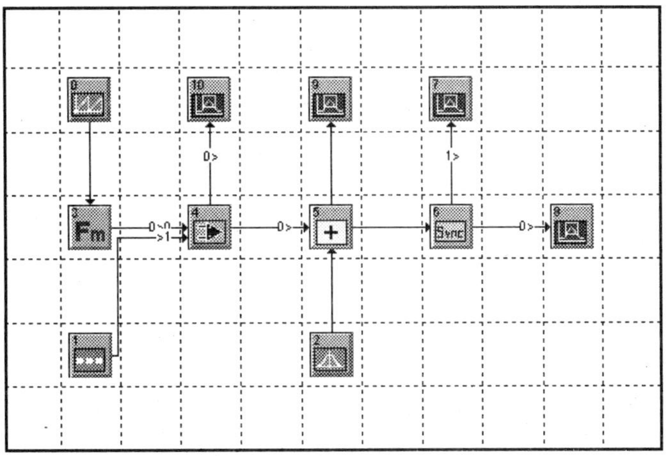

FIGURE 4.9 Early/late gate synchronizer with input jitter.

The output of the Frequency Modulator token is inputted to the Analog Comparator token from the Logic Library, which has another input from a Custom token in the Source Library, with parameters of one output whose algebraic simulation equation is $p(0) = 0$. Since the output frequency of the Frequency Modulator token is variable (500 Hz ± 0.25 V × 1000 Hz/V = 500 Hz ± 250 Hz), the pulse width output signal of the Analog Comparator token has severe induced jitter, as shown in Figure 4.10. For the binary pulses in Figure 4.10, the nominal data rate $r_b = 1$ kb/sec, but varies from 500 b/sec to 1.5 kb/sec because of the sinusoidal input signal to the comparator.

The output of the Analog Comparator token is inputted to the Adder token to also simulate induced noise on the binary pulse output signal. The other input to the Adder token is the Gaussian Noise token from the Source Library, with a standard deviation of $\sigma = 0$ V initially.

The output of the Adder token is inputted to the Bit Synchronizer token from the Communications Library, with parameters of a bit rate $r_b = 1$ kb/sec, a *matched filter* integration time of 1 msec, the early/late delay $d = 0.1$ msec, a loop gain $G = 1$, and a loop filter constant $a = 10^{-3}$. The loop filter has the transfer function $1 + a/s = (s + a)/s = (s + 10^{-3})/s$. The matched filter operation is the correlation process described in Chapter 2.

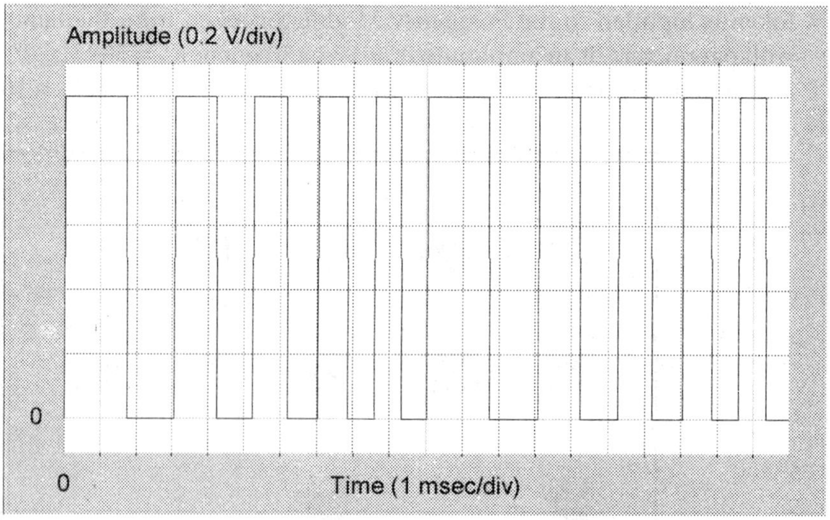

FIGURE 4.10 Binary pulse output signal with induced jitter.

The symbol synchronization signal of the Bit Synchronizer token—a one sample pulse per bit time T_b signal, and the loop error signal—is shown in Figure 4.11. The Connected Points button on the horizontal Analysis toolbar of the SystemVue simulation environment provides a plot that shows the simulation sample points for clarity, as described in Chapter 1.

Figure 4.11 demonstrates the symbol synchronization to the nominal bit time $T_b = 1$ msec or a bit rate $r_b = 1$ kb/sec. Setting the Gaussian Noise token standard deviation parameter $\sigma = 4$ V results in the binary baseband input signal corrupted by both noise and jitter, as shown in Figure 4.12. Compare Figure 4.12 with the noise-free binary baseband input signal with jitter in Figure 4.10.

Figure 4.13 shows the symbol synchronization signal of the Bit Synchronizer token and the loop error signal for the binary pulse output signal with noise and jitter. Remarkably, although the loop error signal is substantially larger, the symbol synchronization signal remains *locked* to the corrupted input signal, with a bit rate $r_b = 1$ kb/sec in Figure 4.12.

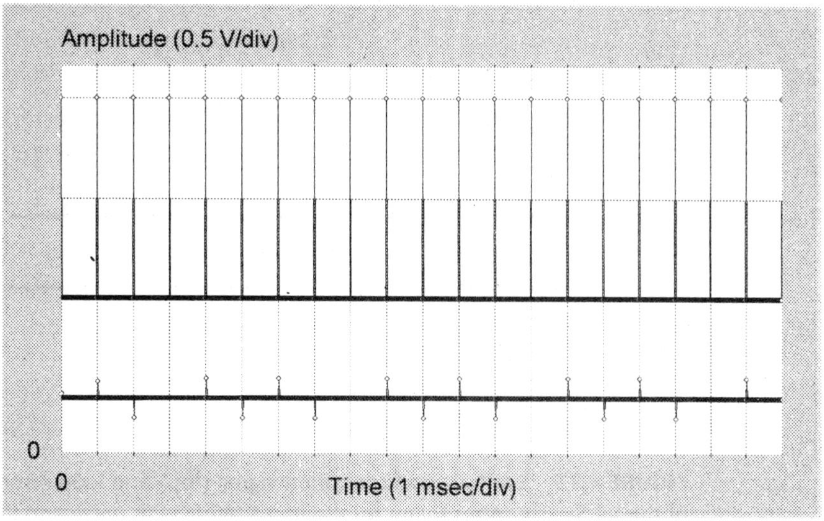

FIGURE 4.11 Symbol synchronization signal (top) and loop error signal of the Bit Synchronizer token for the input signal with jitter.

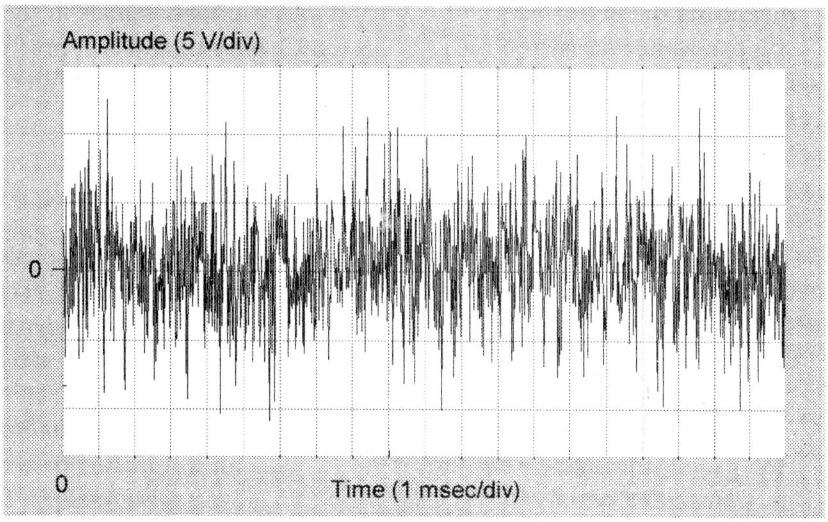

FIGURE 4.12 Binary pulse output signal with noise and jitter.

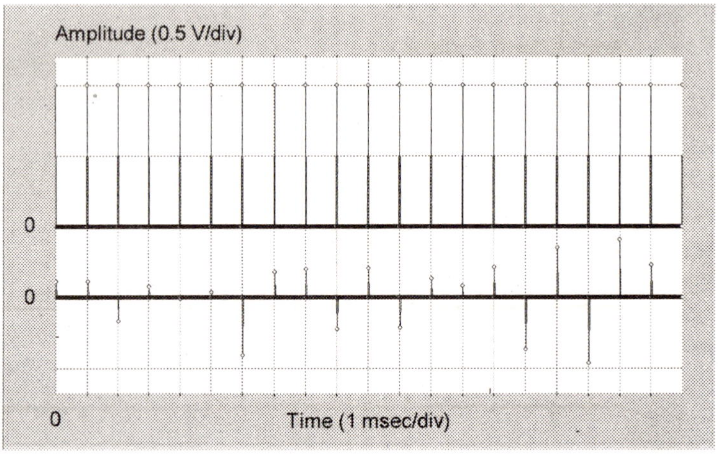

FIGURE 4.13 Symbol synchronization signal (top) and loop error signal of the Bit Synchronizer token for the input signal with noise and jitter.

EQUALIZATION OF BANDLIMITED CHANNELS

Communication channels are not strictly linear and wideband, as utilized when measuring the performance of baseband and bandpass signals in the presence of additive white Gaussian noise (AWGN) in Chapters 2 and 3. A practical communication channel can be characterized as a bandlimited linear filter, as given by Equation 4.1.

$$H(f) = A(f)\exp(j\theta(f)) \quad (4.1)$$

In Equation 4.1, $A(f)$ is the amplitude response and $\theta(f)$ is the phase response of the communication channel as a function of frequency. In the AWGN channels considered previously, the communication channel was ideal (an Adder token in SystemVue) with $A(f) = K = 1$ (a constant, not a function of frequency) and $\theta(f) = k_f = 0 \times f = 0$ (a linear function of frequency).

If $A(f)$ is not constant, then the transmitted signal exhibits *amplitude distortion* after passing through the channel. If $\theta(f)$ is not a linear function of frequency, then the transmitted signal exhibits *delay distortion*. Figure 4.14 (see Fig4-14.svu on the CD-ROM) shows the binary sinc *pulse amplitude modulated* (PAM) transmitter of Figure 2.13 inputted to a Linear System Filter token, which is a 9-pole Butterworth low-pass filter (LPF) with a cutoff frequency $f_{cutoff} = 200$ Hz. The data rate $r_b = 1$ kb/sec and the System Time: Sample Rate is 50 kHz.

ON THE CD

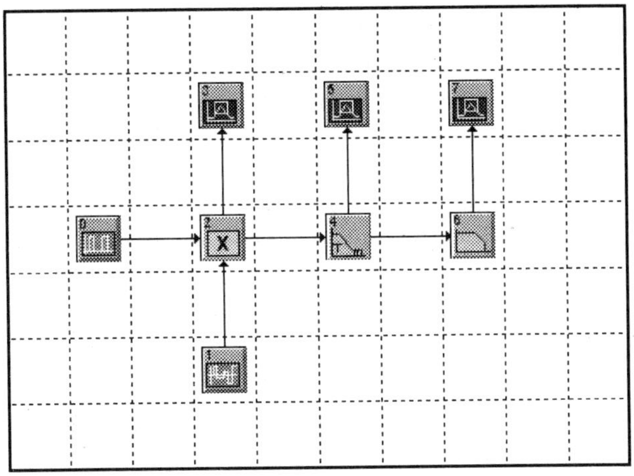

FIGURE 4.14 Binary sinc PAM transmitter and simulated bandlimited channel.

Figure 4.15 shows the transmitted binary sinc signal at 1 kb/sec and the distorted output of the simulated bandlimited channel. The polar sinc PAM signal is sampled every 1 msec (starting at t = 4 msec) and should be either –5 V or +5 V, representing the binary data (0, 1). The output of the simulated bandlimited channel shows no discernable pattern and the original binary data is irretrievably lost.

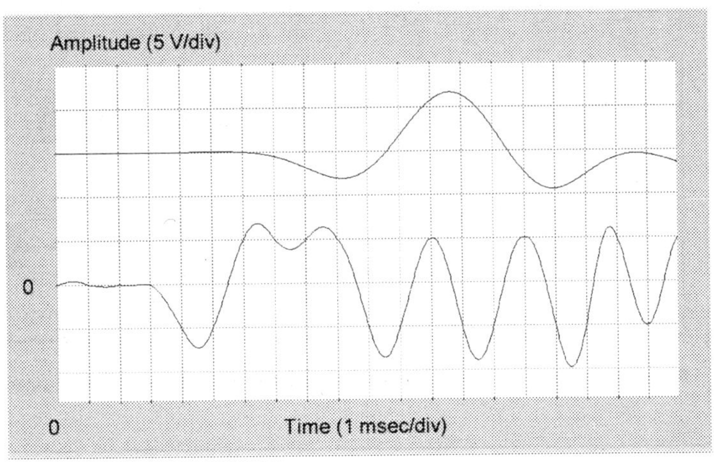

FIGURE 4.15 Distorted output of a simulated bandlimited channel (top) and the binary sinc PAM transmitted signal.

The output has severe delay distortion, in that successive pulses are smeared to the extent that they cannot be distinguished. This condition is *intersymbol interference* (ISI) and the restoration of the ideal characteristics of the channel by additional filtering is *equalization* [Proakis05]. For no ISI to exist, the impulse response of the channel $h(t)$, the inverse Fourier transform of Equation 4.1, must have *zero-crossings* spaced at the bit time $T_b = 1/r_b$ [Stern04]. This requisite is *Nyquist's First Criterion* and is equivalent to a channel with $A(f) = K$ and $\theta(f) = k_f$, but these specifications for a constant amplitude and linear phase response need only apply for $|f| \leq 1/(2T_b)$.

Equalization may be forced to occur by a compensating inverse filter C(*f*) at the input to the receiver, as given by Equation 4.2.

$$H(f)\,C(f) = 1 \quad |f| \leq \frac{1}{2T_b} \tag{4.2}$$

The implementation of the compensating inverse filter can be accomplished by a linear transversal equalizer, as shown in Figure 4.16. In the equalizer, the current and past values of the received signal are weighted (multiplied) by the *equalizer coefficients* c_n, and linearly summed to produce an output sampled at the center of the bit time T_b.

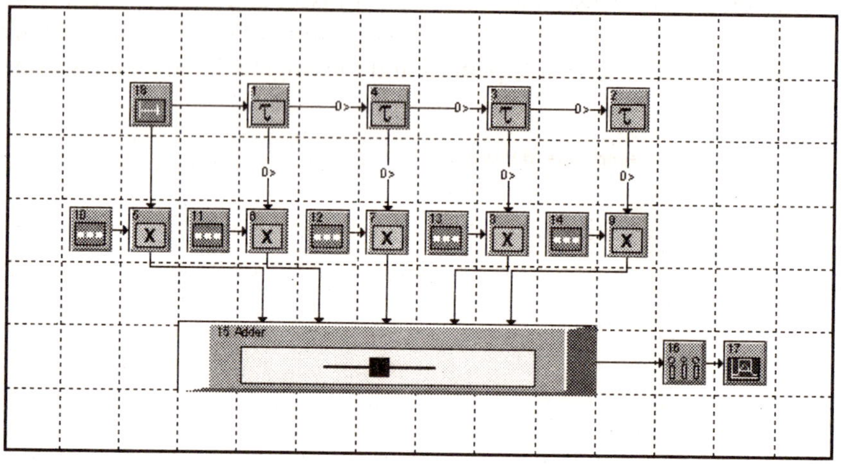

FIGURE 4.16 Linear transversal equalizer (not a SystemVue simulation model).

The coefficients c_n are selected to force the combined channel and equalizer to meet Nyquist's First Criterion. This implementation is the zero-forcing equalizer [Stern04]. The coefficients c_n are obtained by solving N equations simultaneously from the N observations of a possibly known transmitted data signal (a training set).

The zero-forcing equalizer has a finite number of coefficients c_n, and can only approximate the compensating inverse filter C(f). This equalizer may also enhance any AWGN on the communication channel [Sklar01]. However, the least-mean-square (LMS) equalizer determines the coefficients c_n that minimize the sum of the squares of the ISI components and the additive noise power [Proakis05].

Equalizers have been developed for specific types of communication channels. Adaptive equalization compensates for channels with amplitude response and phase response that vary in time (that is, $A(f, t)$ and $\theta(f, t)$). Decision feedback equalizers attempt to compensate for channels that have severe amplitude distortion, including spectral nulls [Sklar01]. However, equalizer design and implementation is intricate and beyond the scope of an introduction to digital communication system simulation.

CHANNEL MODELS

A communications channel can inflict additional distortion by exhibiting the transmission of the signal over multiple paths (multipath) with fading. SystemVue facilitates assessing the performance of a digital communication system under these conditions with channel model tokens. The formulation of a channel model relies heavily upon the concepts of probability and random variables [Xiong00].

The Multipath Channel token from the Communications Library is defined by Equation 4.3.

$$o(t) = c_0\, s(t) + \sum_{k=1}^{N} c_k\, s(t - \tau_k) \tag{4.3}$$

In Equation 4.3, $o(t)$ is the output of the multipath channel model, and $s(t)$ is the input modulated baseband or bandpass signal. The N delays τ_k are random variables that are uniformly distributed from 0 to τ_{max} (maximum delay). The coefficients c_k are Rayleigh distributed random variables with variance σ^2 [Xiong00].

ON THE CD

Figure 4.17 (see Fig4-17.svu on the CD-ROM) shows the binary amplitude shift keyed (ASK) on-off keyed (OOK) coherent digital communication system with bit error rate (BER) analysis, as shown in Figure 3.4, with the Multipath Channel token. The parameters of the Multipath Channel token are $N = 3$, $\tau_{max} = 1$ msec, and K = 1. The K-factor constrains the total power in the channel by the relationship $c_0^2 + N\sigma^2 = 1$. Here $c_0^2 = K/(K+1) = 0.5$, $\sqrt{2} = 1/(N(K+1)) = 0.166$, and $N_0^2 = 0.5$.

To facilitate the simulation in SystemVue, all of the Analysis tokens, except that for the BER token output as the Final Value token, are deleted in the simulation model Fig4-17DT. As described in Chapter 1, the SystemVue Textbook Edition does not permit tokens to be deleted.

ON THE CD

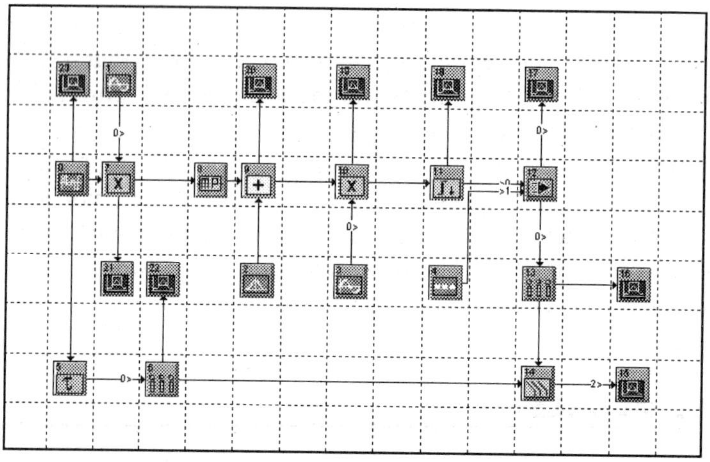

FIGURE 4.17 Binary ASK (OOK) digital communication system with multipath channel distortion.

Figure 4.18 shows the unipolar binary baseband data signal (bottom), the modulated ASK (OOK) signal, and the distortion caused by the multipath channel model (top). The value τ_{max} = 1 msec is excessive, but appropriate for the carrier frequency f_c = 20 kHz and data rate r_b = 1 kb/sec of the ASK (OOK) signal to clearly demonstrate multipath distortion.

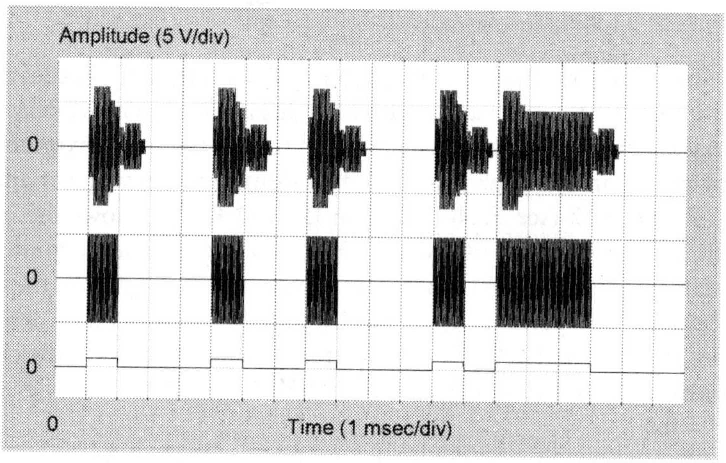

FIGURE 4.18 Binary data signal (bottom), modulated ASK (OOK) signal, and the distortion caused by a multipath channel model.

SystemVue provides other advanced channel models, including the Jakes Mobile Channel token, the Rice Fading Channel token, which models an amplitude fading channel, and the Rummler Fading Channel token, which models a digital microwave link. However, the implementation of these channel models and the compensating receiver structures is intricate and beyond the scope of an introduction to digital communication system simulation.

SUMMARY

In this chapter, SystemVue was used to design and analyze the performance of the phase locked loop (PLL) for carrier, phase, and symbol synchronization in digital communication systems. The performance and configuration of the discrete token PLL, which demonstrated the principle of operation, was simulated and compared to the performance of the SystemVue PLL token. The SystemVue Costas Loop token and the Bit Synchronizer token provided more robust performance in the presence of additive white Gaussian noise and induced jitter. Although beyond the scope of an introduction to digital communication system simulation, channel equalization and channel models were presented to illustrate the degradation in performance that can be inflicted by non-ideal communication channels. In the next chapter, time division multiplexing, frequency division multiplexing, and code division multiplexing are presented as another configuration of a digital communication system.

REFERENCES

[Haykin01] Haykin, Simon, *Communication Systems*. Wiley, 2001.

[Lindsey91] Lindsey, William, et al., *Telecommunication Systems Engineering*. Dover, 1991.

[Proakis01] Proakis, John, *Digital Communications*. McGraw-Hill, 2001.

[Proakis05] Proakis, John, et al., *Fundamentals of Communications Systems*. Prentice Hall, 2005.

[Sklar01] Sklar, Bernard, *Digital Communications*. Prentice Hall, 2001.

[Stern04] Stern, Harold, et al., *Communication Systems Analysis and Design*. Prentice Hall, 2004.

[Xiong00] Xiong, Fuqin, *Digital Modulation Techniques*. Artech House, 2000.

5 Multiplexing

In This Chapter

- Time Division Multiplexing
- Frequency Division Multiplexing
- Code Division Multiplexing
- Direct Sequence Spread Spectrum
- Frequency Hopping Spread Spectrum
- Orthogonal Frequency Division Multiplexing

The baseband and bandpass digital communication systems described in Chapter 2, "Baseband Modulation and Demodulation," and Chapter 3, "Bandpass Modulation and Demodulation," used a single source of information, and transmitted the data across a single communication channel. However, multiple sources of information sharing one or more communication channels is a typical occurrence. *Multiplexing* allows a group of independent sources to share a communication channel without interference, by exploiting the *orthogonality* of the data transmission in the time or frequency signal space.

The independent data sources can be separated in time by transmitting at different intervals using the same baseband or bandpass range of frequencies. This technique is called *time division multiplexing* (TDM). The independent data sources can be separated in frequency by transmissions using non-overlapping ranges of bandpass frequencies. This technique is called *frequency division multiplexing*

(FDM). The independent data also can be transmitted at the same time and in the same range of bandpass frequencies by processing the sources with a unique code to make them orthogonal. This technique is called *code division multiplexing* (CDM) or *code division multiple access* (CDMA).

Finally, *orthogonal frequency division multiplexing* (OFDM) separates a single data source into multiple subchannels, each of which carries only a portion of the total information, but at a lower data rate. OFDM provides an effective technique for the mitigation of multipath fading in wireless communication channels.

ON THE CD The models for these SystemVue simulations in TDM, FDM, CDM, and OFDM are located in the Chapter 5 folder on the CD-ROM, and are identified by the figure number (such as Fig5-1.svu). Appendix A includes a complete description of the contents of the CD-ROM.

TIME DIVISION MULTIPLEXING

Time division multiplexing (TDM) combines a group of low bit rate independent data sources into a high-speed bit stream that can be transmitted as either a modulated baseband or bandpass signal. SystemVue provides the Time Division Multiplexer token and the Time Division Demultiplexer token from the Communications Library, which can simplify the implementation of a TDM system.

The Time Division Multiplexer token buffers N data sources and sequentially outputs each within a specified time of T seconds. The N data sources all have the same data rate r_b b/sec and the output data rate is Nr_b b/sec. The Time Division Demultiplexer token also buffers the multiplexed data input, and outputs the N data sources, each with a data rate of r_b b/sec, with a delay of T seconds.

However, a multiple-token TDM system demonstrates the use of other SystemVue tokens in a simulation, and the actual technique utilized for TDM. The SystemVue simulation of a multiple-token TDM system, where each of the sources have the same data rate, is shown in Figure 5.1 (see Fig5-1.svu on the CD-ROM). ON THE CD The two MetaSystems shown represent any baseband or bandpass modulator and demodulator to complete the digital communication TDM system. Here, the two MetaSystems are merely pass-through, with Meta Input and Meta Output tokens.

The multiplexer shown in Figure 5.1 has four independent unipolar pseudonoise (PN) data sources. They are PN Sequence tokens from the Source Library, all with parameters of a binary amplitude $A = \pm 0.5$ V, a data rate $r_b = 250$ b/sec, and 0.5 V voltage and 0° phase offset. The data sources are time division multiplexed by the Switch token from the Operator Library. The Switch token itself has been expanded to clarify the multiple input connections, as described in Chapter 1, "Communication Simulation Techniques." The parameters of the Switch token are a minimum control input $V_{cmin} = 0$ V and a maximum control input $V_{cmax} = 1$ V.

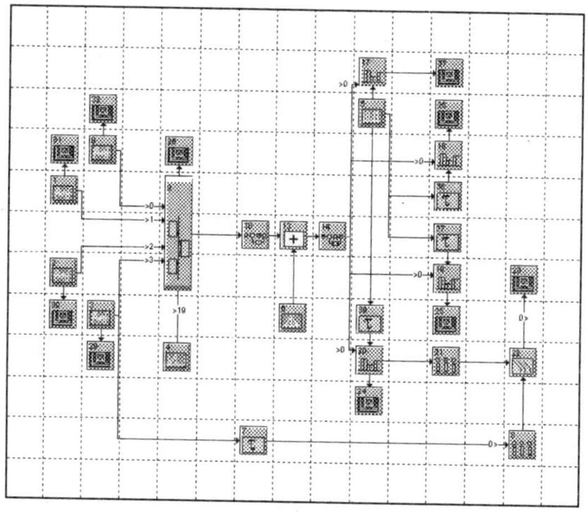

FIGURE 5.1 TDM digital communication system with equal data rate sources.

The control voltage of the Switch token is derived from a Sawtooth token from the Source Library, with parameters of an amplitude $A = 1$ V, a frequency $f_o = 250$ Hz, and 0 V voltage and 0° phase offset. The Switch token responds to the periodic sawtooth control voltage by connecting port N to the output if the control voltage is in the range given by Equation 5.1.

$$\frac{N}{M}\left(V_{cmax} - V_{cmin}\right) < v_{in} < \frac{N+1}{M}\left(V_{cmax} - V_{cmin}\right) \tag{5.1}$$

The number of ports M is four here, the input ports N are either 0, 1, 2, or 3, and the sawtooth input voltage is v_{in}. As the periodic sawtooth source waveform goes from 0 to 1 V in a period of 4 msec, the four input ports are connected in sequence to the output for 1 msec. The TDM output data rate $r_{b\,out}$ then is 4 × 250 b/sec = 1 kb/sec.

The output of the multiplexer is inputted to the modulator MetaSystem, the additive white Gaussian noise (AWGN) communication channel, and the demodulator MetaSystem. The System Time: Sample Rate is set to 50 kHz, well above the nominal TDM data rate of 1 kb/sec.

The output of the demodulator MetaSystem is inputted to the demultiplexer, which consists of four Sample and Hold tokens from the Operator Library with a control threshold voltage of 0.5 V. The Pulse Train token from the Source Library provides the Sample and Hold token control voltage with a single point sample ($T_{system} = 20$ μsec) at a frequency $f_o = 1$ kHz, an amplitude $A = 1$ V, and a 0 V voltage and a 0° phase offset synchronized to the data source.

The control voltage of the Sample and Hold token for the port 0 demultiplexed data source is inputted directly from the Pulse Train token. The control voltages of the Sample and Hold tokens for the port 1, 2, and 3 demultiplexed data sources are delayed by a Delay token from the Operator Library by 1, 2, and 3 msec, respectively.

The two comparison inputs to the BER (bit error rate) token from the Communications Library are single simulation samples per bit time T_b from one of the four data sources. Two Sampler tokens from the Operator Library are used to downsample the 50 kHz System Sampling Rate to the 250 b/sec binary data rate or 250 Hz. The Sampler token parameters are set to a sample rate of 250 Hz and an aperture time of 20 μsec (T_{system}, a single point sample). A Delay token from the Operator Library can be set to compensate for the delay inherent in the baseband or bandpass modulator and demodulator, as appropriate. Here, the delay is set to zero.

The parameters for the BER token are a threshold of 0.5 V and 0 sec offset. The cumulative error output of the BER token verifies the performance of both the TDM system and modulator and demodulator.

The System Stop Time is set to 50 msec for convenience, and the resulting SystemVue simulation of the time division multiplexed unipolar binary data source is shown in Figure 5.2. Port 0 is the uppermost signal, followed by port 1, port 2, and port 3 and the output port. The four demultiplexed data signals replicate the data signals at the four input ports.

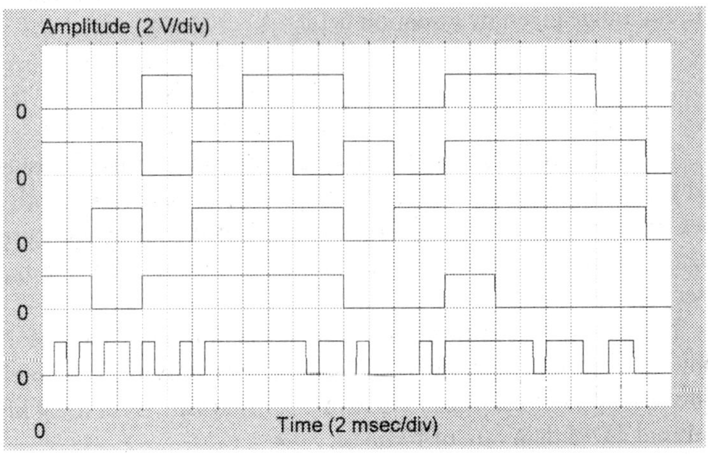

FIGURE 5.2 Four TDM input signals at a data rate $r_{b\ in} = 250$ b/sec and the multiplexed output signal at a data rate $r_{b\ out} = 1$ kb/sec.

This TDM system is uncomplicated when all the input data sources have the same bit rate r_b, as shown in Figure 5.1. However, even when that is not the case, it may be still possible to design a TDM system. A TDM system with two input

ON THE CD sources, both with data rates $r_{b1} = r_{b2} = 250$ b/sec, and a third source with a data rate $r_{b3} = 500$ b/sec is shown in Figure 5.3 (see Fig5-3.svu on the CD-ROM).

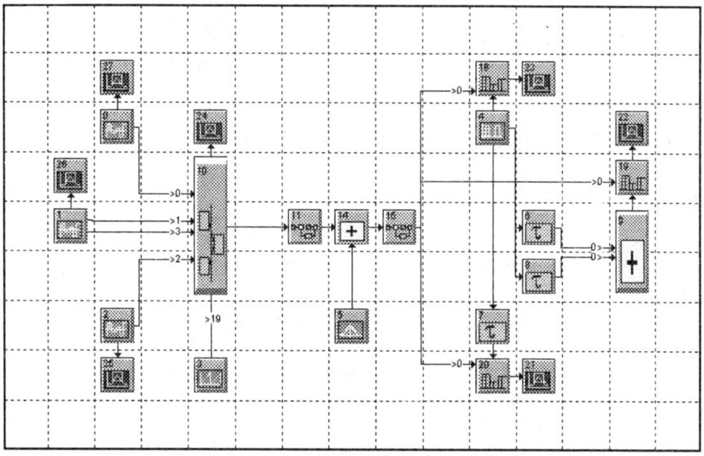

FIGURE 5.3 TDM digital communication system with unequal data rate sources.

The third data source is connected to *both* port 1 and port 3 so that it can be transmitted *twice* during the multiplex time period of 4 msec, at a data rate of 500 Hz or bit time $T_{b3} = 2$ msec. The third data source could also be connected to port 0 and port 2, but not to two adjacent ports (for example, port 0 and port 1). Adjacent ports are sampled 1 msec apart, but the sequence is not repeated until a duration of 3 msec has elapsed. Port 0 and port 2 are sampled 2 msec apart, and the sequence repeats every 2 msec for a data rate $r_{b3} = 500$ b/sec.

Figure 5.4 shows the third data source TDM input as the uppermost signal, the three sources multiplexed at a data rate $r_b = 1$ kb/sec, and the demultiplexed third data source. The demultiplexed third data source has a delay of 1 msec, because it is connected to the TDM input port 1 with a 1 msec delay, and port 2.

The TDM systems remain somewhat uncomplicated if the number of data sources and the data rates are compatible. However if this is not the case, complex binary digital data *first-in first-out* (FIFO) buffers are required, but a reasonable alternative may be to use multiple communication channels. When these incompatible data sources are transmitted by baseband signals, multiple *twisted-pair* or *coaxial* wire channels in parallel can be employed, if obtainable. When they are to be transmitted by bandpass signals, frequency division multiplexing is a reasonable alternative, if the communication channel bandwidth is available.

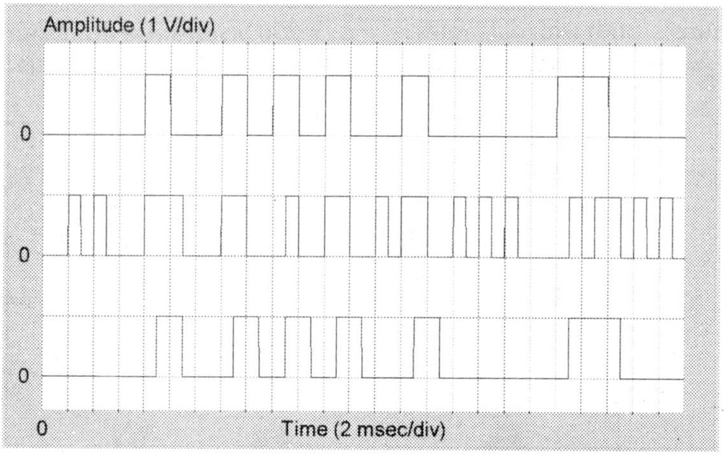

FIGURE 5.4 A TDM input signal at a data rate $r_{b\ in} = 500$ b/sec, the multiplexed output signal at a data rate $r_{b\ out} = 1$ kb/sec, and the demultiplexed output signal.

TDM systems can also transmit more than one bit from each data source at a time. However, the data must also be FIFO buffered here and this produces delay in the transmission. TDM data transmissions using fixed-time allocations (or *slots*) are known as *synchronous TDM*. Alternatively, *statistical TDM* combines data sources with rates that vary by using the average transmission rates, combining them into fixed sized data *blocks*. However, statistical TDM requires a *header* block, which not only contains the overhead of synchronization and address bits, but also provides the advantage of error correction bits [Carlson02].

FREQUENCY DIVISION MULTIPLEXING

Frequency division multiplexing (FDM) transmits a group of independent data sources as modulated bandpass signals that occupy a range of frequencies. A SystemVue simulation of a FDM system is shown in Figure 5.5 (see Fig5-5.svu on the CD-ROM). The six MetaSystems shown represent any bandpass modulator and demodulator to complete the digital communication FDM system. Here, three disparate modulators and demodulators are selected to illustrate the FDM concept. The data sources are PN Sequence tokens from the Source Library, all with a data rate $r_b = 1$ kb/sec, or a bit time $T_b = 1$ msec.

ON THE CD

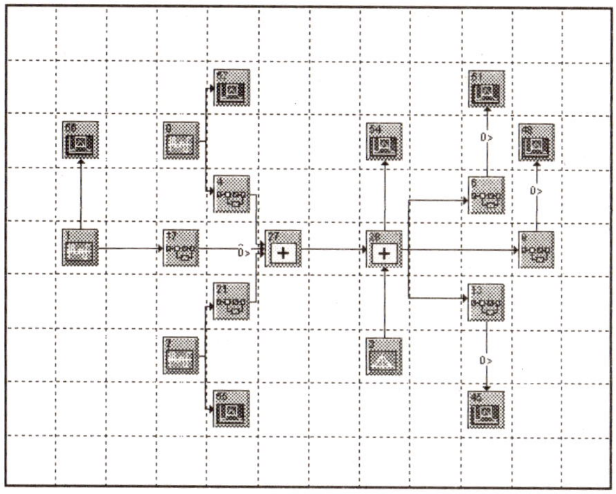

FIGURE 5.5 FDM digital communication system with binary ASK (OOK), binary FSK (MFSK), and binary PSK modulators and demodulation with equal data rate sources.

The first modulator, shown as the uppermost in Figure 5.5, is a binary amplitude shift keying (ASK) digital communication system, as shown in Figure 3.4, with a carrier frequency of 15.5 kHz. The ASK modulator uses on-off keying (OOK) where the PN Sequence token data source has a binary amplitude of $\pm$ 0.5 V, and 0.5 V voltage and 0° phase offset. The ASK (OOK) demodulator is the correlation receiver for asymmetric bandpass signals, also as shown in Figure 3.4. The bandwidth of the binary ASK (OOK) signal is given in Table 3.1, and 95% of the total power transmitted occupies $4/T_b = 4$ kHz, or a range of $\pm$ 2 kHz centered at the 15.5 kHz carrier frequency.

The second modulator is a binary frequency shift keying (FSK) digital communication system, as shown in Figure 3.8 with a carrier frequency of 20 kHz. The PN Sequence token data source has a binary amplitude of $\pm$ 1 V and 0 V voltage and 0° phase offset. The FSK (MFSK) modulator has a modulation gain or *frequency deviation factor* $k_f = 500$ Hz/V, and uses the minimum value of the frequency deviation Δf of $1/(2T_b)$, because $\Delta f = \pm 1 \text{ V} \times 500 \text{ Hz/V} = \pm 500 \text{ Hz}$. The FSK demodulator is the correlation receiver for asymmetric bandpass signals, also as shown in Figure 3.8. The bandwidth of the binary FSK (MFSK) signal is given in Table 3.3 and 95% of the total power transmitted occupies $2\Delta f + 4/T_b = 5$ kHz or a range of $\pm$ 2.5 kHz centered at the 20 kHz carrier frequency.

The third modulator is a binary phase shift keying (PSK) digital communication system, as shown in Figure 3.11, with a carrier frequency of 24.5 kHz. The PN Sequence token data source has a binary amplitude $A = \pm$ 0.5 V and 0.5 V voltage and 0° phase offset. The PSK modulator has a modulation gain or *phase deviation factor* $k_p = 0.5\ 2\pi$ (radians)/V, a 180° phase offset, and produces a phase shift φ of

0 and π radians. The PSK demodulator is the correlation receiver for symmetrical bandpass signals, also as shown in Figure 3.11. The bandwidth of the binary PSK signal is the same as that for binary ASK in Table 3.1, and 95% of the total power transmitted occupies $4/T_b = 4$ kHz, or a range of $\pm$ 2 kHz centered at the 24.5 kHz carrier frequency.

The communication channel is represented by the Adder token and the Gaussian Noise token from the Source Library, with initial parameters of a power spectral density $N_o = 0$ W/Hz. The System Sampling Rate is 500 kHz, and setting the System Time: Number of Samples specification to 524 288 (2^{19}) points results in a spectral resolution of 0.95 Hz. The normalized ($R_L = 1\ \Omega$) power spectral density (PSD) of the Adder token channel output without additive white Gaussian noise (AWGN) is shown scaled and as a single-sided spectrum in Figure 5.6, in dBm/Hz (decibels referenced to 1 milliwatt per Hertz).

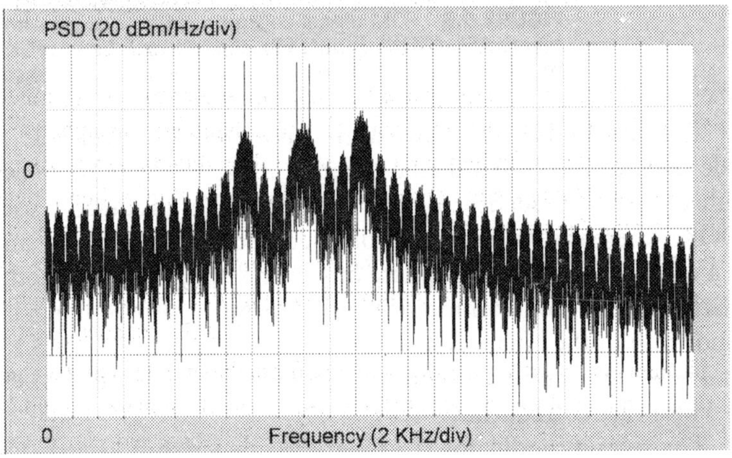

FIGURE 5.6 Power spectral density of a FDM digital communication system with binary ASK (OOK), binary FSK (MFSK), and binary PSK.

The PSD of the FDM digital communication system, as shown in Figure 5.6, is a composite of the PSDs of binary ASK (OOK), as shown in Figure 3.5, binary FSK (MFSK), as shown in Figure 3.10, and binary PSK, as shown in Figure 3.12. The binary ASK (OOK) PSD displays a carrier component at a frequency of 15.5 kHz and approximate nulls in the PSD that occur at the bit rate $r_b = 1/T_b$ (1 kb/sec or 1 kHz) and multiples of the bit rate center about the carrier. Three of the nulls in the binary ASK (OOK) PSD occur at 19.5 kHz, 20.5 kHz, and 24.5 kHz, which are the carrier components of the binary FSK (MFSK) signal and the suppressed carrier component of the binary PSK signal.

There are also approximate spectral nulls of the binary FSK (MFSK) signal centered at the 19.5 kHz and 20 kHz carrier components that occur at the bit rate r_b (1 kb/sec or 1 kHz) and multiples of the bit rate. Two of the nulls in the binary FSK (MFSK) PSD occur at 15.5 kHz, the carrier component of the binary ASK (OOK) signal, and at 24.5 kHz, the suppressed carrier component of the PSK signal.

Finally, the binary PSK PSD displays no distinct carrier component at 24.5 kHz, but also has approximate spectral nulls that occur at the bit rate r_b (1 kb/sec or 1 kHz) and multiples of the bit rate. Three of the nulls in the binary PSK PSD occur at 15.5 kHz, the carrier component of the binary ASK (OOK) signal), and 19.5 kHz and 20.5 kHz, the carrier components of the binary FSK (MFSK) signal.

The appearance of these PSD nulls at the carrier components of these three disparate digital communication systems is a result of the equal bit rates r_b and the selection of the nominal carrier frequencies for the ASK, FSK, and PSK signals here. Obviously, the occurrence of the PSD nulls certainly improves the bit error rate (BER) performance for the FDM digital communication system.

A BER analysis for the three FDM bandpass digital communication systems with AWGN, as shown in Figure 5.5, can be performed with the system shown in Figure 5.7 (see Fig5-7.svu on the CD-ROM). To facilitate the simulation in SystemVue, all of the Analysis tokens, except that for the BER token output as the Final Value token, are deleted in the simulation model Fig5-7DT. As described in Chapter 1, the SystemVue Textbook Edition does not permit tokens to be deleted.

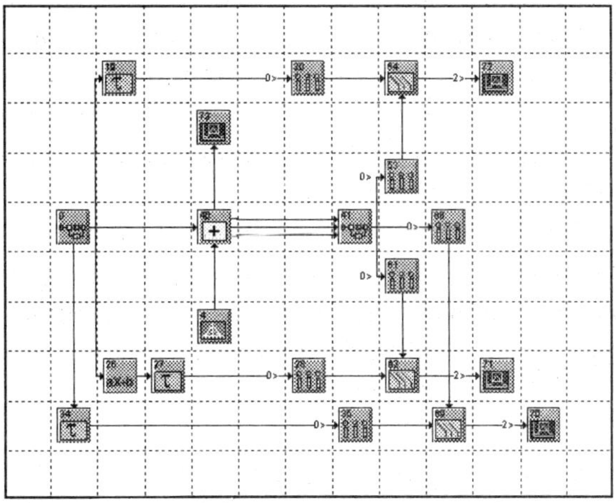

FIGURE 5.7 FDM digital communication system with BER analysis.

The three data sources and the binary ASK (OOK), binary FSK (MFSK), and binary PSK modulators are shown as the MetaSystems in Figure 5.8 to simplify the

SystemVue design window. Figure 5.8 also has MetaSystem tokens within the MetaSystems for the bandpass binary modulators.

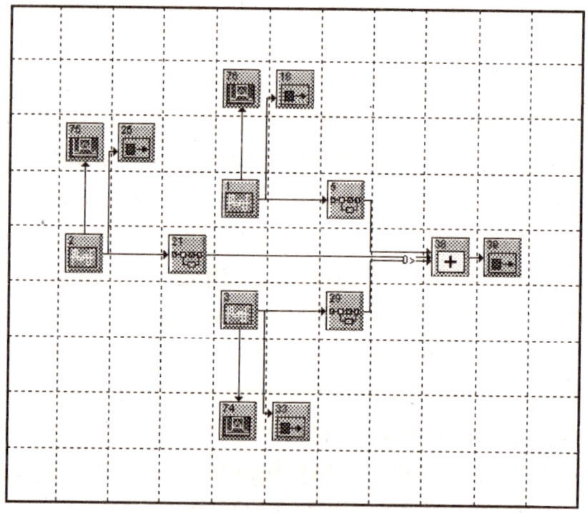

FIGURE 5.8 FDM digital communication system MetaSystem for the pseudonoise data sources and bandpass modulators.

The three demodulators for binary ASK (OOK), binary FSK (MFSK), and binary PSK are shown as the MetaSystems in Figure 5.9, which also has MetaSystem tokens within the MetaSystems for the bandpass binary demodulators. The three MetaSystem Input tokens and the three MetaSystem Output tokens are redundant, as connections could have been made within the demodulator MetaSystems themselves. However, this SystemVue design window construction presents a clear picture of the simulation when MetaSystem tokens are contained within other MetaSystems.

The communication channel is represented by the Adder token and the Gaussian Noise token from the Source Library, with initial parameters of a power spectral density $N_o = 0$ W/Hz. The binary FSK (MFSK) digital communication system can be evaluated for its performance in FDM and the presence of AWGN.

The System Stop Time is set to 10 sec for convenience, which would process 10,000 information bits in possible error. Table 5.1 is a tabular list of the observed BER in a single trial as a function of E_d/N_o to the theoretical probability of bit error P_b for binary FSK (MFSK), as given by Equation 3.11 for asymmetrical signals. The BER performance for binary FSK (MFSK) without the possible interference due to adjacent modulated signals in FDM, as repeated here, is given in Table 3.4.

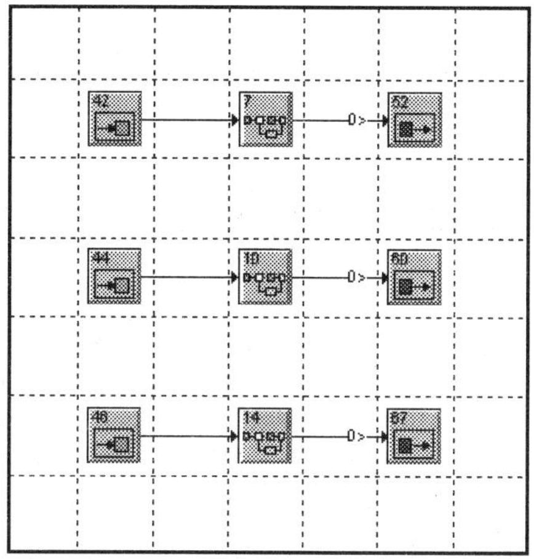

FIGURE 5.9 FDM digital communication system MetaSystems for the bandpass demodulators.

Both binary ASK (OOK), as given in Table 3.2, and binary PSK, as given in Table 3.5, also show no compromise in BER performance in FDM. The results indicate that BER performance is not compromised in FDM with the bandwidth used for each modulated signal and the data rate here.

TABLE 5.1 Observed BER without and with FDM and Theoretical P_b as a Function of E_d/N_o in a Binary FSK (MFSK) Digital Communication System

E_d/N_o dB	N_o V2-sec	BER	BER (FDM)	P_b
∞	0	0		0
12	1.58×10^{-3}	1.7×10^{-3}	2.2×10^{-3}	2.5×10^{-3}
10	2.50×10^{-3}	1.19×10^{-2}	1.21×10^{-2}	1.25×10^{-2}
8	3.96×10^{-3}	3.82×10^{-2}	3.71×10^{-2}	3.75×10^{-2}
6	6.28×10^{-3}	8.19×10^{-2}	7.46×10^{-2}	7.93×10^{-2}
4	9.95×10^{-3}	1.382×10^{-1}	1.343×10^{-1}	1.314×10^{-1}
2	1.58×10^{-2}	1.854×10^{-1}	1.911×10^{-1}	1.872×10^{-1}
0	2.50×10^{-2}	2.363×10^{-1}	2.343×10^{-1}	2.393×10^{-1}

Bandpass filters can also be used with either the coherent or the noncoherent receiver to improve BER performance in FDM and possibly decrease the overall bandwidth. Alternatively, if the data rates are not equal, then the optimal conditions of PSD nulls at adjacent carrier frequencies here are not available and *guardbands*, additional bandwidth separating the modulated signals, are required [Stern04].

CODE DIVISION MULTIPLEXING

Code division multiplexing (CDM) transmits a group of independent data sources at the same time and over the same range of frequencies. The principle of CDM is *spread spectrum*, in which each modulated signal is transmitted with a bandwidth larger than the minimum bandwidth required, as described for bandpass signals in Chapter 3. The bandwidth is increased by the use of pseudorandom sequences or *spreading codes* that are independent of the information, but which also causes the modulated signals to be *orthogonal* to each other. Orthogonality implies that two signals are independent over an interval and is described in Chapter 3 (Equation 3.27), for the bandpass frequency shift keying (FSK) signal.

CDM, or as it is often called *code division multiple access* (CDMA), is implemented by two spread spectrum techniques. Direct sequence spread spectrum (DSSS) processes the information at a data rate r_b b/sec by *exclusive-ORing* it with a spreading code at a higher spread spectrum *chip rate* r_{ss} chips/sec (or Hz). Frequency hopping spread spectrum (FHSS) pseudorandomly changes the carrier frequency under the control of a spreading code during the transmission of data at a *hop rate* r_{hop} Hz. Both DSSS and FHSS increase the bandwidth of the transmitted signal but provide orthogonality with CDM. CDM requires synchronization of both the chip and bit times in the receiver, and this is presented in Chapter 4, "Synchronization and Equalization."

The spreading code is generated by a digital logic circuit consisting of a sequential logic (clock driven) *shift register* and register *taps* that feed back the output to a network of combination logic (event driven). The binary sequence generator is pseudorandom, because the output of a finite digital logic *feedback shift register* (FSR) must repeat after N clock cycles. The desirable properties of the pseudorandom sequence generator are as follows [Sklar01]:

Balance: The number of binary 1s in the finite pseudorandom sequence should differ from the number of binary 0s by no more than 1, so that the spreading code bits are equally probable.

Run Property: One-half of the groups of binary bits should be 1 bit long, one-quarter should be 2 bits long, one-eighth should be 3 bits long, and so forth. The run property is necessary to assure that the sequence is very nearly random and independent.

Correlation: The bit-by-bit XORing of an n-bit binary pseudorandom sequences with the same pseudorandom sequence *rotated* by j-bits should be balanced for all nonzero values of j less than n.

An n-bit binary pseudorandom sequence and the j-bit ($j < n$) rotated sequence is given by Equation 5.2.

$$\begin{aligned} \mathbf{b} &= [b_1\, b_2 b_3\, b_4 \ldots b_n] \\ \mathbf{b^j} &= [b_{(n-j+1)}\, b_{(n-j+2)}\, b_{(n-j+3)} \ldots b_n\; b_1 \ldots b_{n-j}] \end{aligned} \tag{5.2}$$

If the combinational logic of the FSR is a *modulo-2 adder*, then it is called a *linear* FSR. If the FSR consists of M shift registers and the output repeats only after $2^M - 1$ clock inputs, the FSR is called a *maximal length sequence* or an M-sequence [Simon01]. A SystemVue simulation of an FSR is shown in Figure 5.10 (see Fig5-10.svu on the CD-ROM).

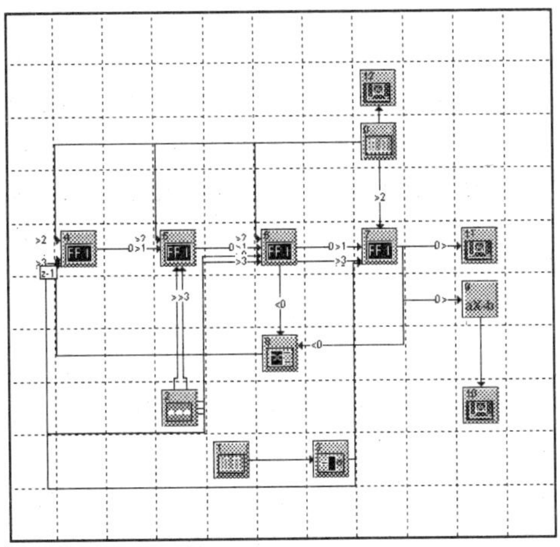

FIGURE 5.10 Finite shift register pseudorandom sequence generator.

The 4-bit shift register consists of four D-Type Flip-Flop tokens from the Logic Library, with parameters of a logic threshold of 0.5 V, a positive true output of 1 V, and a false output of 0 V. The D-Type Flip-Flop token has set and clear inputs that are asserted with a logic 0 (set* and clear*), a positive edge triggered clock input, a data input (D), and both Q and Q* data outputs.

All four D-Type Flip-Flop token clock inputs are provided by a Pulse Train token from the Source Library, with parameters of a frequency $f_o = 20$ kHz ($T_{clock} =$

50 μsec), a pulse width $\tau = 25$ μsec, and a pulse amplitude $A = 1$ V. The D-Type Flip-Flop token set inputs for the first and last tokens are provided by another Pulse Train token from the Source Library, with parameters of a frequency of 0.1 Hz (10 sec period), a pulse width $\tau = 2$ μsec, and a pulse amplitude $A = 1$ V.

The output of the Pulse Train token is logically inverted by an Invert token from the Logic Library to provide the set* (logic 0) signal. The 10 sec period and pulse width equal to the System Time ($T_{system} = 2$ μsec or a 500 kHz System Sampling rate) implies that this Pulse Train token effectively initializes two of the D-Type Flip-Flop tokens with Q = 1. This initialization is the *seed* of the FSR and is 9 or a 4-bit pattern of [1001] for the four registers here. All four of the D-Type Flip-Flop token clear* inputs and two set* inputs are provided with a logical 1 by a Custom token from the Source Library with the output $p(0) = 1$ V.

The feedback shift register taps are obtained from the third and fourth D-Type Flip-Flop token outputs, and processed by an XOR token from the Logic Library, which is essentially a 1-bit modulo-2 adder. The output of the XOR token is fed back to the input of the first D-Type Flip-Flop token to perpetuate the generation of the pseudorandom sequence.

The output of the pseudorandom sequence generator is processed by the Polynomial token from the Function Library to provide a polar + 1 V signal. The parameters of the Polynomial token are that the x^0 (offset) coefficient is –1, the x^1 (linear) coefficient is 2, and all other coefficients are zero.

The System Stop Time is set to 2 msec for convenience and the resulting SystemVue simulation of the FSR in Figure 5.10 is shown in Figure 5.11. The unipolar pseudorandom sequence generator output is the signal at the top, and the 20 kHz clock signal is below.

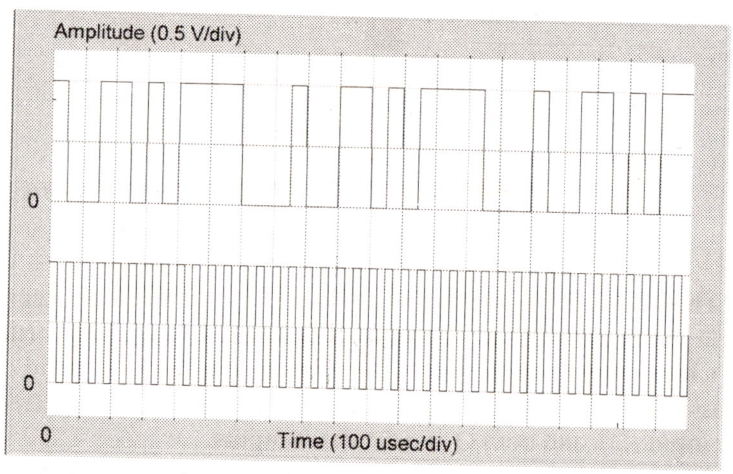

FIGURE 5.11 Finite shift register pseudorandom sequence generator clock and output signals.

The sequence output can be read at the positive edge of the clock signal in Figure 5.11, and is given by Equation 5.3.

$$\mathbf{b} = [100110101111000] \qquad (5.3)$$

The pseudorandom sequence repeats after $N = 15$, as shown in Figure 5.11 with $15 \times 50\ \mu\text{sec} = 750\ \mu\text{sec}$. Because the number of shift registers $M = 4$ and $2^4 - 1 = 15$, the FSR here is a maximal length sequence generator. The sequence has eight 1s and seven 0s and the balance property is nearly satisfied. The sequence has four 1-bit runs (0, 1), two 2-bit runs (00, 11), one 3-bit run (000), and one 4-bit run (1111). The run property is not satisfied because of the small number of shift registers in the FSR.

However, the correlation property is satisfied for all values of the shifting parameter j, $1 \le j \le n\text{–}1 = 14$. A typical bit-by-bit XORing ($\oplus$) of the 15-bit binary pseudorandom sequence with $j = 1$ is given by Equation 5.4. The resulting 15-bit XORed sequence also has eight 1s and seven 0s and is balanced.

$$\begin{aligned} \mathbf{b} \quad &= [100110101111000] \\ \mathbf{b}^1 \quad &= [010011010111100] \\ \mathbf{b} \oplus \mathbf{b}^1 &= [110101111000100] \end{aligned} \qquad (5.4)$$

The period of the pseudorandom sequence or spreading code $c(t)$ is $T_b = nT_c$, where n is the number of bits in the non-repeating spreading code, T_c is the duration of either a 1 or 0 in the spreading code, and T_b is the bit time of the binary data to be code division multiplexed or *spread*.

For a maximal length pseudorandom sequence the number of non-repeating bits is $N = 2^M - 1$. The *discrete* spreading code is $c(jT_c)$ where $0 \le j \le N$ and is periodic with period NT_c. The autocorrelation function $R(kT_c)$ of the polar ($\pm$ 1 V) spreading code $c(jT_c)$ is given by Equation 5.5 [Proakis05].

$$R(kT_c) = \frac{1}{N} \sum_j c(jT_c)\, c(jT_c - kT_c) \quad 0 \le j \le N, 0 \le k \le N \qquad (5.5)$$

From the properties of the pseudorandom sequence, the balanced polar spreading code $c(jT_c)$ has $N/2$ binary 1s (+1 V) and $N/2$ binary 0s (–1 V), and the discrete autocorrelation function $R(kT_c)$ is then given by Equation 5.6.

$$\begin{aligned} R_c(kT_c) &= 1 \qquad & k = 0 \\ R_c(kT_c) &= -\frac{1}{N} \qquad & 1 \le k \le N \end{aligned} \qquad (5.6)$$

The autocorrelation can be computed as the inverse Fourier of the normalized power spectral density (PSD), as described in Chapter 1 [Lathi98]. Setting the System Time: Number of Samples specification to 524 288 (2^{19}) points, the autocorrelation of the finite shift register pseudorandom sequence generator in Figure 5.10 is obtained in the Sink Calculator with the Correlate Convolution primary tab, as shown in Figure 5.12. The Auto Correlate secondary tab is selected with the options of no removal of the mean value before the correlation (this is arbitrary, because the mean here is 0), and a circular, mixed radix fast Fourier transform (FFT) with no zero padding, as described in Chapter 1.

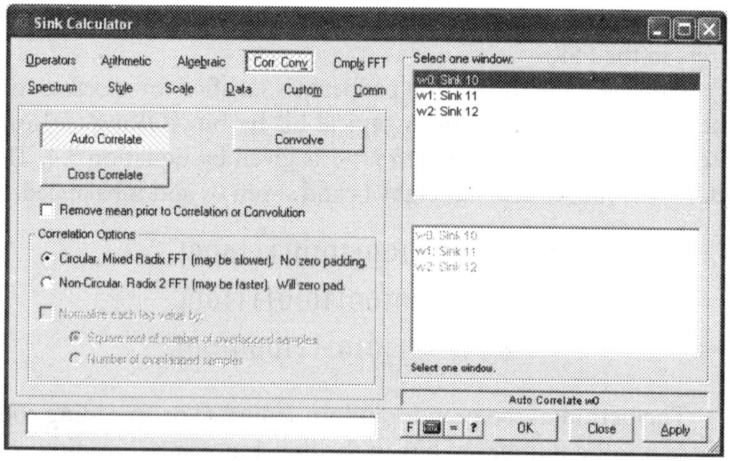

FIGURE 5.12 Sink calculator for the autocorrelation of the pseudorandom sequence.

Figure 5.13 is the scaled autocorrelation of the polar (± 1 V) pseudorandom sequence generator, as given by Equation 5.6. The minimum value of the autocorrelation function here is –1/15 (–0.0666), and the period is $N\ T_c = 750$ μsec, because $N = 15$ and $T_c = 50$ μsec.

The autocorrelation function $R_c(\tau)$ of a periodic continuous spreading code c(t) is given by Equation 5.7. The autocorrelation function here has correlation properties similar to that of white Gaussian noise, as described in Chapter 1. That is, $R_c(0)$ is a maximum value and $R_c(\tau)$ ($\tau \neq 0$) approaches 0. The autocorrelation function $R_c(kT_c)$ of the discrete spreading code $c(jT_c)$, given by Equation 5.6, is a sampled version of Equation 5.7. The autocorrelation of the polar pseudorandom sequence, as shown in Figure 5.13, agrees with Equation 5.7.

$$R_c(\tau) = 1 - \frac{N+1}{N T_c}|\tau| \quad |\tau| \le T_c$$
$$R_c(\tau) = -\frac{1}{N} \qquad \text{otherwise} \tag{5.7}$$

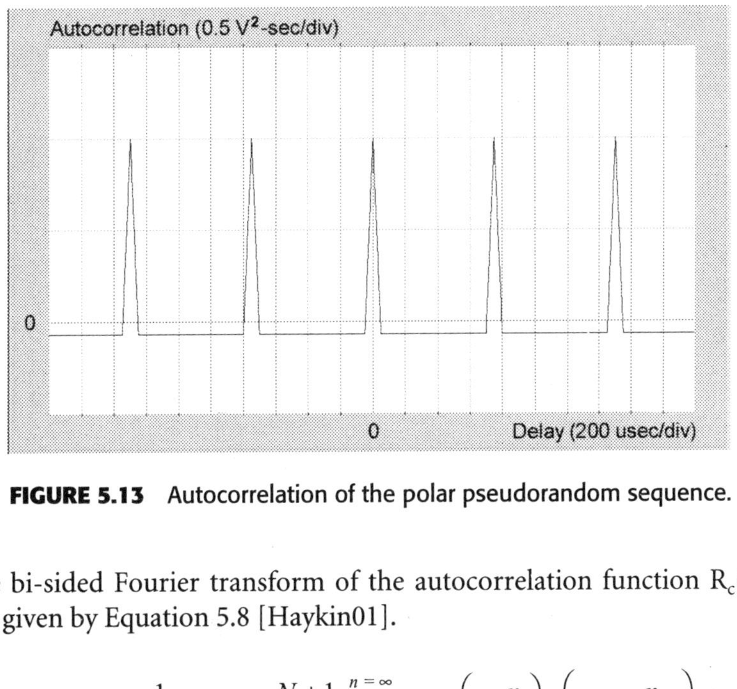

FIGURE 5.13 Autocorrelation of the polar pseudorandom sequence.

The bi-sided Fourier transform of the autocorrelation function $R_c(\tau)$ is the PSD, as given by Equation 5.8 [Haykin01].

$$PSD = \frac{1}{N^2}\delta(f) + \frac{N+1}{N}\sum_{\substack{n=-\infty \\ n\neq 0}}^{n=\infty} \text{sinc}^2\left(\pi\frac{n}{N}\right)\delta\left(f - \frac{n}{NT_c}\right) \tag{5.8}$$

For the pseudorandom sequence here, $N = 15$ and $T_c = 50$ μsec. The impulse terms are discrete components that occur at $f = 0$ Hz and at multiples of $f = 1/(NT_c)$ = 1.333 kHz. The sinc^2 term is zero at $n = \pm N, \pm 2N \ldots$ and at $f = \pm 1/T_c, \pm 2/T_c \ldots$ or multiples of $1/T_c = 20$ kHz. The single-sided PSD of the polar pseudorandom sequence shown in Figure 5.14 agrees with Equation 5.8.

The implementation of the feedback shift register (FSR) pseudorandom sequence generator using Logic Library tokens, even for the small $M = 4$ register size FSR as shown in Figure 5.10, is complicated and inefficient. The PN Sequence Generator token from the Communications Library provides a convenient structure to specify a FSR to a maximum of $M = 33$ registers.

PN implies a pseudonoise source but not a white noise source. Although the autocorrelation of the pseudorandom sequence in Figure 5.13 has the some of the characteristics of white noise, the PSD in Figure 5.14 is not than of a white noise source, as described in Chapter 1. Figure 5.15 shows the parameter window of the PN Sequence Generator token, which implements the FSR shown in Figure 5.10. The register length here is $M = 4$, the seed is 9 (binary 1001), the input clock logic threshold is 0.5 V, the output is polar (± 1 V), and the feedback tap connections are from registers 4 and 3.

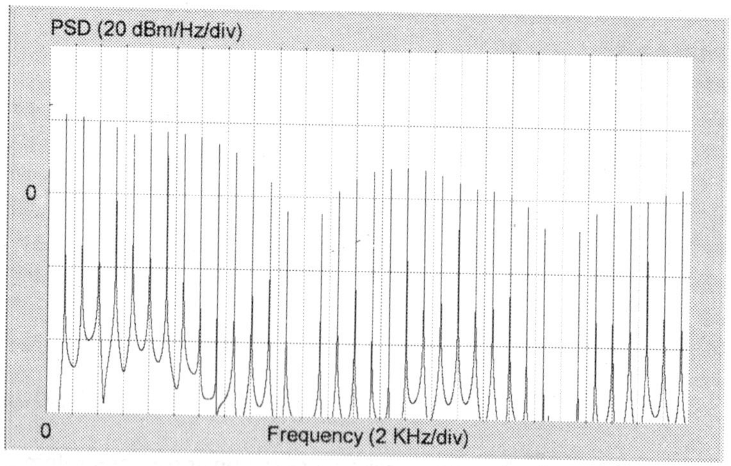

FIGURE 5.14 Power spectral density of the polar pseudorandom sequence.

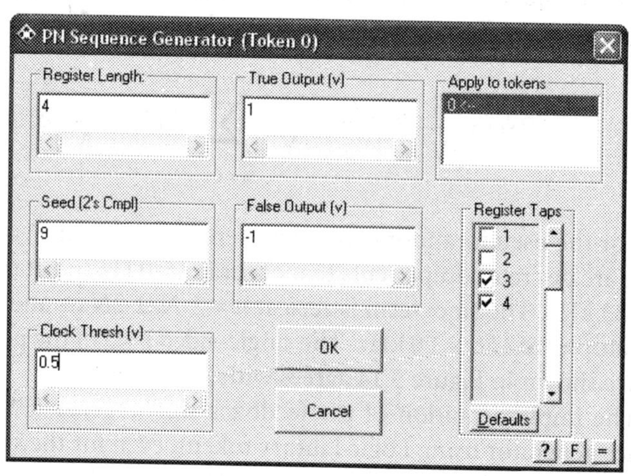

FIGURE 5.15 Parameter window for the PN sequence generator.

The FSR configurations for maximal length sequences are available as extensive tables of the feedback connections [Dixon94]. Table 5.2 lists several of these sets of feedback tap connections for an FSR consisting of M shift registers with a maximal length of $L = 2^M - 1$. An entry [4, 3] for an $M = 4$ FSR, as shown in Figure 5.10 and Figure 5.15, implies that the outputs of register 4 and register 3 are modulo-2 added together, and that output is then inputted to register 1. The pseudorandom sequence is obtained from the output of register 4.

TABLE 5.2 Feedback Tap Connections for Generating Maximal Length Sequences Using *M* Shift Registers and the Maximal Length *L*

M registers	Sets of Feedback Tap Connections	Maximal Length *L*
4	[4, 1] [4, 3]	15
5	[5, 1] [5, 2] [5, 4, 2, 1] [5, 4, 3, 2]	31
6	[6, 1] [6, 5, 2, 1] [6, 5, 3, 2]	63
7	[7, 1] [7, 3] [7, 3, 2, 1] [7, 5, 4, 3, 2, 1]	127
8	[7, 6, 5, 1] [8, 4, 3, 2] [8, 6, 4, 3, 2, 1]	255
16	[16, 14, 5, 1]	65 535
32	[32, 31, 11, 1]	4 294 967 295

For large values of the number of shift registers M, the autocorrelation function $R_c(\tau)$ for $\tau \neq 0$ is small and equal to $1/L = 1/(2^M - 1)$. However, in code division multiplexing (CDM) or code division multiple access (CDMA), each signal is assigned a unique spreading code and these should be uncorrelated so that the signals remain mutually orthogonal. That is, the cross-correlation between the spreading codes should not exhibit a large peak magnitude R_{max}.

ON THE CD

Figure 5.16 shows two maximal length pseudorandom sequence generators with $M = 4$ and feedback tap connections [4, 3] and [4, 1] (see Fig5-16.svu on the CD-ROM). The top PN Sequence Generator token is the FSR, as shown in Figure 5.10. The other PN Sequence Generator is a similar $M = 4$ FSR, but with feedback tap connections at register 4 and register 1. The seed is 9 for both PN Sequence Generator tokens. The clock for both PN Sequence Generator tokens is the Pulse Train token from the Source Library, with parameters of a frequency $f_o = 20$ kHz ($T_{clock} = 50$ μsec), a pulse width $\tau = 25$ μsec, and a pulse amplitude $A = 1$ V.

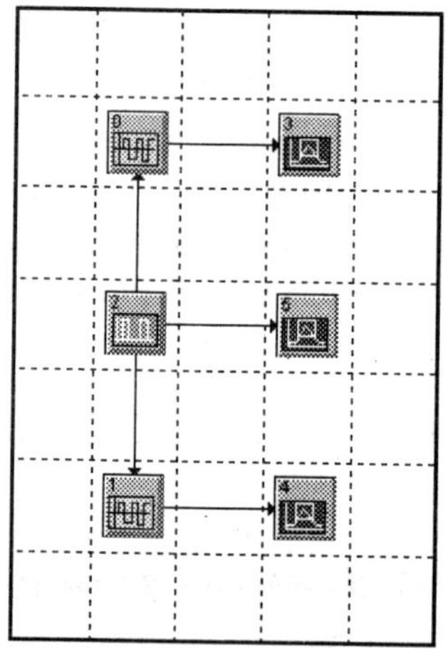

FIGURE 5.16 Two feeback shift register PN sequence generators.

The System Stop Time is set to 2 msec for convenience and the resulting SystemVue simulation of the FSR in Figure 5.15 is shown in Figure 5.17. The polar pseudorandom sequence generator outputs are the two signals at the top and middle, and the 20 kHz clock signal is below. Both of the $M = 4$ maximal length sequences repeat after $L = 2^4 - 1 = 15$, as shown in Figure 5.17 with 15×50 μsec = 750 μsec.

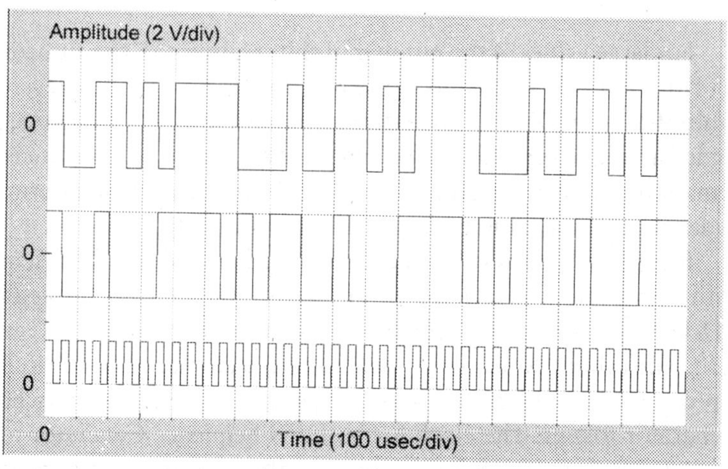

FIGURE 5.17 Two feedback shift register PN sequence generator output signals and common clock.

The cross-correlation can also be computed as the inverse Fourier transform of the normalized cross power spectral density, as described in Chapter 1. Setting the System Time: Number of Samples specification to 524 288 (2^{19}) points, the cross-correlation of the polar (± 1 V) outputs of the two finite shift register pseudorandom sequence generators in Figure 5.16 is obtained in the Sink Calculator with the Correlate Convolution primary tab, as shown in Figure 5.10. However, here the Cross Correlate secondary tab is selected with the options of no removal of the mean value before the correlation (because the mean here is 0), and a circular, mixed radix fast Fourier transform (FFT) with no zero padding, as described in Chapter 1.

The cross-correlation of the two polar $M = 4$ finite shift register pseudorandom sequence generators, both with a seed of 9, is shown in Figure 5.18. The cross-correlation can vary in its offset τ, but not the peak magnitude R_{max}, if the pseudorandom sequences are initialized with other binary bit patterns or seeds.

The ratio of the peak magnitude of the cross-correlation R_{max} to that of the autocorrelation $R(0) = 1$ ($\tau = 0$) here is 0.46, which is large and thus unsuitable for CDMA applications. Improved periodic cross-correlations are provided by the *Gold code sequences*, which are the modulo-2 sum of two *preferred maximum length sequences* [Holmes82]. There are L cyclicly shifted versions of one sequence relative to the other then available for CDMA applications.

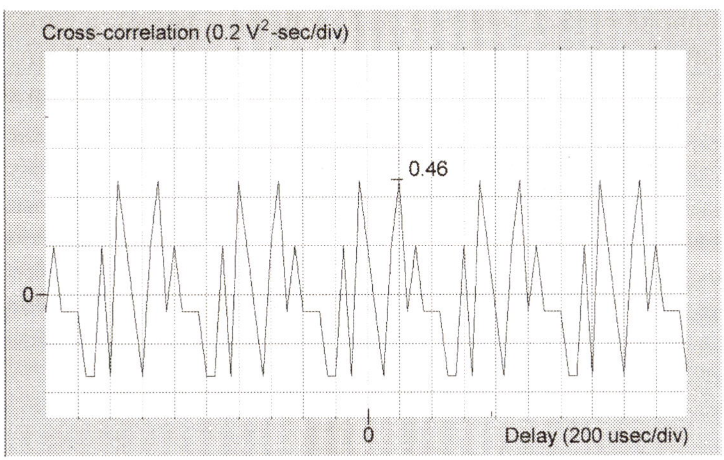

FIGURE 5.18 Cross-correlation of two polar PN sequences.

The Gold Code Generator token from the Communications Library configures the two preferred maximum length sequences. The parameter window for a Gold sequence generator, with an $M = 5$ FSR maximum length sequences and feedback tap connections [5, 2] and [5, 4, 3, 2], and is initialized with the 5-bit patterns

[11101] and [10001], as shown in Figure 5.19. The seed is the registers to be initially set with a binary 1, and the 5-bit pattern [11101] is inputted as [1 2 3 5].

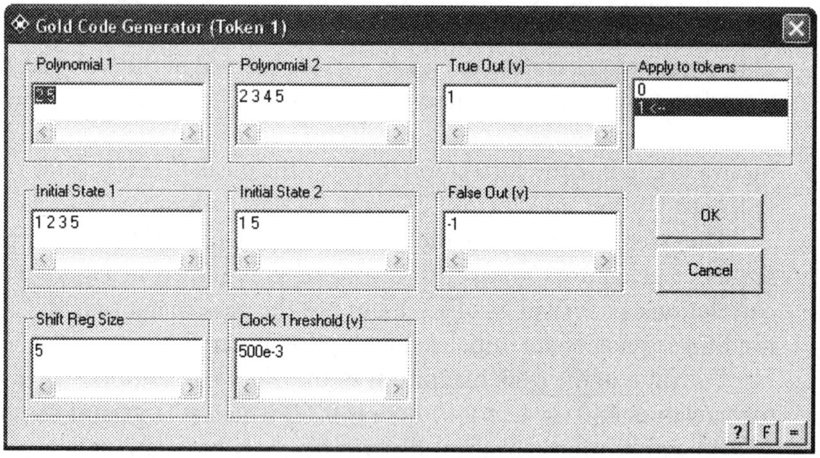

FIGURE 5.19 Parameter window for the Gold code sequence generator.

Figure 5.20 shows two Gold Code Generator tokens, both with feedback tap connections [5, 2] and [5, 4, 3, 2], with the upper token initialized with the 5-bit patterns [11101] and [10001], and the lower token with [10101] and [11100] (see Fig5-20.svu on the CD-ROM). The Pulse Train token from the Source Library clocks the Gold code sequence generator at a 20 kHz rate (T_{clock} = 50 μsec), with a pulse width τ = 5 μsec and a pulse amplitude A = 1 V.

ON THE CD

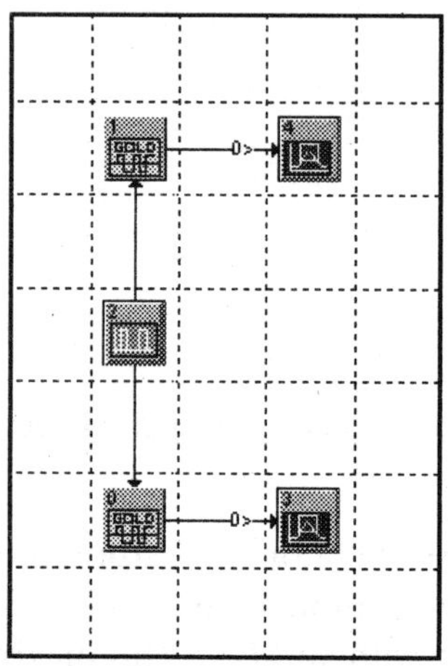

FIGURE 5.20 Two Gold code sequence generators.

The cross-correlation of the two Gold sequence generators in Figure 5.20 is shown in Figure 5.21. The period of both of the Gold sequence generators, with $L = 2^5 - 1 = 31$, is 31×50 μsec = 1550 μsec. The ratio of the peak magnitude of the cross-correlation R_{max} to that of the autocorrelations $R(0) = 1$ ($\tau = 0$) here is 0.233, which is smaller than that for the $M = 4$ maximum length sequence in Figure 5.18 ($R_{max}/R(0) = 0.46$), and is more suitable for CDMA applications.

The ratio $R_{max}/R(0)$ for the PN sequence generators does not significantly improve with an increase in the number of registers M and even for $M = 12$, $R_{max}/R(0) = 0.3$, approximately. However, the ratio $R_{max}/R(0)$ for the Gold sequence generators continues to improve with an increase in M and for $M = 12$, $R_{max}/R(0) = 0.03$ [Proakis05]. A Gold code generator with a large value of the number of shift registers M also has a low autocorrelation function $R_c(\tau)$ for $\tau \neq 0$ equal to $1/L = 1/(2^M - 1)$.

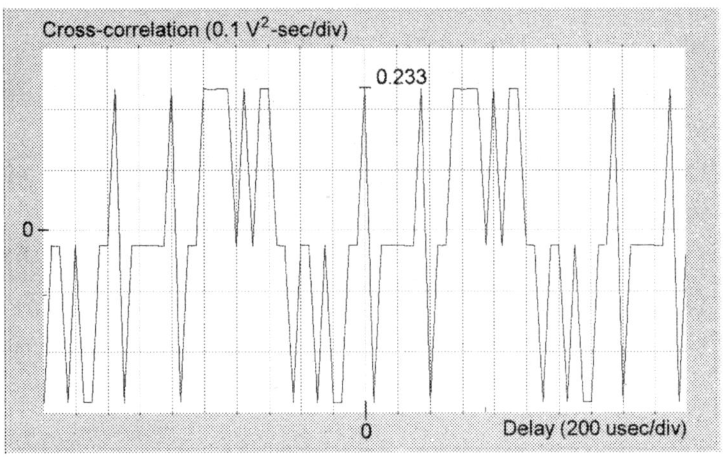

FIGURE 5.21 Cross-correlation of two polar Gold code sequences.

The preferred pairs of maximum length sequences for the Gold code generators are available as tables of the feedback connections [Dixon94]. Table 5.3 lists several of these pairs of FSR feedback tap connections, consisting of M shift registers with a maximal length of $L = 2^M - 1$.

TABLE 5.3 Feedback Tap Connections for the Preferred Pair of Maximal Length Sequences for the Gold Code Generator Using *M* Shift Registers and the Maximal Length *L*

M registers	Pairs of Feedback Tap Connections	Maximal Length *L*
5	[5, 2] [5, 4, 3, 2]	31
6	none	63
7	[7, 3] [7, 3, 2, 1]	127
7	[7, 3, 2, 1] [7, 5, 4, 3, 2, 1]	
8	[8, 7, 6, 5, 2, 1] [8, 7, 6, 1]	255
9	[9, 4] [9, 6, 4, 3]	511
9	[9, 6, 4, 3] [9, 8, 4, 1]	
10	[10, 8, 5, 1] [10, 7, 6, 4, 2, 1]	1023
10	[10, 8, 4, 3] [10, 9, 6, 5, 4, 3]	
11	[11, 2] [11, 8, 5, 2]	2047
11	[11, 8. 5, 2] [11, 10, 3, 2]	
12	[12, 9, 3, 2] [12, 11, 8, 7, 6, 3, 2, 1]	4095
12	[12, 10, 6, 5, 2, 1] [12, 11, 10, 4]	
13	[13, 4, 3, 1] [13, 10, 9, 7, 5, 4]	8191
13	[13, 11, 8, 7, 4, 1] [13, 11, 10, 5, 4, 3, 2, 1]	

DIRECT SEQUENCE SPREAD SPECTRUM

Direct sequence spread spectrum (DSSS) processes the information at a data rate r_b b/sec by exclusive-ORing (XORing) it with an orthogonal spreading code at a higher spread spectrum *chip rate* r_{ss} chips/sec (or Hz). The result is an output rate of r_{ss} b/sec and a wider transmission bandwidth. The bandpass modulation method used in DSSS is usually phase shift keying (PSK) [Dixon94]. The bandwidth required to transmit 90% of the total power in a binary PSK signal without spreading is $2r_b$ Hz, from Chapter 3, but would be $2r_{ss}$ Hz with DSSS. DSSS modulation increases the bandwidth by a factor known as the *processing gain* $G_{pDSSS} = r_s/r_b$.

Because the pseudorandom sequence is essentially uncorrelated, a spreading code supplies an additive white Gaussian noise (AWGN) source, which does not alter the form of noise power spectral density (PSD), as described in Chapter 1. Figure 5.22 shows the SystemVue simulation of a Gold sequence generator added to a Gaussian noise source (see Fig5-22.svu on the CD-ROM).

ON THE CD

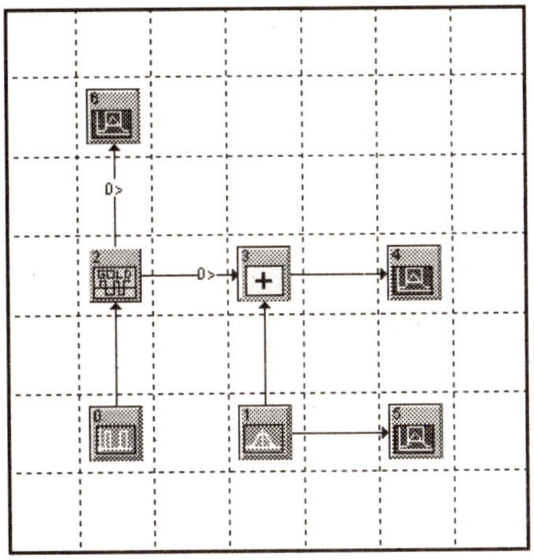

FIGURE 5.22 Polar M = 5 Gold code sequence generator with AWGN.

The Gold Code Generator token has an $M = 5$ FSR maximum length sequences, feedback tap connections [5, 2] and [5, 4, 3, 2], and is initialized with the 5-bit patterns [11101] and [10001]. The Pulse Train token from the Source Library clocks the Gold code sequence generator at a chip rate $r_{ss} =$ 20 kHz ($T_{clock} = 50$ μsec), with a pulse width $\tau = 25$ μsec and a pulse amplitude $A = 1$ V. The System Sampling rate is 500 kHz and the System Time: Number of Samples is 524 288 (2^{19}) points, which provides a spectral resolution of 0.9 Hz.

The single-sided PSD of the polar Gold code sequence shown in Figure 5.23 agrees with Equation 5.8. The impulse terms are discrete components that occur at $f = 0$ Hz and at multiples of $f = 1/(NT_c) \approx 645.2$ Hz, because $N = 31$ and $T_c = 50$ μsec here. The sinc^2 term is zero at $n = \pm N, \pm 2N \ldots$ and correspondingly at $f = \pm 1/T_c, \pm 2/T_c \ldots$ or multiples of $1/T_c = 20$ kHz.

The single-sided PSD of the polar (± 1 V) Gold code sequence with AWGN is shown in Figure 5.24. The Gaussian Noise token from the Source Library has the parameters of zero mean and a standard deviation $\sigma = 1$ V. Although the impulse terms of the polar Gold code sequence remain evident in the PSD, as shown in Figure 5.24, the PSD remains flat, as described in Chapter 1 for white noise.

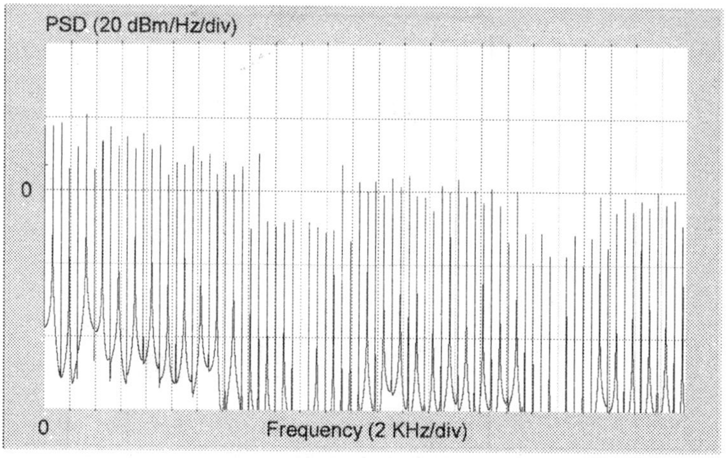

FIGURE 5.23 Power spectral density of the polar M = 5 Gold code sequence.

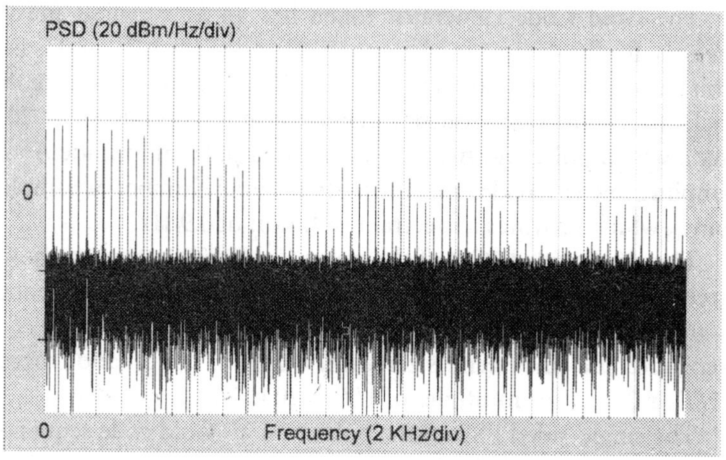

FIGURE 5.24 Power spectral density of the polar M = 5 Gold code sequence with AWGN.

SystemVue Simulation of DSSS

ON THE CD A DSSS transmitter with binary bandpass PSK modulation is shown in Figure 5.25 (see Fig5-25.svu on the CD-ROM). The Gold Code Generator token has the same FSR parameters and initialization patterns as that in Figure 5.22, but with a unipolar (0 V and 1 V) output. The Pulse Train token from the Source Library clocks the Gold code sequence generator at a chip rate r_{ss} = 20 kHz, with a pulse width τ = 25 μsec, and 0 V voltage and 0° phase offset. The data source is the unipolar pseudonoise PN

Sequence token from the Source Library, with parameters of a binary amplitude of 0 V and 1 V, a bit rate $r_b = 1$ kb/sec, and a 0.5 voltage and 0° phase offset.

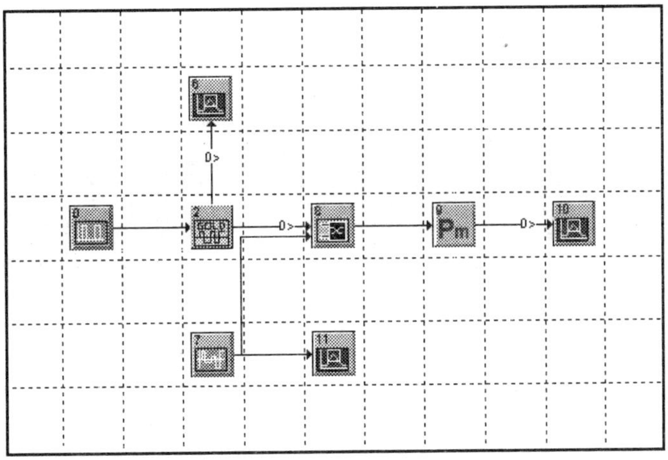

FIGURE 5.25 DSSS transmitter with binary PSK modulation.

The unipolar outputs of the Gold Code Generator and PN Sequence tokens are exclusive-ORed by the XOR token from the Logic Library, with parameters of a threshold of 0.5 V, a logic true output of 1 V, and a logic false output of –1 V. The polar output of the XOR token is inputted to the Phase Modulator token from the Function Library, with parameters of an amplitude $A = 5$ V, a carrier frequency $f_c = 100$ kHz, a modulation gain or phase deviation factor $k_p = 0.5\ 2\pi$ (radians)/V, and a 180° phase offset. The polar input to the Phase Modulator token produces a phase shift φ of 0 and π radians.

With a carrier frequency $f_c = 100$ kHz, the System Sampling rate is increased to 5 MHz and the System Time: Number of Samples is 4 194 304 (2^{22}) points, which provides a spectral resolution of 1.2 Hz. Figure 5.26 shows the normalized ($R_L = 1$ Ω) and scaled PSD of the DSSS binary PSK modulated signal at a chip rate $r_{ss} = 20$ kHz. To facilitate the simulation in SystemVue, all of the Analysis tokens, except that for the Phase Modulator token output, are deleted in the simulation model Fig5-25DT. As described in Chapter 1, the SystemVue Textbook Edition does not permit tokens to be deleted.

The PSD of a binary PSK signal at a carrier frequency $f_c = 20$ kHz and the same data rate $r_b = 1$ kb/sec, is given by Equation 3.36 and shown in Figure 3.12. Both PSDs are similar and exhibit approximate nulls at the chip or bit rate and multiples of the chip or bit rate. Although the data rate r_b is the same in Figure 3.12 and Figure 5.25, the bandwidth equivalent to 90% of the total power for the DSSS binary PSK signal is increased to 2 $r_{ss} = 40$ kHz.

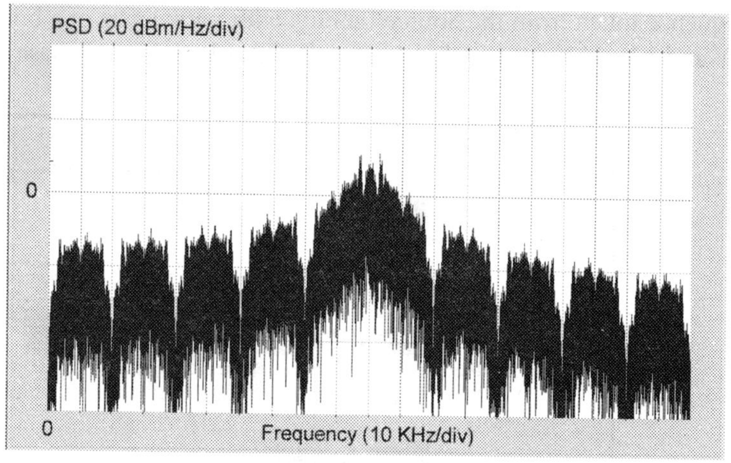

FIGURE 5.26 Power spectral density of the DSSS transmitter with binary PSK modulation.

The DSSS system with binary PSK modulation in Figure 5.25 is the transmitter MetaSystem of the two source DSSS system shown in Figure 5.27 (see Fig5-27.svu on the CD-ROM). The Gold code sequence generators have the preferred pair of feedback tap connections [5, 2] and [5, 4, 3, 2] from Table 5.3, with the first MetaSystem initialized with the arbitrary 5-bit patterns [11101] and [10001], and the second MetaSystem initialized with [10101] and [11100], as in Figure 5.20.

ON THE CD

The communication channel is represented by only the Adder token without AWGN in Figure 5.27, because the other DSSS system represents an *interference* signal. The $M = 5$ Gold code sequence spreading code used here is not ideal and bit errors can occur due to both interference and additive white noise.

The output of the Adder token is inputted to the binary PSK correlation receiver, which is a common MetaSystem for both data sources, is shown in Figure 5.28. The coherent reference signal input to the Multiplier token is a Sinusoidal token from the Source Library, with parameters of an amplitude of 5 V peak, a carrier frequency $f_c = 100$ kHz, and 0° phase offset. The output of the Multiplier is inputted to a lowpass filter (LPF) to attenuate the spectral components that are not at baseband frequencies.

The Linear System Filter token from the Operator Library provides the LPF here, which is selected to be a 9-pole Butterworth filter, with a cutoff frequency $f_{cutoff} = 20$ kHz. The output of the Multiplier token in Figure 5.28 contains spectral components at baseband (near 0 Hz), which are passed by the LPF, and a major bandpass component centered at $2\,f_c = 200$ kHz. The Multiplier token output also has spectral nulls at multiples of 20 kHz, which are a result of the DSSS system chip rate $r_{ss} = 20$ kHz, and not the data bit rate $r_b = 1$ kb/sec or 1 kHz.

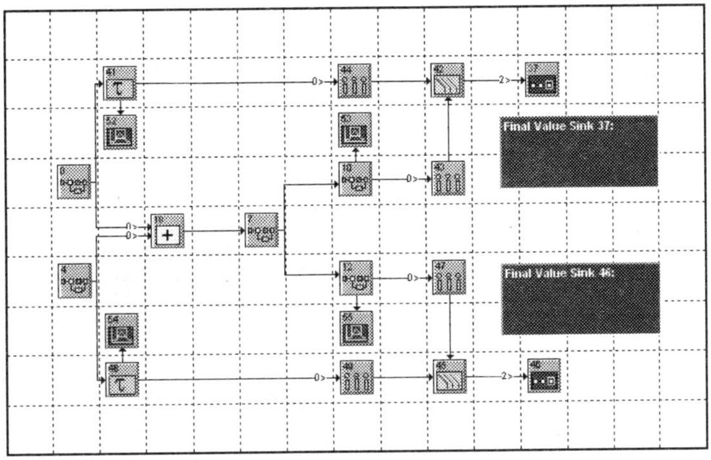

FIGURE 5.27 Two source DSSS system with binary PSK modulation and demodulation.

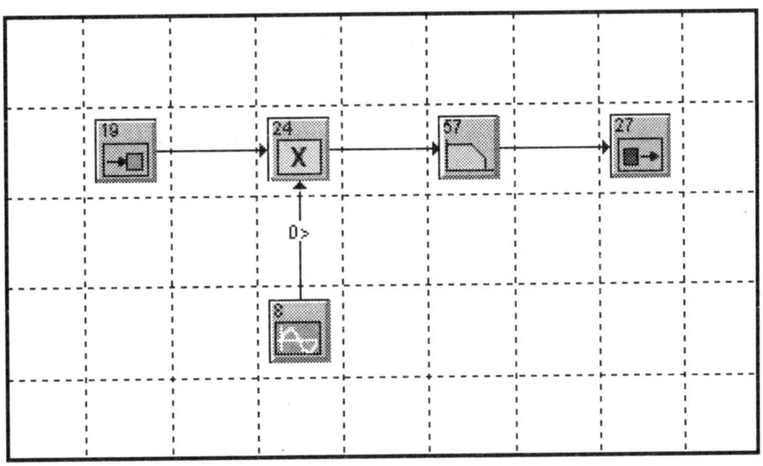

FIGURE 5.28 Binary PSK receiver MetaSystem token of the DSSS system in Figure 5.27.

The output of the binary PSK receiver MetaSystem in Figure 5.28 of the DSSS system in Figure 5.27 is inputted to two *despreading* and demodulating MetaSystems, as shown in Figure 5.29. The Pulse Train token from the Source Library clocks the Gold code sequence generator at a chip rate r_{ss} = 20 kHz. The output of the Gold Code Generator token is delayed by one chip time $1/r_{ss}$ = 50 μsec to despread the DSSS signal.

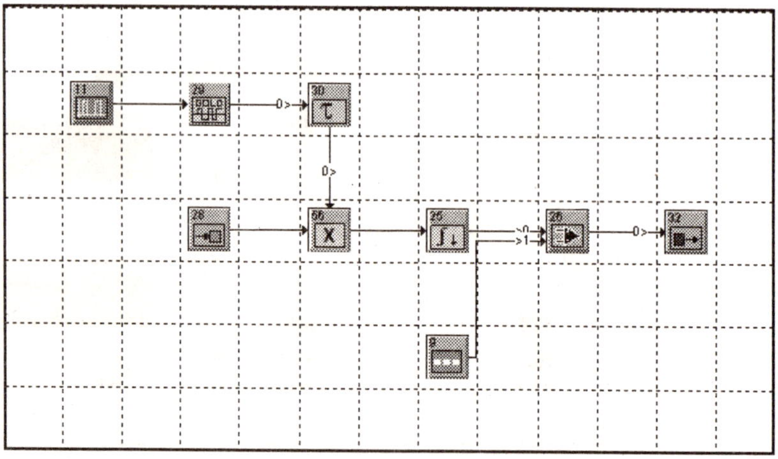

FIGURE 5.29 Despreading and demodulating MetaSystem token of the DSSS system in Figure 5.27.

The Gold code sequence generators in the correlation receiver MetaSystems in Figure 5.29 have the same preferred pair of feedback tap connections, and are initialized with the same arbitrary 5-bit patterns as the DSSS transmitter MetaSystem in Figure 5.25. The output of the Multiplier token as a despreader is inputted to the Integrate and Dump token from the Communications Library synchronized to the input signal. The parameters of the Integrate and Dump token are specified as an integration time equal to the bit time T_b of 1 msec and a continuous output.

The Analog Comparator token comparison input is set to the optimum threshold $\tau_{opt} = 0$ V here, because the bandpass binary PSK signal is symmetrical, as described in Chapter 3. The Custom token from the Source Library sets the threshold.

Two Sampler tokens in Figure 5.27 are used to downsample the 5 MHz System Time rate to the data bit rate of 1 kHz. The Sampler token parameters are set to a sample rate of 1 kHz, and an aperture time of 0.2 µsec (T_{system}, a single point sample). The two PN Sequence token data Sampler inputs are delayed by 1 msec to correlate the data. The BER token parameters are set to output on every trial (number of trials = 1), with a threshold of 0.5 V and a 1 msec offset. The System Stop Time is set to 10 sec for convenience, which would process 10 000 information bits in possible error.

Performance of DSSS with Interference

The DSSS binary PSK transmitter and receiver in Figure 5.27 can be evaluated for its performance in the presence of an interfering DSSS signal. Temporarily disconnecting one of the DSSS systems with binary PSK modulation from the Adder token in Figure 5.27 removes the interfering DSSS signal, which then produces 0 observed errors. Reconnecting the DSSS system is equivalent to adding noise to the desired signal and is an alternative analysis to the additive white Gaussian noise (AWGN) channel. The interfering DSSS signal produces a bit error rate (BER) of approximately 10^{-2}. To facilitate the simulation in SystemVue, all of the Analysis tokens, except that for the BER token output as the Final Value token, are deleted in the simulation model Fig5-27DT. As described in Chapter 1, the SystemVue Textbook Edition does not permit tokens to be deleted.

The observed BER for binary PSK with the optimum correlation receiver in AWGN are given in Table 3.5. The performance of the DSSS system with this short ($M = 5$) Gold code sequence generator and one DSSS system as an interference source is equivalent to an energy per bit E_b to noise PSD N_o rate of approximately 4 dB.

The autocorrelation and cross-correlation performance of Gold code sequence generators with a large number of registers M improves the observed BER. The DSSS system in Figure 5.27 is modified to evaluate the BER performance with the preferred pair of $M = 12$ feedback tap connections [12, 9, 3, 2] and [12, 11, 8, 7, 6, 3, 2, 1] from Table 5.3. The Gold code sequence generator in the first MetaSystem is initialized with the arbitrary 12-bit patterns [010110000101] and [001000111000], and the Gold code sequence generator in the second MetaSystem is initialized with [001100001001] and [100000010110].

The BER for the $M = 12$ Gold code sequence generator is approximately 4×10^{-3} here and, from Table 3.5 for binary PSK with the optimum correlation receiver in AWGN, equivalent to an E_b/N_o rate of 6 dB. This is an improvement of + 2 dB over the $M = 5$ Gold code sequence generator. The BER performance can be improved even further with an increase in the processing gain $G_{pDSSS} = r_{ss}/r_b$, which is $G_{pDSSS} = 20$ here.

The DSSS system in Figure 5.27 with the $M = 5$ Gold code sequence generator is now modified to evaluate the BER performance with an increase in the processing gain G_{pDSSS} by increasing the chip rate. The Pulse Train token from the Source Library clocks the Gold code sequence generators in the DSSS transmitter MetaSystem token in Figure 5.25, and the despreading and demodulating MetaSystem token in Figure 5.29 at a chip rate $r_{ss} =$ 50 kHz, and the resulting processing gain $G_{pDSSS} = r_{ss}/r_b$ is 50.

The Linear System Filter token from the Operator Library is now selected to be a 9-pole Butterworth filter, but with a cutoff frequency $f_{cutoff} = 50$ kHz, which is an appropriate LPF for the chip rate $r_{ss} = 50$ kHz. The output of the Gold Code Generator token in the despreading and demodulating MetaSystem token in Figure 5.29 is also delayed by one chip time $1/r_{ss} = 20$ μsec to despread the DSSS signal here.

The BER performance of the DSSS system with the $M = 5$ Gold code sequence generator, but with a processing gain $G_{pDSSS} = 50$ and one DSSS system as an interference source, is now less than 10^{-4}. From Table 3.5 for binary PSK with the optimum correlation receiver in AWGN, this BER is equivalent to an E_b/N_o ratio of approximately 10 dB, and is an improvement of + 6 dB over the same DSSS system simulation, but with a processing gain $G_{pDSSS} = 20$.

Signal to Noise Ratio in DSSS

For a DSSS system with J transmitters using CDM with an average normalized (R_L = 1 Ω) transmitted power P and an AWGN channel with a bi-sided normalized PSD of $N_o/2$ W/Hz, the average normalized power $P_{received}$ in the received signal is given by Equation 5.9.

$$P_{received} = J\,P + \left[\frac{N_o}{2}\right]\mathrm{BW}_{\mathrm{DSSS}} \tag{5.9}$$

The bandwidth of the bandpass DSSS signal is $\mathrm{BW}_{\mathrm{DSSS}}$ Hz. The first term JP in Equation 5.9 is the power from all the DSSS transmitters and the second term is the bandpass noise power from the AWGN channel. The resulting signal-to-noise ratio (SNR) at each DSSS receiver before despreading then is given by Equation 5.10.

$$\mathrm{SNR}_{\mathrm{DSSS}} = \frac{P}{(J-1)P + \left[\frac{N_o}{2}\right]\mathrm{BW}_{\mathrm{DSSS}}} \tag{5.10}$$

The first denominator term $(J - 1)P$ in Equation 5.10 is the interference from all of the other DSSS transmitters and the second term is the bandpass noise. Despreading reduces the DSSS signal bandwidth by the inverse of the processing gain $\mathrm{G}_{\mathrm{pDSSS}}$, and the resulting SNR is given by Equation 5.11.

$$\mathrm{SNR} = \frac{P}{\frac{1}{\mathrm{G}_{\mathrm{pDSSS}}}\left\{(J-1)P + \left[\frac{N_o}{2}\right]\mathrm{BW}_{\mathrm{DSSS}}\right\}} \tag{5.11}$$

The concept of the processing gain $\mathrm{G}_{\mathrm{pDSSS}}$ now is evident from Equation 5.11, because the SNR is directly improved by increasing $\mathrm{G}_{\mathrm{pDSSS}}$. Usually the power in all the interfering DSSS transmitted signals $(J - 1)P$ is much greater than the bandpass noise from the AWGN channel $[N_o/2]$ $\mathrm{BW}_{\mathrm{DSSS}}$, and the SNR is then given by Equation 5.12.

$$\mathrm{SNR} = \frac{\mathrm{G}_{\mathrm{pDSSS}}\,P}{(J-1)P} = \frac{\mathrm{G}_{\mathrm{pDSSS}}}{J-1} \tag{5.12}$$

Probability of Bit Error Bit in Multiple Access Interference in DSSS

The probability of bit error P_b in a DSSS system with J transmitters, but neglecting the contribution to P_b of the noise from the AWGN channel, is due to the interference caused by the remaining J–1 transmitters. The interference DSSS signal $r_k(t)$ from the kth transmitter at the input of the first receiver is given by Equation 5.13.

$$r_k(t) = A_k\, m_k(t-\tau_k)\, c_k(t-\tau_k) \cos(2\pi f_c(t-\tau_k)+\theta_k) \tag{5.13}$$

In Equation 5.13, $m(t)$ is the data signal, $c(t)$ is the pseudorandom spreading signal, A_k is the amplitude, and τ_k is the relative delay of the kth transmitted signal at the input of the first receiver. The carrier frequency for the J transmitters is f_c Hz, but each with an arbitrary phase τ_k. In the SystemVue simulation of a DSSS system in Figure 5.27, J = 2, all the received amplitudes A_k are equal, and all the received phases τ_k and relative delays τ_k are 0.

If the received amplitudes A_k are all equal, the interference signal I_k output of the despreading and demodulating MetaSystem token of the DSSS system in Figure 5.29 from the kth transmitted signal for the first receiver I_k is given by Equation 5.14 [Lathi98].

$$I_k = \cos(\theta_k - 2\pi f_c \pi_k - \theta_l) \int_0^{T_b} m_k(t-\tau_k)\, c_k(t-\tau_k)\, c_l(t)\, dt \tag{5.14}$$

The data signal $m_k(t)$ is a polar (± 1 V) signal, but it is not synchronized to the first data signal $m_1(t)$ during the bit time T_b. Therefore, a possible change of amplitude of $m_k(t)$ can occur at $t = \tau_k$. If the amplitude of $m_k(t)$ prior to $t = \tau_k$ is a_0 and afterwards is a_1, then the interference signal output from Equation 5.14 is given by Equation 5.15.

$$I_k = \cos(\theta_k - 2\pi f_c \tau_k - \theta_l)\left[a_0 \int_0^{\tau_k} c_k(t-\tau_k)\, c_l(t)\, dt + a_l \int_0^{T_b} c_k(t-\tau_k)\, c_l(t)\, dt \right] \tag{5.15}$$

Equation 5.15 for the interference I_k from the kth transmitter at the despread and demodulated output of the first receiver shows the implicit dependence of the interference on the cross-correlation of the pseudorandom spreading codes $c_k(t)$ in CDM. The M = 4 FSR sequence generators have a maximum cross-correlation to autocorrelation ratio $R_{max}/R(0) = 0.46$, while the M = 12 Gold code sequence generators have $R_{max}/R(0) = 0.03$ [Proakis05].

The *multiple access interference* (MAI) given by Equation 5.15 can be modeled as a sum of independent *Bernoulli trials.* A simple approach is to assume that the MAI is the sum of the interferences I_k, as given by Equation 5.16.

$$\text{MAI} = \sum_{k=2}^{J} I_k \tag{5.16}$$

MAI then can be approximated by the *Central Limit Theorem* as a Gaussian random variable, and P_b is given by Equation 5.17 [Lathi98].

$$P_b = Q\left(\left(\sqrt{\frac{J-1}{3G_{\text{pDSSS}}} + \frac{N_o}{2E_b}}\right)^{-1}\right) \tag{5.17}$$

In Equation 5.17, G_{pDSSS} is the processing gain, $N_o/2$ is the bi-sided normalized PSD of the noise, E_b is the energy per bit, and the Q function is the complementary error function, as described in Chapter 2. If $J = 1$, Equation 5.17 reduces as expected to Equation 3.7 for a symmetrical binary PSK signal in AWGN. If $J > 1$ and the signal-to-noise ratio (SNR) E_b/N_o is very high, then P_b is an irreducible *error floor* in MAI given by Equation 5.18.

$$P_b = Q\left(\sqrt{\frac{3G_{\text{pDSSS}}}{J-1}}\right) \qquad \frac{E_b}{N_o} \to \infty \tag{5.18}$$

FREQUENCY HOPPING SPREAD SPECTRUM

Frequency hopping spread spectrum (FHSS) changes the carrier frequency under the control of a pseudorandom sequence or spreading code during the transmission of data. The data can either be transmitted as binary with a bit time T_b, or multiple level (M-ary) with a symbol time T_s as described in Chapter 3 for bandpass digital communication systems. If the hop rate r_{hop} is less than one change per bit or symbol time, the FHSS system is *slow hopping*, or if greater, the FHSS system is *fast hopping*. Although more complicated and prone to carrier and bit or symbol time synchronization concerns, fast hopping FHSS provides more immunity than direct sequence spread spectrum to channel distortions and fading, as described in Chapter 4 [Dixon94].

For the multiplexing of several FHSS signals over the same communication channel, orthogonality requires that the hopping period in a slow hopping system is an integral multiple of the bit or symbol period. For a fast hopping FHSS system, the bit or symbol period must be an integral multiple of the hop time T_{hop}.

The bandpass modulation method used in the FHSS system is usually frequency shift keying (FSK). Phase shift keying (PSK) is difficult to employ in FHSS because of synchronization problems that arise in maintaining phase coherence as the carrier frequency is changed. Bandpass amplitude shift keying is seldom used in FHSS because of channel distortions and fading [Dixon94].

In an FHSS system, the processing gain G_{pFHSS} is defined as the ratio of the bandwidth in Hz of the spread spectrum signal to the bandwidth in Hz of the original signal. The definition for the processing gain G_{pFHSS} in FHSS is consistent with that of the processing gain G_{pDSSS} in DSSS, although the terms are defined differently. G_{pDSSS} is the ratio of the bit rate r_b b/sec to the chip rate r_{ss} Hz.

SystemVue Simulation of FHSS

ON THE CD

An FHSS binary FSK transmitter is shown in Figure 5.30 (see Fig5-30.svu on the CD-ROM). To facilitate the simulation in SystemVue, all of the Analysis tokens, except that for the BER token output as the Final Value token, are deleted in the simulation model Fig5-30DT. As described in Chapter 1, the SystemVue Textbook Edition does not permit tokens to be deleted.

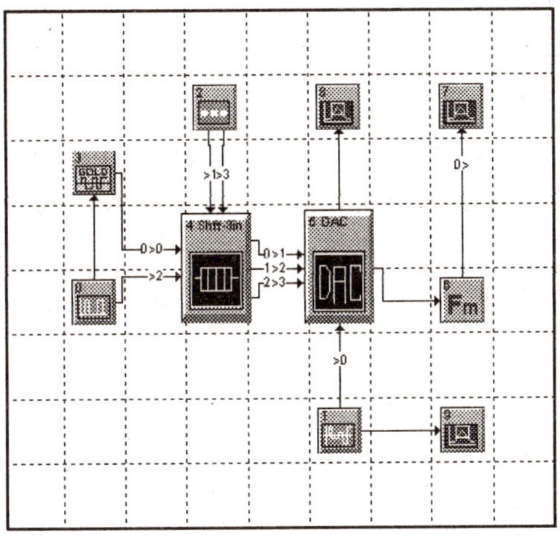

FIGURE 5.30 FHSS transmitter with binary FSK modulation.

The Gold Code Generator token has the same $M = 5$ FSR parameters and initialization patterns as that in Figure 5.22, but with a unipolar (0 and 1 V) output. The Pulse Train token from the Source Library clocks the Gold Code Generator at the hop rate $r_{hop} = 10$ kHz, with parameters of an amplitude $A = 1$ V, a pulse width $\tau = 50$ Msec, and 0 V voltage and 0° phase offset.

The output of the Gold Code Generator token is inputted as data to the 8-bit Shift Register token from the Logic Library, which is the behavioral synthesis of the MSI 74164 IC. The 8-bit Shift Register token Clock input is derived from the same Pulse Train token from the Source Library that clocks the Gold Code Generator. The B Input and active low Master Reset control signal of the 8-bit Shift Register token is provided by the Custom token from the Source Library, and one output with an algebraic simulation equation that is $p(0) = 1$ V.

The three least significant bits (LSB) of the output of the 8-bit Shift Register token are inputted to the three most significant bits (MSB) of the 4-bit Digital to Analog Converter (DAC) token from the Logic Library. The parameters of the DAC token are 4 bits of resolution, $V_{maxp} = 7$ V, $V_{maxn} = -8$ V, a voltage step output of 1 V, and a logic threshold of 0.5 V. The LSB of the DAC token is inputted from the unipolar pseudonoise PN Sequence token data source from the Source Library, with parameters of a binary amplitude of 0 V and 1 V, a bit rate $r_b = 1$ kb/sec, and a 0.5 voltage and 0° phase offset.

The voltage output of the DAC token is inputted to the Frequency Modulator token from the Function Library, with parameters of a carrier amplitude $A_c = 5$ V, a carrier frequency $f_c = 100$ kHz, and a modulation gain or frequency deviation factor $k_f = 5$ kHz/V. The minimum and maximum frequency deviation in the FHSS system due to the center frequency f_c, the DAC token voltage output and frequency deviation k_f are $(100 - 8 \times 5)$ kHz = 60 kHz and $(100 + 7 \times 5)$ kHz = 135 kHz, respectively. The minimum frequency hop is 5 kHz.

There are 16 possible frequencies outputted then in the FHSS system representing a 4-bit logic signal, the LSB of which is the binary data source, and the 3 MSBs are the spreading code. Because the data rate $r_b = 1$ kb/sec and the hop rate $r_{hop} = 10$ kHz, the FHSS system here is fast hopping.

With a center frequency $f_c = 100$ kHz, the System Sampling rate is increased to 5 MHz and the System Time: Number of Samples is 4 194 304 (2^{22}) points, which provides a spectral resolution of 1.2 Hz. Figure 5.31 shows the normalized ($R_L = 1\ \Omega$) and scaled power spectral density (PSD) of the FHSS M-ary (multiple level) FSK modulated signal at a hop rate $r_{hop} = 10$ kHz. To facilitate the simulation in SystemVue, all of the Analysis tokens, except the Frequency Modulator token output, are deleted in the simulation model Fig5-30DT. As described in Chapter 1, the SystemVue Textbook Edition does not permit tokens to be deleted.

ON THE CD

For each of the 8 effective binary FSK signals, the DAC token output voltage due to the data source LSB shifts the frequency by ± 2.5 kHz about the 8 carrier frequencies, which are spaced 10 kHz apart starting at 62.5 kHz and ending at 132.5 kHz. Each of the FSK signals for the FHSS system have a bandwidth for 95% of the total power equal to $2\Delta f + 4r_b = 2(2.5) + 4(1)$ kHz = 9 kHz, as described in Chapter 3.

The binary FSK signals in the FHSS system then have a *guard band* of 1 kHz for the 98% total power bandwidth. The bandwidth of the FHSS signal extends from a lowest frequency of $62.5 - 2.5 - 2 = 58$ kHz to the high frequency of $132.5 + 2.5 + 2 = 137$ kHz. The processing gain G_{pFHSS} for the FHSS system here, for the bandwidth that contains 95% of the total power, is approximately 8.8 (79 kHz/9 kHz).

The FHSS transmitter with binary FSK modulation in Figure 5.30 is the transmitter MetaSystem of the complete FHSS system with bit error rate (BER) analysis

ON THE CD shown in Figure 5.32 (see Fig5-32.svu on the CD-ROM). To facilitate the simulation in SystemVue, all of the Analysis tokens, except that for the BER token output as the Final Value token, are deleted in the simulation model Fig5-32DT.

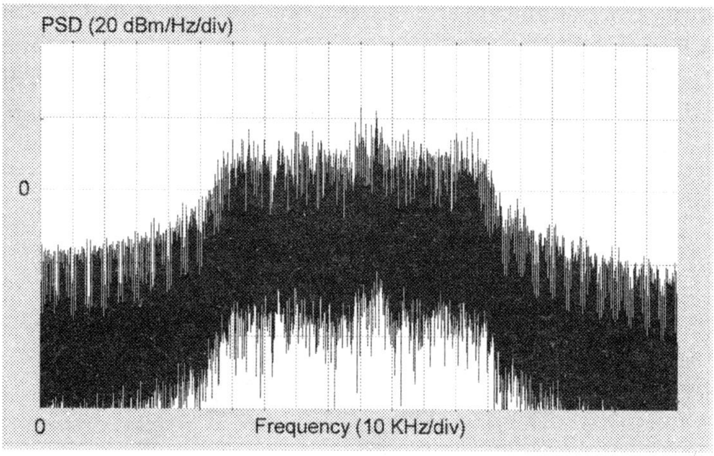

FIGURE 5.31 Power spectral density of the FHSS transmitter with M-ary FSK modulation.

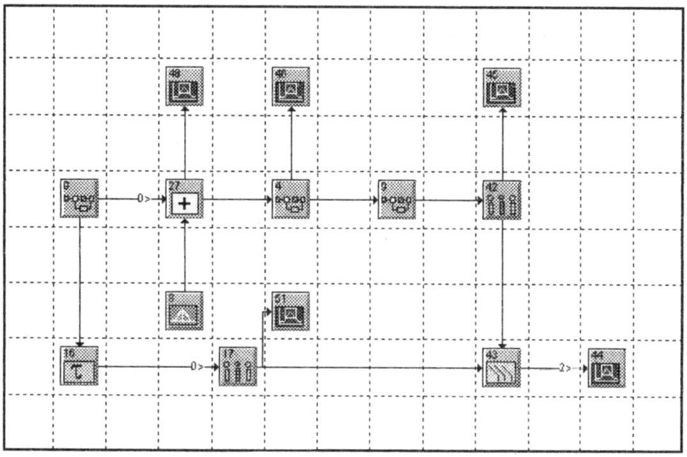

FIGURE 5.32 FHSS system with binary FSK modulation and demodulation and BER analysis.

The communication channel is represented by the Adder token and the Gaussian Noise token from the Source Library, with initial parameters of a power spectral density (PSD) $N_o = 0$ W/Hz. Rather than assess the effect of a second FHSS system as an interfering signal here, a conventional BER analysis is performed in an additive white Gaussian noise (AWGN) channel.

The FHSS receiver binary FSK demodulator and despreader MetaSystem of Figure 5.32 is shown in Figure 5.33. The Gold Code Generator token has the same $M = 5$ FSR parameters and initialization patterns as that for the FHSS transmitter in Figure 5.30. The Pulse Train token from the Source Library clocks the Gold Code Generator at a hop rate r_{hop} = 10 kHz, with parameters of an amplitude A = 1 V, a pulse width τ = 50 μsec, and 0 V voltage and 0° phase offset.

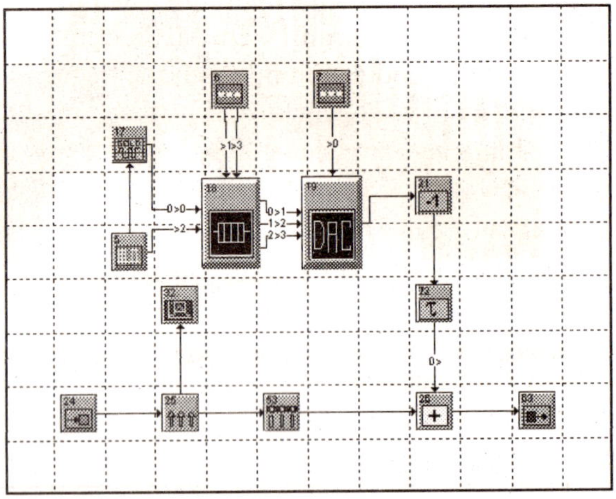

FIGURE 5.33 FHSS receiver binary FSK demodulator and despreader MetaSystem.

The output of the Gold Code Generator token is inputted as data to the 8-bit Shift Register token from the Logic Library. The 8-bit Shift Register token Clock input is derived from the same Pulse Train token that clocks the Gold Code Generator. The B Input and active low Master Reset control signal of the 8-bit Shift Register token is provided by the Custom token, and one output with an algebraic simulation equation that is $p(0)$ = 1 V.

The three LSBs of the output of the 8-bit Shift Register token are inputted to the 3 MSBs of the 4-bit DAC token. The parameters of the DAC token are 4 bits of resolution, V_{maxp} = 15 V, V_{maxn} = 0 V, a voltage step output of 1 V and a logic threshold of 0.5 V. The LSB of the DAC token is inputted from another Custom token, and one output with an algebraic simulation equation that is $p(0)$ = 0 V. The 8 level voltage output of the DAC token then is from 0 V to 14 V in steps of 2 V, because the LSB = 0.

The output of the DAC token is processed by the Negate and Delay tokens from the Operator Library. The delay parameter is 100 μsec, which is the hop time T_{hop}.

The output of the Delay token ranges from 0 V to –14 V with a voltage step size of 2 V, because the LSB of the DAC token is always 0.

The FHSS system multiple FSK (MFSK) received signal is inputted to the MFSK Demodulator token from the Communications Library with parameters of 16 tones, a rate of 10 kHz, the lowest frequency of 60 kHz (100 – 8 × 5 kHz), and a tone spacing of 5 kHz. Figure 3.34 shows a noncoherent binary FSK demodulator MetaSystem, but implemented with bandpass filter, rectifier, and lowpass filter tokens for only two tones ($f_c \pm \Delta f$), to illustrate the processing concept in Chapter 3.

However, the MFSK Demodulator token is more efficient, because 16 tones are required here, and it implements the optimum maximum likelihood correlation detector for multiple FSK signals [Sklar01]. The MFSK Demodulator token is normally used for *multitone* (MT) demodulation in which the frequencies represent a multiple bit data symbol and are orthogonally spaced by the symbol rate r_s Hz apart. The Communications Library also provides an MFSK Modulator token for MT transmission.

The output of the MFSK Demodulator token ranges from 0 V to 15 V, with a voltage step size of 1 V for each demodulated tone at the hop rate r_{hop} = 10 kHz or hop time T_{hop} = 100 μsec. The Resampler token from the Operator Library upsamples the MFSK Demodulator token output, which is at the hop rate r_{hop} = 10 kHz, to the System Time: Sample Rate = 5 MHz (T_{system} = 0.2 μsec). The Adder token adds the Delay token voltage output (0 V to –14 V in steps of 2 V) and Resampler token voltage output to produce a binary 0 V and 1 V data source signal.

The FHSS receiver data detector of Figure 5.32 is shown in Figure 5.34. The output of the demodulation and despreading MetaSystem of Figure 5.33 is inputted to the AND token from the Logic Library. The output of the Pulse Train token from the Source Library, with parameters of a frequency of the data rate r_{hop} = 10 kHz, an amplitude A = 1 V, a pulse width τ = 99.8 μsec ($T_{hop} - T_{system}$), and 0 V voltage and 0° phase offset, is the other input to the AND token. The detected data signal is thus interrupted producing a clock transition once per hop time T_{hop} for a string of binary 1s, while a string of binary 0s would produce no clock transition.

The output of the AND token is inputted to the positive edge trigger Clock input of the 4-bit Counter token from the Logic Library, which counts the number of binary 1s demodulated once per hop time T_{hop} = 0.1 msec during the bit time T_b = 1 msec here. The maximum number of demodulated binary 1s then is T_b/T_{hop} = 10. The 4-bit Counter token is reset every T_b seconds and the *majority rule* is applied to the resulting count of the number of demodulated binary 1s. Here, 4-bit counts of 5 (0101) through 10 (1010) are used to estimate if a binary 1 is received.

The 4-bit Counter token is the behavioral synthesis of the MSI 74161 IC. The Pulse Train token from the Source Library, with parameters of a frequency of the data rate r_b = 1 kHz, an amplitude A = 1 V, a pulse width τ = 0.2 μsec (T_{system}), and 0 V voltage and 0° phase offset, the Delay token from the Operator Library, with a delay $\tau = 0.2$ μsec (T_{system}), and the Inverter token from the Logic Library provides the active low Master Reset signal for the 4-bit Counter token.

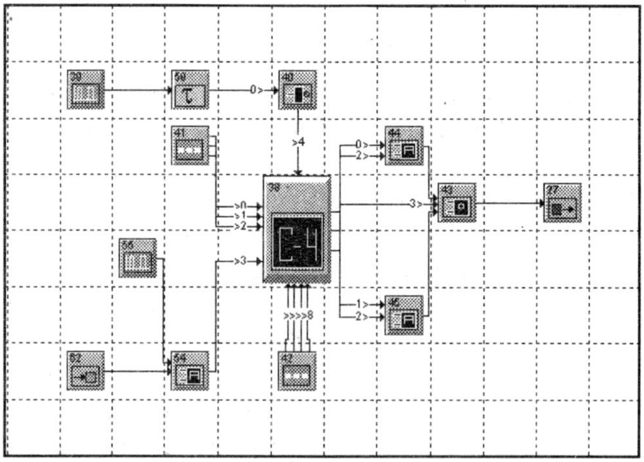

FIGURE 5.34 FHSS receiver data detector MetaSystem.

The active low Count Enable Parallel Input, Count Enable Trickle Input, and Parallel Enable inputs are disabled by the output of the Custom token, and one output with an algebraic simulation equation that is $p(0) = 1$ V. The four Parallel inputs are set low by another Custom token, and one output with an algebraic simulation equation that is $p(0) = 0$ V.

The output of the 4-bit Counter token is processed by a combinational logic circuit consisting of two 2-input AND tokens and a 3-input OR token from the Logic Library. A binary 1 is detected by majority rule from the Boolean logic equation $C3 + C2 \cdot C1 + C2 \cdot C0$, where $C3$ is the MSB and $C0$ is the LSB of the 4-bit counter. This equation is true (1) for all of the 4-bit binary counter outputs from 0101 (5) to 1010 (10).

The output of the FHSS receiver data detector MetaSystem in Figure 5.34 is downsampled by the Decimator token from the Operator Library, with a decimation of 5000 to reduce the output from the System Time: Sample Rate = 5 MHz to the binary data source rate $r_b = 1$ kHz. The output of the PN Sequence token data source in the FHSS system transmitter MetaSystem is delayed by 1 msec to correlate the data for the BER analysis by the Delay token from the Operator Library. The output of the Delay token is also downsampled by the Decimator token from the Operator Library, with a decimation of 5000 to reduce the output from the System Time: Sample Rate = 5 MHz to the binary data source rate $r_b = 1$ kHz.

Performance of FHSS in AWGN

The FHSS binary FSK transmitter and receiver in Figure 5.32 can be evaluated for its performance in the presence of AWGN. The BER token parameters are set to output on every trial (number of trials = 1), with a threshold of 0.5 V, and a 1 msec offset. The System Stop Time is set to 10 sec for convenience, which would process 10 000 information bits in possible error. Setting the Gaussian noise source PSD to 0 W/Hz should produce 0 observed errors.

To facilitate the simulation in SystemVue, all of the Analysis tokens, except that for the BER token are deleted from the FHSS system with binary FSK modulation ON THE CD and demodulation in Figure 5.32, as in the simulation model Fig5-32DT. The Final Value token from the Sink Library records the total number of errors here.

Table 5.4 is a tabular list of the observed BER in a single trial of 10,000 information bits as a function of $E_{b\,hop}/N_o$, to the theoretical probability of bit error (P_b) for noncoherent binary FSK (BFSK) for the same E_b/N_o ratio, as given by Equation 3.93 and Table 3.17. The amplitude is $A = 5$ V, the effective bit time is the hop time $T_{hop} = 0.1$ msec, and the normalized energy per bit in a hop time is $E_{b\,hop} = 1.25 \times 10^{-3}$ V^2-sec, as given by Equations 3.4 and 3.21. The attenuation of the communication channel is taken to be zero ($\gamma = 1$).

A comparison of Table 5.4 with Table 3.17 indicates that the FHSS system seems to perform significantly better than noncoherent BFSK for E_b/N_o ratios greater than 0 dB. However, the observed BER performance is somewhat biased because, although the $E_{b\,hop}$ is less than the E_b for noncoherent BFSK by a factor of 10 here, the FHSS data detector uses 10 estimates of the transmitted data bit.

However, if an analysis with equal energy per bit is used, a comparison of Table 5.4 with Table 3.16 shows an approximate –10 dB decrease in BER performance. The *performance penalty* is due to the smaller signal processing time provided by the hop time T_{hop}, during which a bit must be detected.

TABLE 5.4 Observed BER as a Function of $E_{b\,hop}/N_o$ in a FHSS Digital Communication System and the Theoretical P_b for Noncoherent BFSK

$E_{b\,hop}/N_o$ dB	N_o V^2-sec	BER	P_b (Noncoherent BFSK)
∞	0	0	0
10	1.25×10^{-4}	0	4.10×10^{-2}
8	1.98×10^{-4}	$1.2 \times 10-^{-3}$	1.031×10^{-1}
6	3.13×10^{-4}	2.35×10^{-2}	1.842×10^{-1}
4	4.97×10^{-4}	9.96×10^{-2}	2.666×10^{-1}
2	7.89×10^{-4}	2.051×10^{-1}	3.365×10^{-1}
0	1.25×10^{-3}	3.089×10^{-1}	3.894×10^{-1}

Jamming in FHSS

Jamming is the deliberate interference to a modulated signal to interrupt communication. An FHSS system can be subjected to a variety of jamming procedures. Broadband noise jamming is similar to AWGN, but it is difficult to deploy a large amount of power over the large bandwidth of the FHSS signal. Partial noise jamming attempts to solve this problem by concentrating the available noise power over a smaller bandwidth. Multiple-tone jamming deploys several equal power, random phase, unmodulated (also know as continuous wave or CW) signals within the bandwidth of the FHSS signal.

ON THE CD

The multiple-tone jamming of the FHSS system of Figure 5.32 is shown in Figure 5.35 (see Fig5-35.svu on the CD-ROM). The FHSS system utilizes 8 carrier frequencies for binary FSK modulation that are spaced 10 kHz apart, starting at 62.5 kHz and ending at 132.5 kHz. The two CW jamming signals have an amplitude of 5 V, the same as the carrier frequencies of the FHSS signal, and are arbitrarily centered at two of the carrier frequencies, 82.5 kHz and 112.5 kHz.

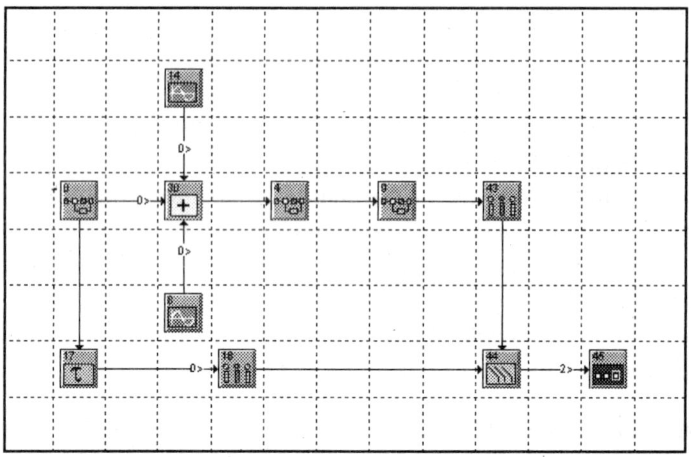

FIGURE 5.35 FHSS system with CW jamming signals.

The observed bit error rate (BER) for the two jamming signals in a trial of 10,000 information bits is 1.115×10^{-1}. Both lowering the amplitude of the two jamming signals below 5 V, or a single jamming signal with an amplitude of 5 V and a frequency of either 82.5 kHz or 112.5 kHz, produces no errors.

This indicates that the MFSK Demodulator token provides a spurious output when two or more jamming signals with amplitude equal to the binary FSK carrier frequencies are present. Although the MFSK Demodulator token implements the optimum maximum likelihood correlation detector, multiple noncoherent binary

FSK demodulators, implemented with bandpass filters, rectifiers, and lowpass filter tokens as in Figure 3.34, would perform better in the presence of CW jamming signals [Dixon94].

ORTHOGONAL FREQUENCY DIVISION MULTIPLEXING

Orthogonal frequency division multiplexing (OFDM) is related to multitone modulation and is used for digital data transmission over wireless communication channels [Haykin01]. Such communication channels are often nonideal over the required bandwidth, and distort the data transmission by causing intersymbol interference (ISI), as described in Chapter 4. However, the channel distortion can be obviated by dividing the available channel bandwidth B Hz into equal bandwidth subchannels, where the subchannel bandwidth Δf Hz is relatively narrow and nearly ideal [Proakis05].

OFDM uses $K = B/\Delta f$ subchannels and transmits different data symbols simultaneously in the K subchannels. Each subchannel $s_k(t)$ has a carrier frequency f_k, as given by Equation 5.19.

$$s_k(t) = A\cos 2\pi f_k t \qquad k = 1, 2, \ldots, K \tag{5.19}$$

The carrier frequencies f_k are the mid-frequencies of the bandwidth Δf of the kth subchannel and are orthogonal if $\Delta f = r_{sc}$, where r_{sc} is the subchannel symbol rate and $1/r_{sc} = T_{sc}$ is the subchannel symbol time. The OFDM system has a symbol rate $r_s = Kr_{sc}$, and a symbol time $T_{sc} = KT_s$. If K is selected to be large enough so that $T_{sc} = KT_s$ is much greater than the channel dispersion time τ_{dis}, as described in Chapter 4, then ISI is negligible [Proakis05].

Each subcarrier is modulated by multilevel (M-ary) quadrature amplitude modulation (QAM), as described in Chapter 3. The M-ary QAM signal is given by Equation 3.68 and for the kth subcarrier $s_k(t)$ is given by Equation 5.20.

$$s_k(t) = A_I^k \cos(2\pi f_k t) + A_Q^k \sin(2\pi f_k t) \quad (i-1)T_s \le t \le iT_s \quad k = 1, 2, \ldots, K \tag{5.20}$$

The signals transmitted on any pair of the K subchannels, as given by Equation 5.20, are orthogonal over the symbol interval T_s. The in-phase A_I^k and quadrature A_Q^k carrier amplitudes determine the resulting complex phasor constellation plot for the QAM signal, as in Chapter 3. The complex constellation point P_k then is given by Equation 5.21.

$$P_k = A_k \exp[j\theta_k]$$
$$A_k = \sqrt{\left(A_I^k\right)^2 + \left(A_Q^k\right)^2} \qquad \theta_k = \tan^{-1}\left[\frac{A_Q^k}{A_I^k}\right] \tag{5.21}$$

If the number of subchannels is large ($K >> 1$), each subchannel $C(f_k)$ is approximately linear and can be approximated as complex-valued, with magnitude and phase as given by Equation 5.22.

$$C(f_k) = C_k \exp(j\varphi_k) \tag{5.22}$$

The OFDM received complex constellation point R_k with no noise from Equation 5.21 and Equation 5.22, is given by Equation 5.23.

$$R_k = C_k \exp(j\varphi_k)\, A_k \exp(j\theta_k) = C_k A_k \exp(j(\theta_k + \varphi_k)) \tag{5.23}$$

The real part of the complex phase R_k from Equation 5.23 and the additive noise in the kth subchannel $n_k(t)$ produce the received OFDM signal $r_k(t)$ in the kth subchannel, as given by Equation 5.24 [Proakis05].

$$\begin{gathered} r_k(t) = C_k\, A_I^k \cos(2\pi f_k t + \varphi_k) + C_k\, A_Q^k \sin(2\pi f_k t + \varphi_k) + n_k(t) \\ (i-1)T_s \le t \le iT_s \quad k = 1,2,\ldots,K \end{gathered} \tag{5.24}$$

The additive noise $n_k(t)$ is assumed to be zero-mean Gaussian with a power spectral density (PSD), which is flat across the bandwidth Δf of the kth subchannel.

The reception of an OFDM signal utilizes two correlators, or matched filters, for each of the K subchannels. The $2K$ filters used to generate the OFDM signal and the $2K$ correlators to demodulate the signal are equivalent to the computation of the discrete Fourier transform (DFT) and its inverse. An efficient implementation of the DFT is the fast Fourier transform (FFT) if $K = 2^n$. OFDM digital communication systems usually employ the FFT in the receiver and the inverse FFT (IFFT) in the transmitter [Proakis05].

OFDM digital communication systems demonstrate robustness against multipath delay spread, as described in Chapter 4, which is achieved by having a long effective subchannel symbol time T_{sc}. The subchannel symbol time T_{sc} can be increased even more by the addition of a guard time between transmitted symbols. The guard time obviates ISI to a greater extent by allowing sufficient time for multipath interference signals to effectively decay. The design tradeoff is the increased complexity of the OFDM digital communication system.

An OFDM baseband digital communication system is shown in Figure 5.36 (see Fig5-36.svu on the CD-ROM). A further extension to the OFDM bandpass modulation system is not shown to reduce the complexity of the system, and to illustrate only the principle of OFDM. The data source is a pseudonoise PN Sequence token from the Source Library, with parameters of a binary amplitude $A = 0.5$ V, a data rate $r_b = 1.024$ kb/sec, and 0.5 V voltage and 0° phase offset.

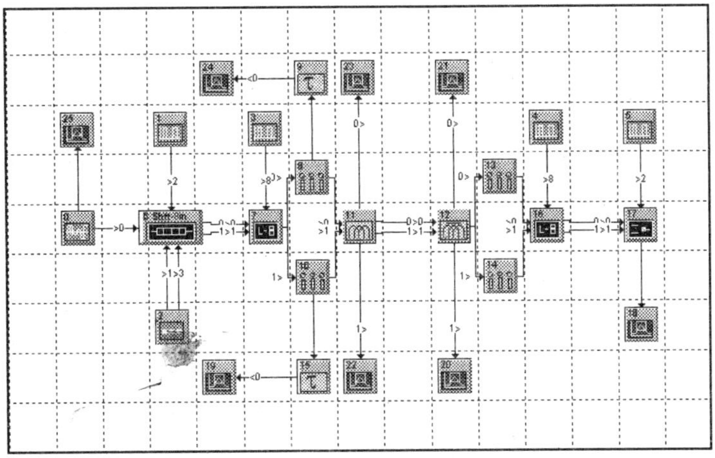

FIGURE 5.36 OFDM baseband digital communication system.

The unipolar output of the PN Sequence token is inputted to the 8-bit Shift Register token of the Logic Library. The clock signal of the 8-bit Shift Register token is derived from a Pulse Train token from the Source Library, with a frequency $f_{o1} = 1.024$ kHz and a pulse with $\tau_1 \approx 488$ μsec (a square wave). The output of the 8-bit Shift Register token is inputted to the 8-bit Latch token of the Logic Library. The clock signal of the 8-bit Latch token is derived from a second Pulse Train token from the Source Library, with a frequency $f_{o2} = 512$ Hz, and a pulse with $\tau_2 \approx 976$ μsec (a square wave). Only the two least significant bits (LSB) of the 8-bit Shift Register token and the 8-bit Latch token are used here.

The two LSB bit outputs of the 8-bit Latch token represent the in-phase (I) and quadrature (Q) data components to be transmitted. The I and Q data signals have a symbol rate of 512 Hz. Two Sampler tokens from the Operator Library down-sample the I and Q data signals from the System Sampling Rate of 8.192 kHz to the symbol rate $r_s = 512$ Hz. The outputs of the two Sampler tokens are inputted to the OFDM Modulator token of the Communications Library.

The OFDM Modulator token has parameters of an FFT sample per block of 512 ($K = 2^9$), a subchannel symbol time $T_{sc} = 1$ sec, and a guard time of 0 sec. The OFDM Modulator token operates on a block-by-block basis. The outputs of the

OFDM Modulator token are the in-phase A_I^k and quadrature A_Q^k carrier amplitudes from the IFFT of the input data block. The A_I^k and A_Q^k components are available for the extension to a bandpass digital communication system, which for simplicity is not shown here.

Rather, the outputs of the OFDM Modulator token are directly inputted to the OFDM Demodulator token of the Communications Library, with the same parameters of an FFT sample per block of 512, a symbol time $T_s = 1$ sec, and a guard time of 0 sec. The outputs of the OFDM Demodulator token are the I and Q data signals derived from the FFT, and are upsampled by two Sampler tokens from the Operator Library to return the symbol rate $r_s = 512$ Hz to the System Sampling Rate of 8.192 kHz for further processing.

The outputs of the OFDM Demodulator token are inputted to another 8-bit Latch token of the Logic Library. The clock signal of the 8-bit Latch token is derived from a third Pulse Train token from the Source Library, with a frequency $f_{o3} = 512$ Hz and a pulse with $\tau_3 \approx 976$ °sec (a square wave). Only the two least significant bits (LSB) of the 8-bit Latch token are used here, which are outputted by the SPDT Switch token of the Logic Library. Finally, the clock signal of the 8-bit Latch token is derived from a fourth Pulse Train token from the Source Library, with a frequency $f_{o4} = 1.024$ kHz and a pulse with $\tau_4 \approx 488$ μsec (a square wave).

The OFDM digital communication system, as shown in Figure 5.36, is simulated in SystemVue with a System Sampling Rate of 8.192 kHz, and a System Time of 4 seconds, which produces four blocks of data. Figure 5.37 shows the initial portion of the transmitted I and Q data signals and their reception at the output of the OFDM demodulator. From the top to the bottom, the plots are the received Q data signal, the transmitted Q data signal, the received I data signal, and the transmitted I data signal, demonstrating recovery of the original information.

Figure 5.38 shows the $K = 512$ in-phase A_I^k (bottom) and quadrature A_Q^k (top) carrier amplitude components for complex phasor constellation plot for the QAM signal. The $K = 512$ components appear sequentially as a block of data outputted from the OFDM Modulator token in $T_{sc} = 1$ sec here.

The OFDM digital bandpass communication system utilizes $K = 512$ independent QAM subchannels with a symbol time for each subchannel $T_{sc} = 1$ sec. A lower value for the number of subchannels K produces a higher subchannel symbol rate r_{sc} and a resulting higher bandwidth Δf of the kth subchannel because for orthogonality $\Delta f = r_{sc}$.

The OFDM digital communication system in Figure 5.26 can be simulated in SystemVue with $K = 16$ and $T_{sc} = 31.25$ msec, for example, by changing the parameters of the OFDM Modulator token and the OFDM Demodulator token. The received I and Q data signals would now appear with a 2 $T_{sc} = 62.5$ msec delay, because processing requires one subchannel symbol time for the transmit IFFT, and one for the receive FFT of the OFDM system.

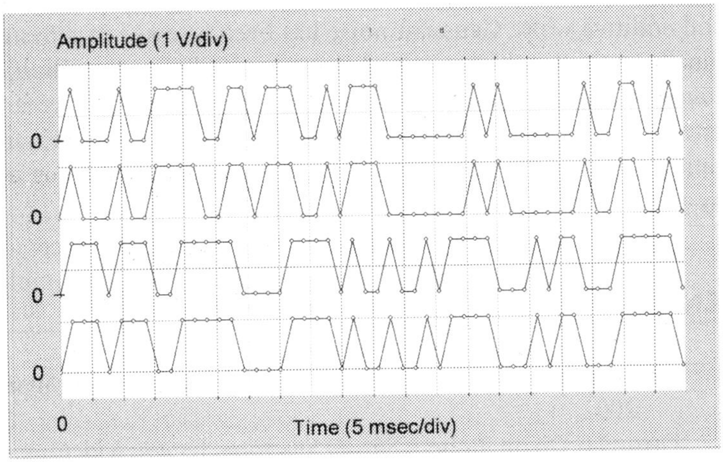

FIGURE 5.37 Received (top) and transmitted Q data signal and the received and transmitted I data signal for the OFDM digital communication system.

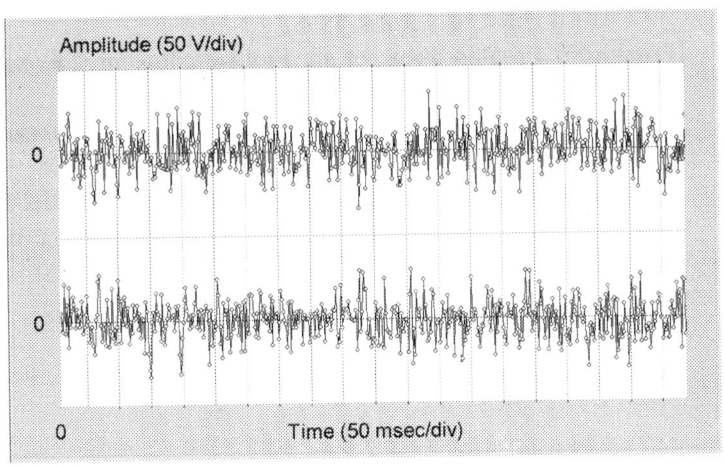

FIGURE 5.38 In-phase A_I^k (bottom) and quadrature A_Q^k (top) carrier amplitude components for the OFDM digital communication system.

SUMMARY

In this chapter, SystemVue was used to design and analyze time division multiplexing, frequency division multiplexing, and code division multiplexing as methods for facilitating the transmission of multiple sources of information over a common communication channel. The performance of direct sequence spread spectrum and frequency hopping spread spectrum in the presence of interference and additive white Gaussian noise has been assessed in a SystemVue simulation. Finally, orthogonal frequency division multiplexing uses multiple subchannels to transmit a single source of information to improve performance with multipath interference in wireless communication. In the next chapter, sampling and quantization is presented as the method by which continuous analog signals are processed by a digital communication system.

REFERENCES

[Carlson02] Carlson, A. Bruce, et al., *Communication Systems*. McGraw-Hill, 2002.

[Dixon94] Dixon, Robert, *Spread Spectrum Systems with Commercial Applications*. Wiley, 1994.

[Haykin01] Haykin, Simon, *Communication Systems*. Wiley, 2001.

[Holmes82] Holmes, John, *Coherent Spread Spectrum Systems*. Wiley, 1982.

[Lathi98] Lathi, B.P., *Modern Digital and Analog Communication Systems*. Oxford University Press, 1998.

[Proakis05] Proakis, John, et al., *Fundamentals of Communications Systems*. Prentice Hall, 2005.

[Simon01] Simon, Marvin, et al., *Spread Spectrum Communication Handbook*. McGraw-Hill, 2001.

[Sklar01] Sklar, Bernard, *Digital Communications*. Prentice Hall, 2001.

[Stern04] Stern, Harold, et al., *Communication Systems Analysis and Design*. Prentice Hall, 2004.

6 Sampling and Quantization

In This Chapter

- Sampling Baseband Analog Signals
- Companding
- Pulse Code Modulation
- Line Codes
- Differential Pulse Code Modulation
- Sampling Bandpass Analog Signals

Analog sources of information are often derived from transducers that provide continuous electrical (usually) voltage signals from physical phenomenon such as light, pressure, temperature, vibration, and acceleration. These analog baseband signals are bandlimited to a maximum frequency. Analog bandpass signals are modulated sources of information that are transmitted on a communication channel, and are substantially higher in frequency. Both analog baseband and bandpass signals are continuous in time and amplitude, and are *sampled* and *quantized* for digital signal processing.

Analog signals are first sampled at discrete intervals of time, but continuous in amplitude. Quantization then is the roundoff of the continuous amplitude sample to a discrete preset value, represented as a binary number or binary bit pattern. The preset values are equally spaced in *uniform quantization*, and the total number of binary bits is the resolution. In *nonuniform quantization*, the preset values are not equally spaced, and the resolution varies.

nonuniform quantization is often used to improve the perceived quality of a sampled speech, where nonlinear compression is used at the transmitter, and nonlinear expansion is used at the receiver. The procedure for nonlinear compressing and expanding of a signal is referred to as *companding.*

Pulse code modulation (PCM) represents the sampled and quantized analog baseband signal as a sequence of encoded pulses. The PCM receiver regenerates, decodes, and reconstructs the sequence of quantized samples of the original analog signal. The *line codes* used here for the baseband signals are an extension of the concepts of baseband modulation in Chapter 2, "Baseband Modulation and Demodulation," and the precepts of bit time synchronization in Chapter 4, "Synchronization and Equalization."

Differential PCM (DPCM) is a continuation of the techniques of baseband *delta modulation* in Chapter 2. The source files for these SystemVue simulations in sampling, quantization, PCM, DPCM, and line codes are located in the Chapter 6 folder on the CD-ROM, and are identified by the figure number (such as Fig6-1.svu). Appendix A includes a complete description of the contents of the CD-ROM.

ON THE CD

SAMPLING BASEBAND ANALOG SIGNALS

A periodic baseband analog signal can be implemented as a sum of sinusoids, which will have a discrete voltage (or *line*) spectra, and a limit to the frequency content. The SystemVue simulation sampling time is set arbitrarily high to approximate an analog signal. Figure 6.1 shows a periodic baseband signal source, ideal discrete sampler, discrete sample and hold, discrete analog-to-digital converter (ADC) to digital-to-analog converter (DAC), and continuous quantizer (see Fig6-1.svu on the CD-ROM). The source is formed by combining three Sinusoid tokens from the Source Library, with parameters of an amplitude and frequency of ± 1 V at 500 Hz, ± 0.5 V at 1.5 kHz, and ± 0.2 V at 2.5 kHz. The System Sampling Rate is set to 5 MHz or T_{system} = 0.2 μsec, which is 2000 times higher than the highest frequency in the periodic baseband source. This is known as *oversampling.*

ON THE CD

The Adder token sums the output from the three Sinusoid tokens to form the composite baseband signal, as shown in Figure 6.2. The composite baseband signal has a fundamental period of 2 msec or a fundamental frequency of 500 Hz. The other two sinusoids have frequencies (1.5 kHz and 2.5 kHz) that are integer multiples of the fundamental frequency here. The summated peak amplitude is approximately 1.15 V at 0.166 msec.

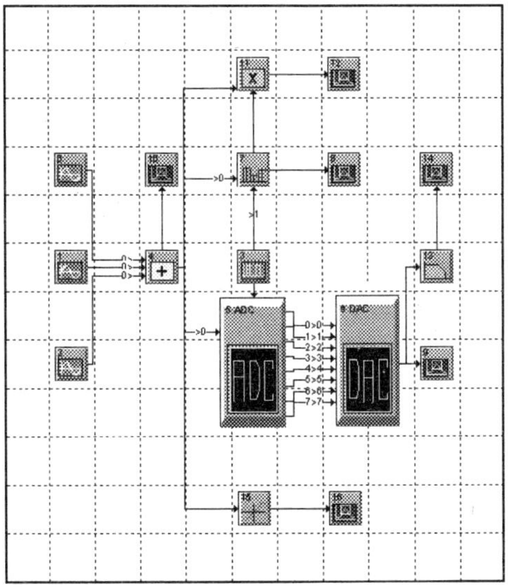

FIGURE 6.1 Periodic baseband analog signal source, ideal sampler, sample and hold, ADC-DAC, and quantizer.

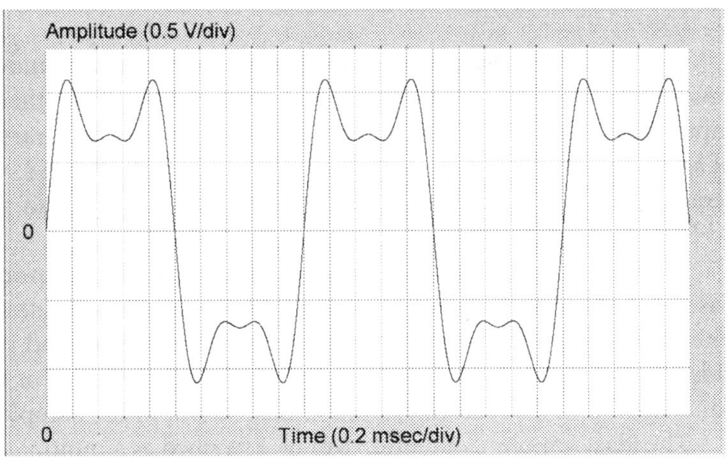

FIGURE 6.2 Periodic baseband analog signals formed by summing three sinusoids.

Setting the System Time: Number of Samples specification to 4 194 304 (2^{22}) points results in a spectral resolution of 1.19 Hz. The normalized ($R_L = 1\ \Omega$) power spectral density (PSD) of the Adder token output is shown scaled and as a single-sided spectrum in Figure 6.3, in dBm/Hz (decibels referenced to 1 milliwatt per Hertz). The discrete spectral components of the three sinusoids at 500 Hz, 1.5 kHz, and 2.5 kHz are evident. However, the PSD is not an exact-line spectra, because of its computation in SystemVue as a discrete (and not continuous) Fourier transform (DFT) [Carlson02].

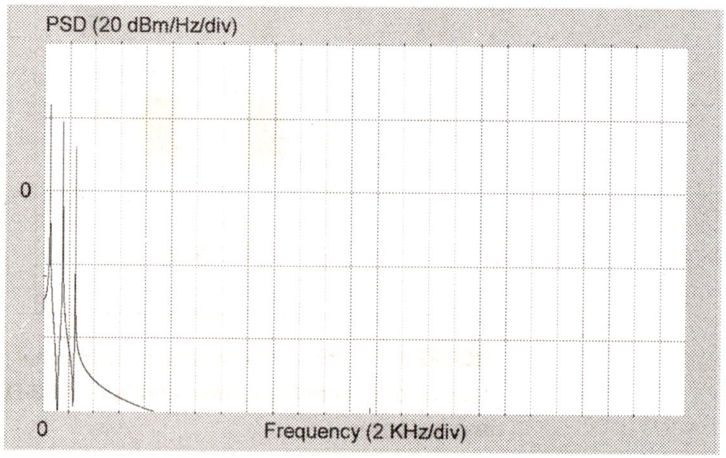

FIGURE 6.3 PSD of the periodic baseband analog signal.

The Multiplier token ideally processes the periodic baseband analog signal, and produces a single sample at discrete intervals of time. The other Multiplier token input is provided by a Pulse Train token from the Source Library, with parameters of an amplitude $A = 1$ V, a frequency $f_o = 8$ kHz, a pulse width ? = 0.2 ?sec (the System Time, T_{system}), and 0 V voltage and 0° phase offset. The sampling rate f_s then is 8 kHz.

The Sample and Hold token from the Operator Library processes the periodic baseband analog signal, and produces a continuous amplitude sample at discrete intervals of time. The control voltage input of the Sample and Hold token is provided by the Pulse Train token. Figure 6.4 shows an overlay of the periodic baseband signal and the continuous amplitude output of the Sample and Hold token.

The ideal sampling operation can be described as a multiplication of the baseband analog signal $x(t)$ by a periodic series of unit impulse functions ?(t ? nT_s), where T_s is the sampling interval. The ideal sampling process is given by Equation 6.1.

$$x(nT_s) = \sum_n x(t)\ \delta(t - nT_s) \tag{6.1}$$

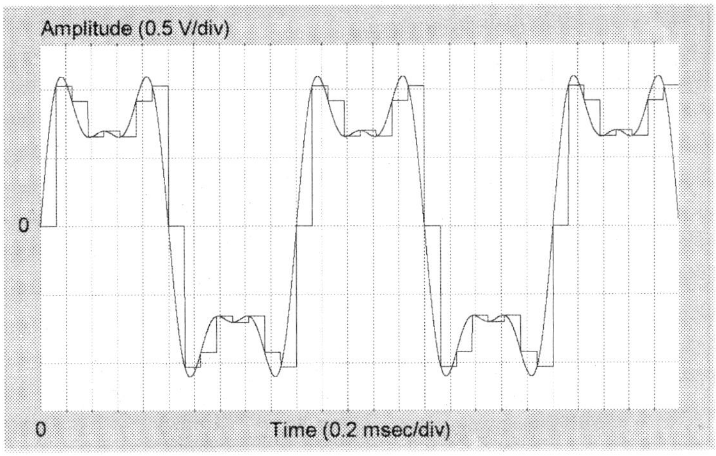

FIGURE 6.4 Periodic baseband analog signal and continuous amplitude sample and hold signal.

The power spectral density (PSD) of the ideal sampling operation is given by Equation 6.2 [Lathi98].

$$PSD = f_s^2 \sum_k \left| X(f - k\,f_s) \right|^2 \tag{6.2}$$

The Fourier transform of the baseband analog signal $x(t)$ is $X(f)$, and the sampling rate $f_s = 1/T_s = 8$ kHz here. The normalized PSD of $x(t)$ is $|X(f)|^2$ and is shown in Figure 6.3. The PSD of the ideal sampling operation of the Multiplier token output in Figure 6.1 is shown scaled, and as a single-sided spectrum in Figure 6.5, in dBm/Hz.

From Equation 6.2 and Figure 6.3, the single-sided PSD of the ideal sampling operation in Figure 6.5, as expected, shows three discrete line spectra at 500 Hz, 1.5 kHz, and 2.5 kHz. These line spectra are centered about 0 Hz ($k = 0$), a periodic replication of six discrete line spectra, centered about $k \times 8$ kHz ($k = 1, 2, 3\ldots$), and spaced ± 500 Hz, ± 1.5 kHz and ± 2.5 kHz about the center frequency.

The continuous amplitude sample and hold operation is a first order process given by Equation 6.3.

$$\begin{aligned} y_{s-h}(t) &= \sum_n x(nT_s)\,h(t - nT_s) \quad nT_s \le t < (n+1)T_s \\ \text{where} \quad h(t) &= 1 \quad 0 \le t < T_s \\ h(t) &= 0 \quad \text{otherwise} \end{aligned} \tag{6.3}$$

From Equations 6.1 and 6.3, the power spectral density PSD_{s-h} of the sample and hold operation is given by Equation 6.4 [Haykin01].

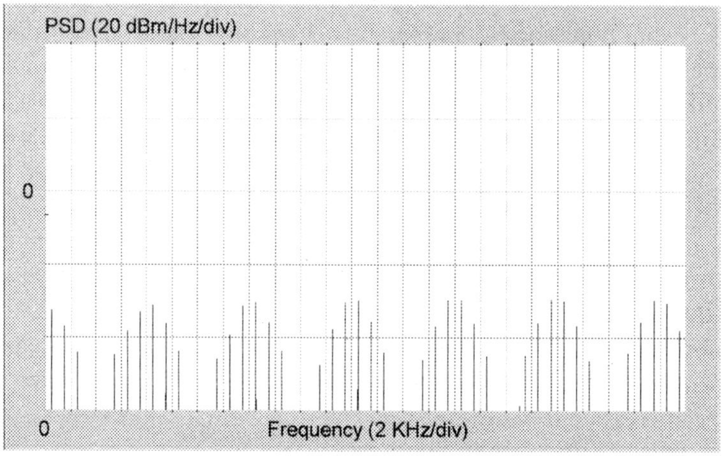

FIGURE 6.5 PSD of the ideally sampled signal.

$$PSD_{s-h} =$$
$$f_s^2 \sum_k \left| X(f - k f_s) \right|^2 T_s^2 \operatorname{sinc}^2(2\pi f T_s) = \sum_k \left| X(f - k f_s) \right|^2 \operatorname{sinc}^2(2\pi f T_s) \tag{6.4}$$

The PSD of the continuous amplitude Sample and Hold token output is shown scaled and as a single-sided spectrum in Figure 6.6, in dBm/Hz.

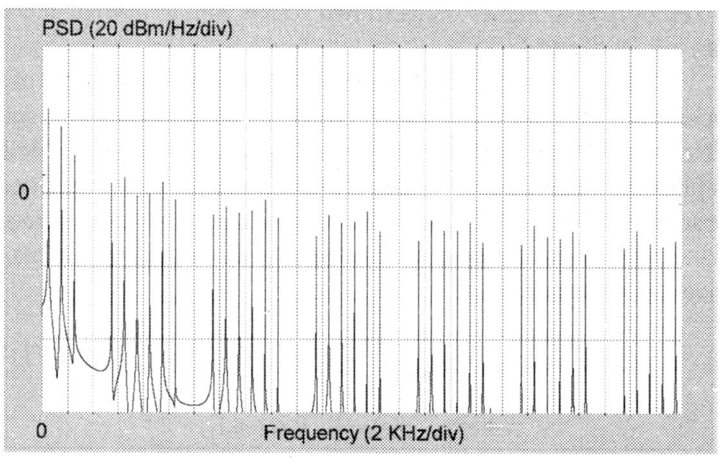

FIGURE 6.6 PSD of the continuous amplitude sample and hold signal (fs = 8 kHz).

From Equation 6.3 and Figure 6.3, the single-sided PSD of the sample and hold operation in Figure 6.6 again shows three discrete line spectra at 500 Hz, 1.5 kHz, and 2.5 kHz. These line spectra are centered about 0 Hz ($k = 0$), a periodic replication of six discrete line spectra, centered about $k \times 8$ kHz ($k = 1, 2, 3\ldots$), and spaced $\pm$ 500 Hz, $\pm$ 1.5 kHz, and $\pm$ 2.5 kHz about the center frequency, but here with a decreasing PSD amplitude because of the sinc^2 term.

The Analog to Digital Converter (ADC) token from the Logic Library uniformly converts the analog baseband signal to a binary number with parameters of 8 bits of resolution, *two's complement signed integer* parallel binary output, a maximum positive input voltage V_{maxp} = 1.27 V, a maximum negative input voltage $V_{maxn} = -1.28$ V, and a clock threshold of 0.5 V. The ADC token output can also be specified as an *unsigned integer*. The ADC token sample clock input is provided by the Pulse Train token in Figure 6.1.

With the voltage range and number of bits specified, the actual uniform voltage step size Δ_{ADC} for the ADC token per bit of resolution here is given by Equation 6.5.

$$\Delta_{ADC} = \frac{1.27-(-1.28)}{2^8-1} = 10 \text{ m} \tag{6.5}$$

The 8 bits of output of the ADC token are connected in parallel to the 8 bits of input of the Digital to Analog Converter (DAC) token from the Logic Library.

The parameters of the DAC token are the same as that for the ADC token, with 8 bits of resolution, V_{maxp} = 1.27 V, V_{maxn} = ?1.28 V, and a logic threshold of 0.5 V. The difference between the Sample and Hold token continuous amplitude output and the voltage step output of the ADC-DAC token system with 8 bits of resolution is a maximum of 4 mV, and shown in Figure 6.7.

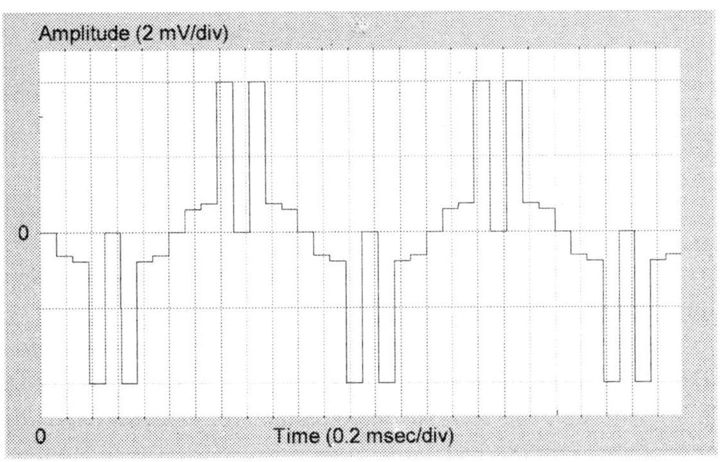

FIGURE 6.7 Voltage difference between the Sample and Hold token signal and the ADC-DAC token system signal.

The PSD of the voltage step output of the ADC-DAC token is very similar to that of the Sample and Hold token signal in Figure 6.6, because the temporal voltage difference between these signals is small. Comparing Figure 6.6 for a close replication of the PSD of the voltage step output of the ADC-DAC token system signal, and Figure 6.3 for the PSD of the periodic baseband analog signal, indicates that a low-pass filter (LPF) can closely recover the original signal.

The Linear System Filter token from the Operator Library provides the LPF in Figure 6.1, which is selected to be a 9-pole Butterworth filter, with a cutoff frequency f_{cutoff} of 3 kHz. The discrete frequencies in the periodic baseband analog signal are 500 Hz, 1.5 kHz, and 2.5 kHz. The nonideal LPF here is selected to pass the highest frequency f_{max} in the original periodic signal. Figure 6.8 shows the Butterworth LPF reconstructed signal with f_{cutoff} = 3 kHz from the voltage step output of the ADC-DAC token system and the original periodic baseband analog signal.

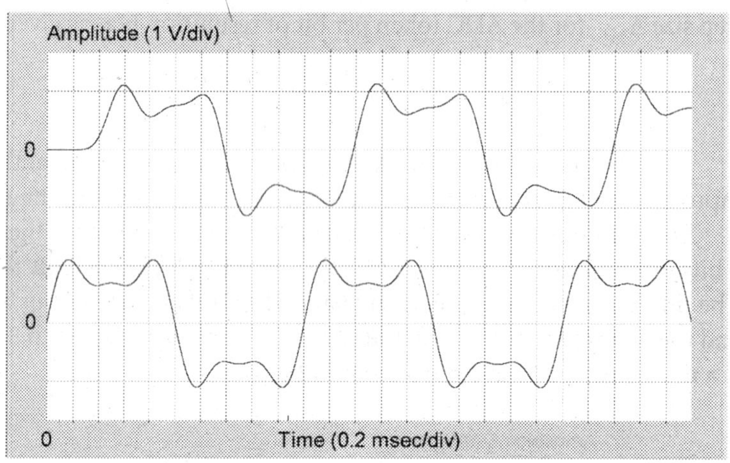

FIGURE 6.8 Butterworth LPF (f_{cutoff} = 3 kHz) ADC-DAC token system signal (top) and periodic baseband analog signal (bottom).

The reconstructed signal in Figure 6.8 is not exact, and exhibits delay and distortion because the LPF is not ideal. A nearly ideal linear phase LPF filter with 0.05 dB of ripple in the response, which can be implemented as a digital filter, produces a more exact reconstruction.

Figure 6.9 shows the linear phase digital LPF reconstructed signal with f_{cutoff} = 3 kHz from the voltage step output of the ADC-DAC token and the original analog signal. Although the reconstructed signal still exhibits delay, distortion from the linear phase LPF is reduced.

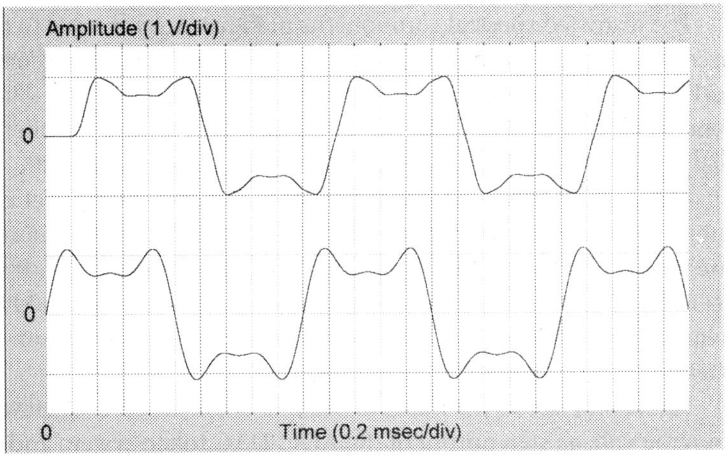

FIGURE 6.9 Linear phase LPF (fcutoff = 3 kHz) ADC-DAC token system signal (top) and periodic baseband analog signal (bottom).

An analog baseband signal can be nearly ideally reconstructed only if the sampling rate is at least twice the highest frequency in the analog. This Nyquist sampling rate then is $f_s > 2f_{max}$, and here 8 kHz > 2 × 2.5 kHz = 5 kHz. If the sampling rate for the sample and hold output of the periodic baseband analog signal for the system in Figure 6.1 is lowered to 3.5 kHz, below the Nyquist rate of 5 kHz, the PSD in Figure 6.10 results. A comparison of Figure 6.6, where the analog signal was sampled at 8 kHz, and Figure 6.10 shows not only the original spectral components at 500 Hz, 1.5 kHz, and 2.5 kHz, but also distortion components at 1 kHz, 2 kHz, and 3 kHz.

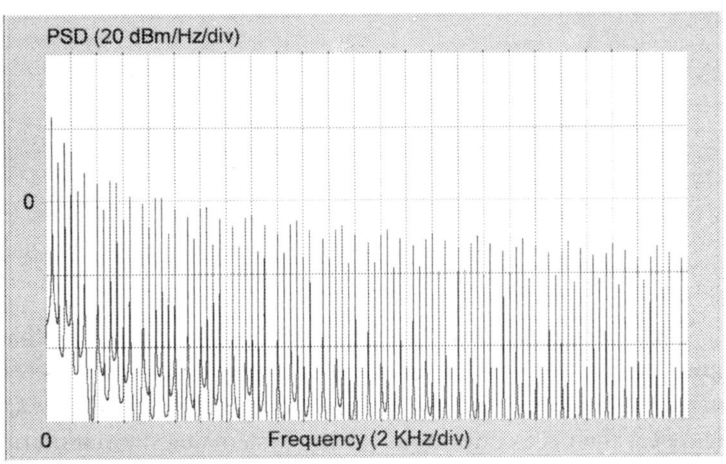

FIGURE 6.10 PSD of the continuous amplitude sample and hold signal (f_s = 3.5 kHz < 2 f_{max}).

The distorted spectral components in Figure 6.10 are due to the frequency difference between the sampling rate and the original frequency components (that is, 3.5 kHz – 2.5 kHz, 3.5 kHz – 1.5 kHz, and 3.5 kHz – 500 Hz). These spectral components illustrate the *spectral folding* or *aliasing* that occurs when signals are sampled at rates below the theoretical Nyquist minimum [Lathi98].

Although the Nyquist sampling rate represents the theoretical lower limit for sampling of baseband signal, in practice the sampling rate is increased to compensate for the baseband rolloff characteristics of a nonideal LPF. If the sampling rate for LPF ADC-DAC token system in Figure 6.1 is raised to 20 kHz, well above the Nyquist rate of 5 kHz, then the cutoff frequency f_{cutoff} of the nonideal Butterworth filter can also be raised to at least 15 kHz.

Figure 6.11 shows the Butterworth LPF reconstructed signal with $f_{cutoff} = 15$ kHz from the voltage step output of the ADC-DAC token system and the original periodic baseband analog signal. A comparison of Figures 6.8 and 6.11 shows that although the reconstructed signal again still exhibits delay, distortion from the nonideal Butterworth LPF is now reduced.

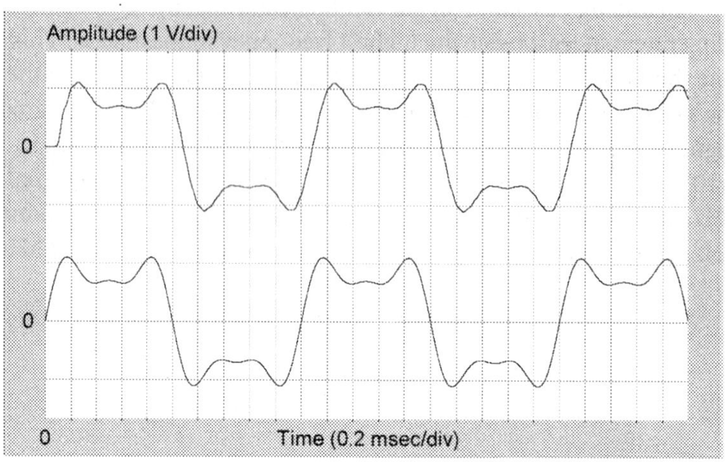

FIGURE 6.11 Butterworth LPF ($f_{cutoff} = 15$ kHz) ADC-DAC token system signal (top) and periodic baseband analog signal (bottom).

The Quantizer token from the Function Library does not have a sample clock input, and provides a discrete amplitude output quantized to N bits of resolution, but with continuous and not sampled time. The output of the Quantizer token in Figure 6.1 then is a continuous (to the SystemVue simulation time T_{system}) voltage signal in either floating point or an N-bit two's complement signed integer format.

Figure 6.12 shows the continuous time response of the output of the 8-bit Quantizer token and the discrete time response of the output of the Sample and

Hold token with a sampling rate of 8 kHz. The Quantizer token output responses *asynchronously* as the input voltage level changes, while the Sample and Hold token output is synchronous with the clock input signal.

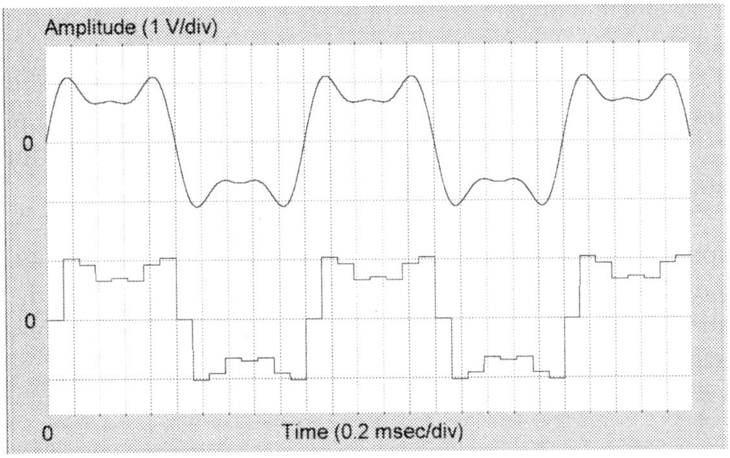

FIGURE 6.12 Quantizer token output (top) and Sample and Hold token output (bottom) with a sampling rate of 8 kHz.

The ideal uniform quantizer voltage step size ? is given by Equation 6.6.

$$\Delta = \frac{2V_{max}}{L} \tag{6.6}$$

V_{max} is the equal positive and negative maximum input voltage, and $L = 2^n$ is the number of levels in the output of the ideal uniform quantizer and n is the number of bits. The maximum *quantization error q* that can occur in the sampled output of the ideal uniform quantizer is $\pm\ \Delta/2$ V. This assumes that all values of quantization error within the range $+\Delta/2$ to $-\Delta/2$ are equally likely, and from Equation 6.6, the mean square quantizing error E_q is given by Equation 6.7 [Haykin01].

$$E_q = \frac{1}{\Delta}\int_{-\Delta/2}^{\Delta/2} q^2\, dq = \frac{\Delta^2}{12} = \frac{V_{max}^2}{3\,L^2} \tag{6.7}$$

E_q is also the normalized power in the quantizing noise, similar to N_o for additive channel noise. The root mean square (RMS) quantizing noise is $\Delta/\sqrt{12} = \Delta/3.464$. If the normalized power in the signal is S_o then, from Equations 6.6 and 6.7, the signal to quantization noise ratio (SNR_q) is given by Equation 6.8.

$$\mathrm{SNR}_q = \frac{12\, S_o}{\Delta^2} = 3\, L^2 \frac{S_o}{V_{max}^2} \tag{6.8}$$

SNR_q is a linear function of the normalized power in the signal S_o, and a second order function of the number of levels $L = 2^n$ of the ideal uniform quantizer. If S_o and V_{max} remain constant, but the number of bits n increases to $n + 1$ (the number of levels L doubles), SNR_q quadruples or increases by +6 dB ($10 \log_{10} 4$). For a sinusoidal input signal with a positive and negative maximum input voltage equal to V_{max}, the normalized power in the signal S_o is $V_{max}^2/2$ V^2 and the signal to quantization noise ratio SNR_q is $1.5\, L^2$.

For the ADC specified here $q \approx \Delta_{ADC} = 10$ mV, $L = 2^8 = 256$ and, for a sinusoidal input signal with an amplitude of V_{max}, $\mathrm{SNR}_q = 1.5$? $256^2 = 49.93$ dB. The maximum quantization error $q \approx \Delta_{ADC}/2 = 10/2$ mV $= 5$ mV, but the observed quantization error, shown as the voltage difference in Figure 6.7, is only 4 mV, because the peak voltage of the analog signal is 1.1 V here.

COMPANDING

Nonuniform quantization is preferable for speech signal processing in telephony, because the ratio of peak voltages in a loud passages to those of a weak passage could be 60 dB [Stern04]. nonuniform quantization increases the perceived quality of the telephony signal. The voltage step size Δ is not given by Equation 6.6, but increases as the absolute input voltage deviates from 0 to the absolute maximum value V_{max}.

The procedure for nonlinear compressing a signal at the transmitter and expanding a signal at the receiver is referred to as *companding*. Two companding standards have been accepted by the CCITT (*Comité Consultatif International Télé-phonique et Télégraphique*, an international standards organization). The European standard, used in most of the world, is the A-law, where the absolute value of the output voltage V_{out} is given by Equation 6.9.

$$\begin{aligned} |V_{out}| &= \frac{A}{1+\ln A}\left(\left|\frac{V_{in}}{V_{max}}\right|\right)|V_{max}| && 0 \le \left|\frac{V_{in}}{V_{max}}\right| \le \frac{1}{A} \\ |V_{out}| &= \frac{|V_{max}|}{1+\ln A}\left(1+\ln\left(A\left|\frac{V_{in}}{V_{max}}\right|\right)\right) && \frac{1}{A} \le \left|\frac{V_{in}}{V_{max}}\right| \le 1 \end{aligned} \tag{6.9}$$

The input voltage is V_{in} and A is a constant equal to 87.6 [Haykin01]. The North American standard, used in the United States and Japan, is the μ-law, where the absolute value of the output voltage V_{out} is given by Equation 6.10.

$$\left|V_{out}\right| = \frac{\ln\left(1+\mu\left|\dfrac{V_{in}}{V_{max}}\right|\right)}{\ln\left(1+\mu\right)}\left|V_{max}\right| \qquad 0 \le \left|\frac{V_{in}}{V_{max}}\right| \le 1 \tag{6.10}$$

The input voltage is V_{in} and μ is a constant equal to 255 [Haykin01]. Figure 6.13 is a SystemVue simulation test system that generates the polar voltage transfer functions for the μ-law and *A*-law companders (see Fig6-13.svu on the CD-ROM)

ON THE CD

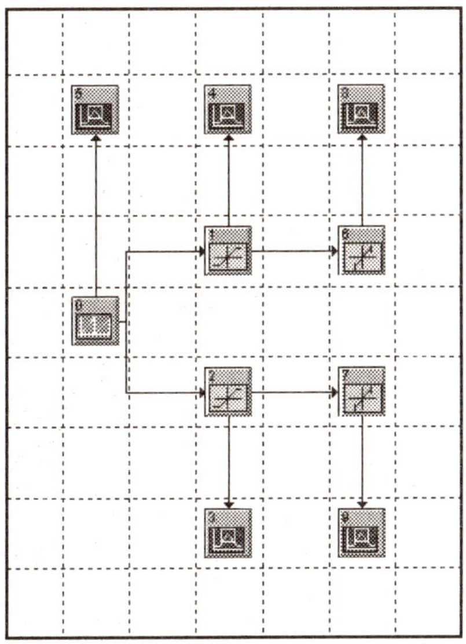

FIGURE 6.13 μ-law and *A*-law compression and expansion test system.

The Sawtooth token from the Source Library provides the input voltage as a polar (± 1 V) linear ramp voltage input V_{in} with parameters of an amplitude $A = 2$ V, –1 V offset, and a frequency of 10 Hz and 0° phase offset. The μ-law and *A*-law Compander and Decompander tokens are from the Communications Library, with parameters of a maximum input amplitude $V_{max} = \pm 1$ V. The linear ramp frequency is 10 Hz, the System Sampling Rate is 10 kHz, and the System Stop Time is 0.1 sec (one period of the linear ramp), although these parameters are arbitrary for the test system here.

Figure 6.14 shows the μ-law compander transfer function as a scatter plot of the linear input voltage (x axis) against the output voltage (y axis). A scatter plot of *A*-law compressor transfer function would show a nearly identical response. A scatter plot of the difference between either the μ-law or *A*-law compandor input and output voltages would also be nearly error-free.

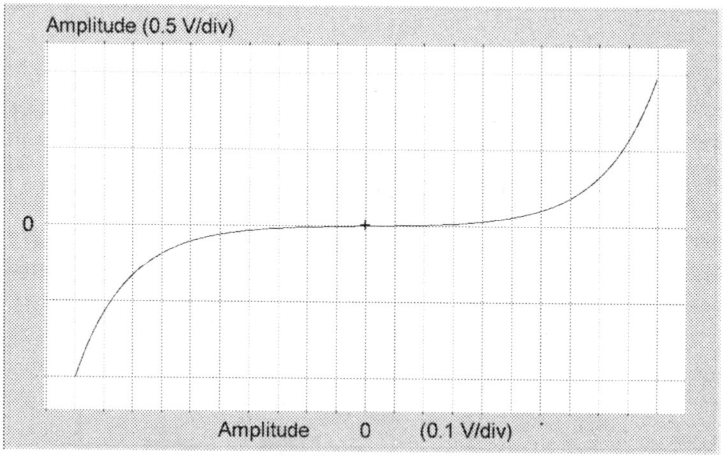

FIGURE 6.14 μ-law compandor voltage transfer function.

As described in Chapter 1, "Communication Simulation Techniques," SystemVue can import a pseudo-analog audio file (.wav) to illustrate the overall performance of a digital communication system. Figure 6.15 shows a companding analog-to-digital (ADC) and digital-to-analog (DAC) SystemVue simulation test system (see Fig6-15.svu on the CD-ROM). The Wave external file token from the Source Library inputs the audio file SVAudioIn.wav, which pronounces "SystemVue" at 8 bits of resolution with an initial 8-kHz sampling rate, or a data rate of 64 kb/sec.

ON THE CD

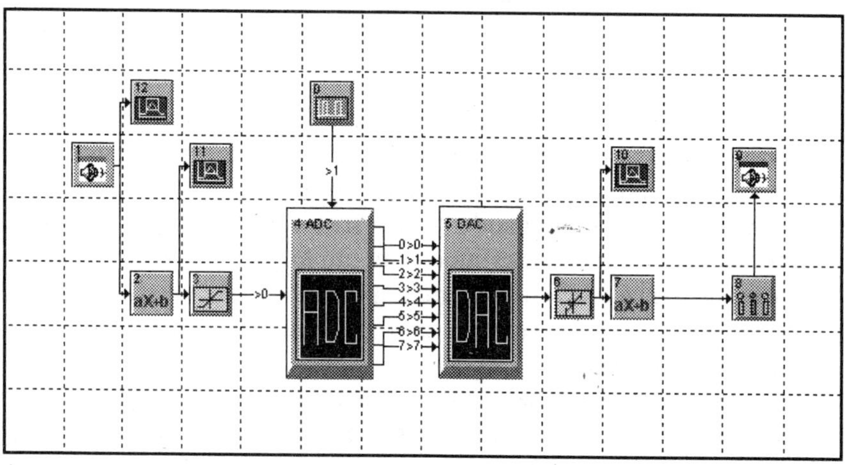

FIGURE 6.15 Companding ADC and DAC test system.

However, the Wave external input file token is specified to be upsampled by a factor of 10, padding the input sample with nine of the previous values. This is set in the Wave external file token parameter window as shown in Figure 6.16, which increases the audio file sample rate to the System Sampling Rate of 80 kHz (T_{system} = 12.5 μsec) here. The System Time is 620 msec, which is the length of the external input audio file here. The Launch Audio Player box is also selected in Figure 6.16, which runs the default audio player at the end of the simulation.

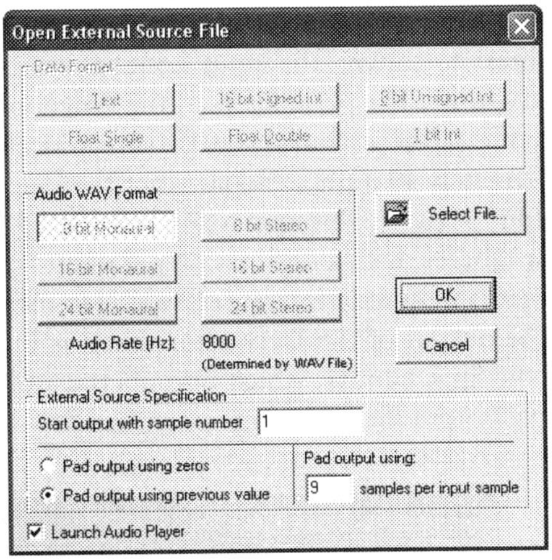

FIGURE 6.16 Wave external input file token | parameter window.

The audio file is scaled by the Polynomial token from the Function Library to convert the input from 8-bit arbitrary units to approximately ± 1.28 V. The parameters of the Polynomial token are that the x^0 (offset) coefficient is 0, the x^1 (linear) coefficient is 0.032 (the peak value of the audio file is approximately 40 units), and all other coefficients are zero. The output of the Polynomial token is inputted to the μ-law Compander token from the Communications Library with a parameter V_{max} = 1.28 V.

The output of the μ-law Compander token is inputted to the ADC token from the Logic Library with parameters of 8 bits of resolution, two's complement signed integer parallel binary output, a maximum positive input voltage V_{maxp} = 1.27 V, a maximum negative input voltage V_{maxn} = −1.28 V, and a clock threshold of 0.5 V. The ADC token sample clock input is provided by the Pulse Train token with parameters of an amplitude A = 1 V, frequency f_o = 8 kHz, pulse width τ = 12.5 μsec (the System Time, T_{system}), and 0 V voltage and 0° phase offset. The resolution of the output of the ADC token and the audio file is the same, of course, but the input can be replaced with any signal for a general analysis of companding.

The 8 bits of output of the ADC token are connected in parallel to the 8 bits of input of the Digital to Analog Converter (DAC) token from the Logic Library. The parameters of the DAC token are 8 bits of resolution, a maximum positive output voltage $V_{maxp} = 1.27$ V, a maximum negative output voltage $V_{maxn} = -1.28$ V, and a logic threshold of 0.5 V. The output of the DAC token is inputted to the μ-law Decompander token from the Communications Library with a parameter $V_{max} = 1.28$ V.

The output of the DAC token is inputted to the Polynomial token from the Function Library to convert the input from approximately ± 1.28 V to the original 8-bit arbitrary units. The parameters of the Polynomial token are that the x^0 (offset) coefficient is 0, the x^1 (linear) coefficient is 31.25 (the peak value of the audio file is approximately 40 units), and all other coefficients are zero.

The output of the Polynomial token is inputted to the Decimator token from the Operator Library, with a decimation parameter of 10, which reduces the signal from the SystemVue simulation rate of 80 kHz to the original audio file sample rate of 8 kHz. The output of the Decimator token is input to the Wave external file token from the Sink Library, which outputs the audio file SVAudioOut.wav. The Wave external output file token parameter window is shown in Figure 6.17.

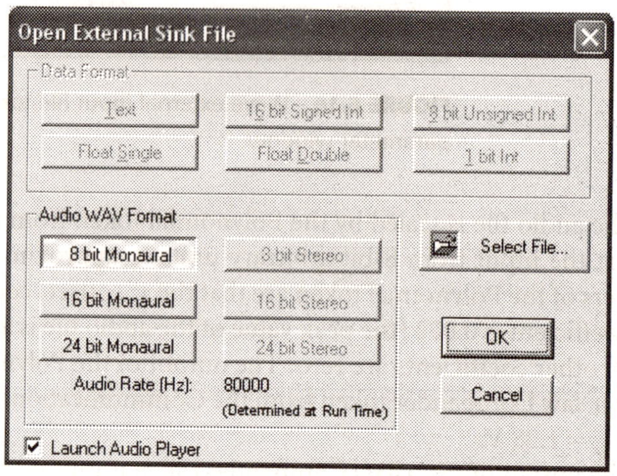

FIGURE 6.17 Wave external output file token parameter window.

Note that from Figures 6.16 and 6.17, the external input and output audio filenames and folder locations can be changed as required. The default audio player is automatically launched at the end of the simulation of the companding ADC and DAC test system, as shown in Figure 6.18.

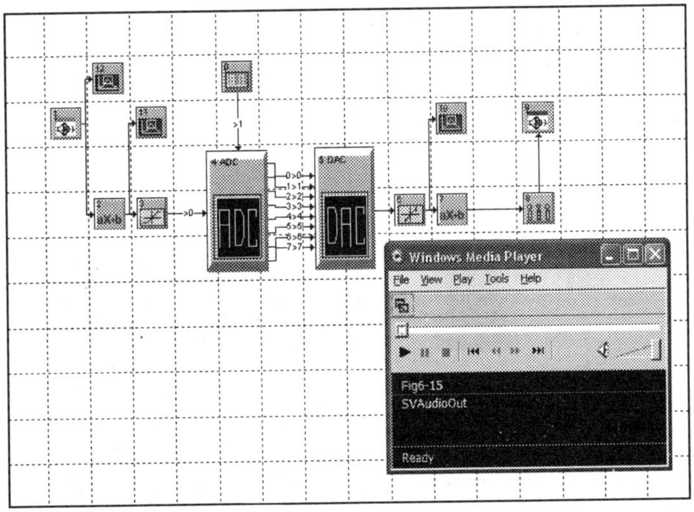

FIGURE 6.18 Companding ADC and DAC test system with default audio player for the external input and output audio files.

PULSE CODE MODULATION

Pulse code modulation (PCM) represents the sampled and quantized analog baseband signal as a sequence of encoded baseband pulses using line codes. A PCM system, which uses the external audio file as an input source, is shown in Figure 6.19 (see Fig6-19.svu on the CD-ROM). The Wave external file token from the Source Library inputs the audio file SVAudioIn.wav, which pronounces "SystemVue" at 8 bits of resolution with an initial 8-kHz sampling rate, or a data rate of 64 kb/sec.

ON THE CD

The Wave external input file token is specified to be upsampled by a factor of 80 here, padding the input sample with 79 of the previous values. This is set in the Wave external file token parameter window, similar to that shown in Figure 6.16, which increases the audio file sample rate to the System Sampling Rate of 640 kHz (T_{system} = 1.5625 μsec) here. The System Time is 620 msec, which is the length of the external input audio file here. The Launch Audio Player box is also selected, which runs the default audio player at the end of the simulation.

The PCM system of Figure 6.19 uses the companding analog-to-digital (ADC) and digital-to-analog (DAC) test system of Figure 6.15 as the source input and output. The serial data transmitter MetaSystem, shown in Figure 6.20, inputs 8 parallel data bits from the ADC token at 8000 8-bit samples/sec and transmits the data serially, least significant bit (LSB) first, at 64 kb/sec.

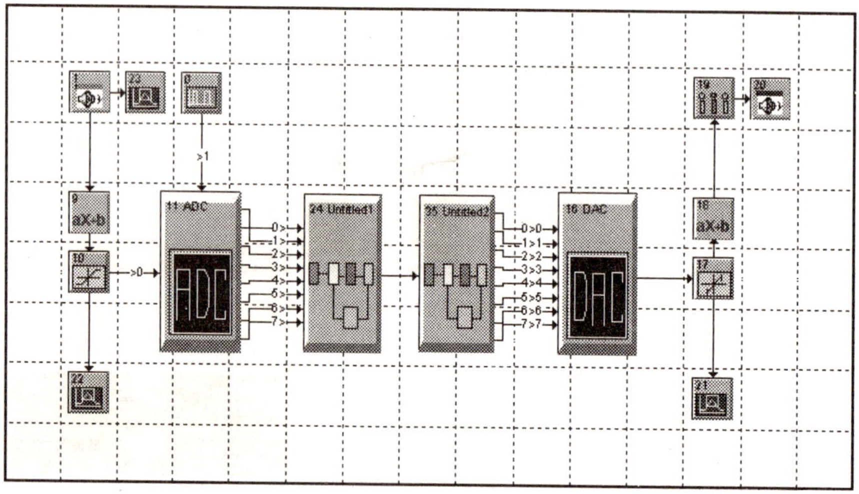

FIGURE 6.19 PCM system with external audio file input and output.

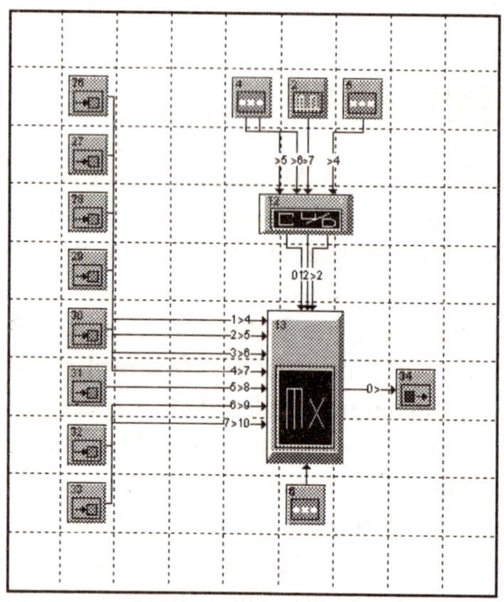

FIGURE 6.20 Serial data transmitter MetaSystem of the PCM system.

The 8-bit Mux (multiplexer) token from the Logic Library selects one of eight input signals and is a behavioral synthesis of the medium scale integration (MSI) 74151 integrated circuit (IC), as described in Chapter 1. The active low Enable control signal of the 8-bit Mux token is provided by a Custom token from the Source Library, and one output with an algebraic simulation equation that is $p(0) = 0$ V.

The 8-bit Mux token 3-bit data selection control signal is derived from the 3 LSBs of the 4-bit Up/Down Binary Counter token from the Logic Library, which is the behavioral synthesis of the MSI 74191 IC, as described in Chapter 1. The Up/Down Binary Counter token Clock input is derived from the Pulse Train token from the Source Library, with parameters of an amplitude $A = 1$ V, a frequency $f_o = 64$ kHz, a pulse width $\tau = 1.5625$ μsec (the System Time, T_{system}), and 0 V voltage and 0° phase offset.

The active low Up and Count Enable control signals of the Up/Down Binary Counter token is provided by a Custom token from the Source Library, and one output with an algebraic simulation equation that is $p(0) = 0$ V. The inactive high Parallel Load control signal is provided by another Custom token and one output with an algebraic simulation equation that is $p(0) = 1$ V.

The serial data receiver MetaSystem, shown in Figure 6.21, inputs a serial data bit stream, LSB first, at 64 kb/sec, and outputs 8 parallel data bits at 8000 samples/sec. The serial data is inputted to the 8-bit Shift Register token from the Logic Library, which is the behavioral synthesis of the MSI 74164 IC, as described in Chapter 1. The 8-bit Shift Register token Clock input is derived from the Pulse Train token from the Source Library, with parameters of an amplitude $A = 1$ V, a frequency $f_o = 64$ kHz (the data rate), a pulse width $\tau = 1.5625$ μsec (the System Time, T_{system}), and 0 V voltage and 0° phase offset. The B Input and the active low Master Reset control signal of the 8-bit Shift Register token is provided by a Custom token from the Source Library, and one output with an algebraic simulation equation that is $p(0) = 1$ V.

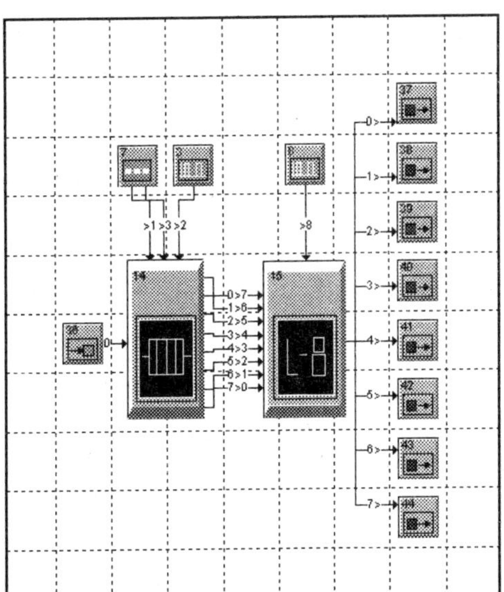

FIGURE 6.21 Serial data receiver MetaSystem of the PCM system.

The output of the 8-bit Shift Register token is inputted to the 8-bit Latch token, which is the behavioral synthesis of the MSI 74573 IC, and holds the parallel 8-bit data signal for each sample, as described in Chapter 1. The 8-bit Latch token clock input is derived from the Pulse Train token from the Source Library, with parameters of an amplitude $A = 1$ V, a frequency $f_o = 8$ kHz (the sampling rate), a pulse width $\tau = 1.5625$ μsec (the System Time, T_{system}), and 0 V voltage and 0° phase offset.

The active high Latch Enable control signal of the 8-bit Latch token is provided by a Custom token from the Source Library, and one output with an algebraic simulation equation that is $p(0) = 1$ V. Note that because the LSB is transmitted first here, bit 0 of the 8-bit Shift Register token is inputted to bit 7 of the 8-bit Latch token, and likewise in reverse order for all the 8 data bits.

The output of the serial data receiver MetaSystem of Figure 6.21 is inputted to the DAC token of the companding ADC and DAC system of Figure 6.15 and the PCM system of Figure 6.19. The output of the Polynomial token is inputted to the Decimator token from the Operator Library, with a decimation parameter of 80 here, which reduces the signal from the SystemVue simulation rate of 640 kHz to the original audio file sample rate of 8 kHz.

LINE CODES

Binary baseband data streams are encoded for the representation of information by pulses that extend either for a single bit time T_b or, if extending for more than one bit time, do not interfere substantially with adjacent symbol pulses (*intersymbol interference*). Polar rectangular, sinc, raised cosine, and duobinary pulses for baseband modulation are discussed in Chapter 2. However, additional line codes are often employed which feature some or all of the characteristics that are deemed useful. The desirable properties of a line code for baseband digital data communication are as follows [Lathi98]:

Transmission Bandwidth: A line code should exhibit a minimum transmission bandwidth.

Power Efficiency: For a given transmission bandwidth and probability of bit error P_b, the transmitted power should be as small as possible.

Error Detection and Correction: A line code should enable the detection, and preferably the correction, of errors in transmission.

Desirable PSD: The PSD of a line code should be zero at a frequency of 0 Hz (DC), because many baseband digital data communication systems use AC coupling and magnetic transformers.

Sufficient Timing Information: A line code should enable the extraction of timing and synchronization information, as described in Chapter 4.

Transparency: A line code should be transparent so that a digital signal is transmitted correctly regardless of the source information bit pattern, including long strings of binary 1s or 0s.

Line codes often utilize rectangular pulses to encode or *map* the input binary data. Line codes with pulses that maintain constant amplitude throughout the bit time T_b use rectangular *nonreturn-to-zero* (NRZ) signaling. Line codes with pulses that maintain constant amplitude for a portion of the bit time, usually $T_b/2$, use rectangular *return-to-zero* (RZ) signaling. Finally, line codes with pulses that maintain constant amplitude for a portion of the bit time, usually $T_b/2$, then reverse amplitude for the remainder of the bit time use *split-phase* signaling.

Power Spectral Density of Line Codes

Although the power spectral density (PSD) of a line code can be developed from a single pulse $p(t)$, as in Chapter 2, for rectangular, sinc, raised cosine, and duobinary pulses, a general procedure would facilitate the analysis for complex line codes. An arbitrary line code can be generated by a linear time invariant (LTI) system with an impulse response $h(t) = p(t)$. The impulse response $p(t)$ is non-zero only over the interval $0 \leq t \leq T_b$, where T_b is the bit time.

The LTI system is excited by an *impulse train* as given by Equation 6.11.

$$x(t) = \sum_k a_k\, \delta\left(t - kT_b\right) \tag{6.11}$$

The impulse function is $\delta(t)$ and the strengths of the impulse are a_k, which are related to the input binary data $b_k = 0, 1$. The output of the LTI system excited by $x(t)$ then is a pulse amplitude modulated (PAM) signal $y(t)$ as given by Equation 6.12.

$$y(t) = \sum_k a_k\, p(t - kT_b) \tag{6.12}$$

The *autocorrelation* $R_x(\tau)$ of the impulse train $x(t)$ in Equation 6.11 is given by Equation 6.13 [Lathi98].

$$R_x(\tau) = \frac{1}{T_b} \sum_{n=-\infty}^{\infty} R_n\, \delta(\tau - nT_b) \tag{6.13}$$

The coefficients R_n of the autocorrelation $R_x(\tau)$ are the *time average* of the product $a_k\, a_{k+n}$ as given by Equation 6.14.

$$\mathrm{R}_n = \lim \frac{1}{N}\sum_k a_k a_{k+n} \qquad N \to \infty \tag{6.14}$$

The normalized ($R_L = 1\ \Omega$) power spectral density $S_x(f)$ of the impulse train $x(t)$ then is given by Equation 6.15 [Lathi98].

$$S_x(f) = \frac{1}{T_b}\sum_{n=-\infty}^{\infty} \mathrm{R}_n \exp(-\mathrm{j}2\pi f n T_b) \tag{6.15}$$

The impulse response of the LTI system is $p(t)$ and its Fourier transform is $P(f)$. The normalized power spectral density $S_y(f)$ of the output of the LTI system $y(t)$ in Equation 6.12 then is given by Equation 6.16 [Haykin01].

$$S_y(f) = |P(f)|^2 S_x(f) \tag{6.16}$$

Substituting Equation 6.15 into Equation 6.16 results in the normalized PSD of an arbitrary line code in terms of the coefficients R_n of the autocorrelation $\mathrm{R}_x(\tau)$ and the magnitude $|P(f)|$ of the Fourier transform of the pulse $p(t)$, as given by Equation 6.17.

$$S_y(f) = \frac{|P(f)|^2}{T_b}\sum_{n=-\infty}^{\infty} \mathrm{R}_n \exp(-\mathrm{j}2\pi f n T_b) \tag{6.17}$$

Equation 6.17 can also be rendered with an algebraic manipulation by Equation 6.18.

$$\begin{aligned} S_y(f) &= \frac{|P(f)|^2}{T_b}\left[\mathrm{R}_0 + \sum_{n=-\infty,\, n\neq 0}^{\infty} \mathrm{R}_n \exp(-\mathrm{j}2\pi f n T_b)\right] \\ S_y(f) &= \frac{|P(f)|^2}{T_b}\left[\frac{\mathrm{R}_0}{2} + \sum_{n=-\infty}^{\infty} \mathrm{R}_n \exp(-\mathrm{j}2\pi f n T_b)\right] \end{aligned} \tag{6.18}$$

If all the R_n are equal to some constant K for $n \neq 0$ then the *Poisson sum* formula can be used in Equation 6.18 to substitute an impulse sum for the exponential sum. The result is given by Equation 6.19 for the normalized PSD of the line code [Xiong00].

$$S_y(f)=\frac{|P(f)|^2}{T_b}\left[\frac{R_o}{2}+R_n\sum_{n=-\infty}^{\infty}\exp(-j2\pi f n T_b)\right] \quad R_n = K,\ n\neq 0$$
$$S_y(f)=\frac{|P(f)|^2}{T_b}\left[\frac{R_o}{2}+\frac{R_n}{T_b}\sum_{n=-\infty}^{\infty}\delta(f-n/T_b)\right] \tag{6.19}$$

Equation 6.19 implies that if the R_n are all equal and non-zero, then the normalized PSD of an arbitrary line code has both a continuous spectrum, which is a scaled version of the magnitude $|P(f)|$ of the Fourier transform of its pulse $p(t)$, and a discrete spectrum at multiples of the data rate $r_b = 1/T_b$.

If the pulse $p(t)$ is rectangular with constant amplitude A V throughout the bit time T_b or NRZ, the magnitude of the Fourier transform $|P_{rectNRZ}(f)|$ is given by Equation 6.20, which is the same as Equation 2.1.

$$\left|P_{rect\ NRZ}(f)\right| = AT_b\,\text{sinc}(2\pi(T_b/2)f) \tag{6.20}$$

If the pulse $p(t)$ is rectangular with constant amplitude A V throughout the first half of the bit time T_b, then zero for the remainder of T_b or RZ, the magnitude of the Fourier transform $|P_{rectRZ}(f)|$ is given by Equation 6.21. This equation is the same as Equation 6.20, but with T_b replaced by $T_b/2$.

$$\left|P_{rect\ RZ}(f)\right| = \frac{AT_b}{2}\text{sinc}(2\pi(T_b/4)f) \tag{6.21}$$

Finally, if the pulse has constant positive amplitude A V for the first half of the bit time T_b, then has a constant negative amplitude ?A V for the remainder of T_b or split-phase, the magnitude of the Fourier transform $|P_{SplitPhase}(f)|$ is given by Equation 6.22 [Xiong00].

$$\left|P_{Split\ Phase}(f)\right| = AT_b\ \text{sinc}(2\pi(T_b/2)f)\sin(2\pi(T_b/2)f) \tag{6.22}$$

Polar NRZ Line Code

The polar NRZ line code maps the input binary data $b_k = 0$ to $a_k = -1$ and $b_k = 1$ to $a_k = +1$. If the probability of occurrence P of binary 0 and 1 are equally likely ($P(b_k = 0) = P(b_k = 1) = 0.5$), then $P(a_k = -1) = P(a_k = +1) = 0.5$ and $a_k^2 = -1^2$ for $N/2$ terms and $a_k^2 = 1^2$ for $N/2$ terms. From Equation 6.14, R_0 is given by Equation 6.23.

$$R_0 = \lim\frac{1}{N}\sum_k a_k^2 = \lim\frac{1}{N}\left[\frac{N}{2}(-1)^2+\frac{N}{2}(1)^2\right]=1 \qquad \lim N\to\infty \tag{6.23}$$

For N terms in the summation of Equation 6.14, the product $a_k a_{k+1}$ is either −1 on average for $N/2$ terms or +1 for $N/2$ terms, and R_1 (or R_{-1}) is given by Equation 6.24.

$$R_1 = \lim \frac{1}{N}\left[\frac{N}{2}(1)+\frac{N}{2}(-1)\right]=0 \qquad \lim N \to \infty \tag{6.24}$$

Similar reasoning implies that $R_n = 0$ for $|n| \geq 1$. The magnitude $|P(f)|$ of the Fourier transform of the rectangular NRZ pulse $p(t)$, the impulse response of the LTI system here, is given by Equation 6.20. From Equation 6.17, the normalized power spectral density $S_{polar\ NRF}(f)$ of the polar rectangular NRZ line code is given by Equation 6.25, which is the same as Equation 2.3.

$$\begin{aligned} S_{polar\ NRZ}(f) &= \frac{|P_{rect\ NRZ}(f)|^2}{T_b}(1) \\ S_{polar\ NRZ}(f) &= A^2\, T_b\, \text{sinc}^2(2\pi(T_b/2)f) \end{aligned} \tag{6.25}$$

Unipolar NRZ Line Code

The unipolar NRZ rectangular line code maps the input binary data $b_k = 0$ to $a_k = 0$ and $b_k = 1$ to $a_k = +1$. If the probability of occurrence P of binary 0 and 1 are equally likely ($P(b_k = 0) = P(b_k = 1) = 0.5$), then $P(a_k = 0) = P(a_k = +1) = 0.5$ and $a_k^2 = 0$ for $N/2$ terms and $a_k^2 = 1$ for $N/2$ terms. From Equation 6.14, R_0 is given by Equation 6.26.

$$R_0 = \lim \frac{1}{N}\sum_k a_k^2 = \lim \frac{1}{N}\left[\frac{N}{2}(0)+\frac{N}{2}(1)^2\right]=\frac{1}{2} \qquad \lim N \to \infty \tag{6.26}$$

Because a_k and a_{k+n} are equally likely to be 0 or 1, the product $a_k a_{k+n}$ is equally likely to be 0 ? 0, 0 ? 1, 1 ? 0, or 1 ? 1 and is 0 for 3 $N/4$ terms and 1 for $N/4$ terms. The autocorrelation R_n for $|n| \geq 1$ then is given by Equation 6.27.

$$R_n = \lim \frac{1}{N}\left[\frac{3N}{4}(0)+\frac{N}{4}(1)^2\right]=\frac{1}{4} \qquad \lim N \to \infty,\ |n| \geq 1 \tag{6.27}$$

The magnitude $|P(f)|$ of the Fourier transform of the rectangular NRZ pulse $p(t)$, the impulse response of the LTI system here, is given by Equation 6.20. From Equation 6.19, the normalized PSD of the unipolar rectangular NRZ line code is given by Equation 6.28.

$$S_{unipolar\ NRZ}(f) = \frac{|P_{rect\ NRZ}(f)|^2}{T_b}\left[\frac{R_0}{2} + \frac{R_n}{T_b}\sum_{n=-\infty}^{\infty}\delta(f - n/T_b)\right]$$

$$S_{unipolar\ NRZ}(f) = A^2\, T_b \operatorname{sinc}^2(2\pi(T_b/2)f)\left[\frac{1}{4} + \frac{1}{4T_b}\sum_{n=-\infty}^{\infty}\delta(f - n/T_b)\right] \tag{6.28}$$

Although Equation 6.28 seems to indicate that the normalized PSD of the unipolar rectangular NRZ line code has components at multiples of the bit rate $r_b = 1/T_b$, the sinc² term is zero at the same frequency n/T_b Hz, $n \neq 0$. The unipolar rectangular NRZ line code does not enable the extraction of timing and synchronization information, as described in Chapter 4. The resulting normalized PSD then is given by Equation 6.29.

$$S_{unipolar\ NRZ}(f) = A^2\, T_b \operatorname{sinc}^2(2\pi(T_b/2)f)\left[\frac{1}{4} + \frac{1}{4T_b}\delta(f)\right] \tag{6.29}$$

Alternate Mark Inversion NRZ Line Code

The alternate mark inversion (AMI) is a *pseudoternary* (three-level) NRZ code that maps the input binary data $b_k = 0$ to $a_k = 0$ and $b_k = 1$ to either $a_k = -1$ if $a_{k-j} = +1$ or $a_k = +1$ if $a_{k-j} = -1$, where the $b_{k-j} = 1$ is the immediate past binary 1 inputted. AMI is also called *bipolar* signaling. If the probability of occurrence P of binary 0 and 1 is equally likely $(P(b_k = 0) = P(b_k = 1) = 0.5)$, then $P(a_k = 0) = 0.5$ and $P(a_k = -1$ or $a_k = +1) = 0.5$ and $a_k^2 = 0$ for $N/2$ terms, and $a_k^2 = 1$ for $N/2$ terms. From Equation 6.14, R_0 is given by Equation 6.30.

$$R_0 = \lim \frac{1}{N}\sum_k a_k^2 = \lim \frac{1}{N}\left[\frac{N}{2}(0) + \frac{N}{2}(\pm 1)^2\right] = \frac{1}{2} \qquad \lim N \to \infty \tag{6.30}$$

For R_1 there are four equally likely sequences of two consecutive input binary data $b_k\, b_{k+1}$, that is 00, 01, 10, and 11. However, because $b_k = 0$ maps to $a_k = 0$, the product $a_k\, a_{k+1} = 0$ for $3N/4$ of the terms. Furthermore, because $b_k = 1$ maps to either $a_k = -1$ or $a_k = +1$, the product $a_k\, a_{k+1} = -1$ for $N/4$ of the terms and R_1 (or R_{-1}) is given by Equation 6.31.

$$R_1 = \lim \frac{1}{N}\left[\frac{N}{4}(-1) + \frac{3N}{4}(0)\right] = -\frac{1}{4} \qquad \lim N \to \infty \tag{6.31}$$

Similarly for R_2 there are eight equally likely sequences of three consecutive input binary data $b_k\, b_{k+1}\, b_{k+2}$, that is 000, 001, 010, 011, 100, 101, 110, and 111. However, because $b_k = 0$ maps to $a_k = 0$, the product $a_k\, a_{k+2} = 0$ for $6N/8$ or $3N/4$ of the terms because either the first bit, the last bit, or both input data bits are 0. Furthermore, because $b_k = 1$ maps to either $a_k = -1$ or $a_k = +1$ by the alternate rule, the product $a_k\, a_{k+2} = -1$ for $b_k\, b_{k+1}\, b_{k+2} = 101$ and $N/8$ of the terms, but the product $a_k\, a_{k+2} = +1$ for $b_k\, b_{k+1}\, b_{k+2} = 111$ for $N/8$ of the terms. The correlation R_2 (or R_{-2}) then is given by Equation 6.32.

$$R_2 = \lim \frac{1}{N}\left[\frac{3N}{4}(0) + \frac{N}{8}(-1) + \frac{N}{8}(1)\right] = 0 \qquad \lim N \to \infty \tag{6.32}$$

Similar reasoning implies that $R_n = 0$ for $|\,n\,| \geq 2$. The magnitude $|\,P(f)\,|$ of the Fourier transform of the AMI rectangular NRZ pulse $p(t)$, the impulse response of the LTI system here, is given by Equation 6.20. From Equation 6.18, the normalized power spectral density $S_{AMI\ NRZ}(f)$ of the AMI line code is given by Equation 6.33.

$$\begin{aligned}
S_{AMI\ NRZ}(f) &= \frac{|P_{rect\ NRZ}(f)|^2}{T_b}\left[\frac{R_o}{2} + \sum_{n=-\infty}^{\infty} R_n \exp(-\,j2\pi f n T_b)\right] \\
S_{AMI\ NRZ}(f) &= \frac{|P_{rect\ NRZ}(f)|^2}{T_b}\left[\frac{1}{2} - \frac{1}{4}\left(\exp(j2\pi f T_b) + \exp(-\,j2\pi f T_b)\right)\right] \\
S_{AMI\ NRZ}(f) &= \frac{A^2\, T_b}{2}\,\mathrm{sinc}^2(2\pi (T_b/2) f)\left[1 - \cos 2\pi f T_b\right] \\
S_{AMI\ NRZ}(f) &= A^2\, T_b\,\mathrm{sinc}^2(2\pi (T_b/2) f)\ \sin^2\left(\pi f T_b\right)
\end{aligned} \tag{6.33}$$

Euler's Identity for $\cos x$ and the trigonometric identity $\sin^2 x = (1 - \cos 2x)/2$ is used to simplify the PSD in Equation 6.29.

The normalized PSD of the AMI line code is desirable in that it has no component at $f = 0$ Hz (DC), regardless of the shape of the pulse $p(t)$, because the $\sin^2(0) = 0$. In addition, the $\sin^2$ term assures that the AMI line code has a minimum transmission bandwidth with the first null occurring at $f = 1/T_b = r_b$ Hz, again regardless of the shape of the pulse $p(t)$. Finally, the AMI line code is partially transparent, because at least a long string of binary 1s would not imply the loss of timing and synchronization information, as described in Chapter 4.

However, unlike the unipolar rectangular NRZ line code, the PSD of the AMI rectangular NRZ line code does not have a component at the bit rate $r_b = 1/T_b$ to enable the extraction of timing and synchronization information, because the sinc^2

term is zero at $f = 1/T_b$. Regardless, because of its other desirable properties, the AMI line code is often used in pulse code modulation (PCM) [Carlson02].

Split-Phase NRZ Line Code

The split-phase or Manchester NRZ line code uses a pulse $p(t)$, which is a positive constant amplitude for the first half, and an equal negative constant amplitude for the second half of the bit time T_b. The split-phase code maps the input binary data $b_k = 0$ to $a_k = -1$ and $b_k = 1$ to $a_k = +1$. If the probability of occurrence P of binary 0 and 1 is equally likely ($P(b_k = 0) = P(b_k = 1) = 0.5$), then $P(a_k = -1) = 0.5$ and $P(a_k = +1) = 0.5$.

This distribution of probabilities and the resulting analysis for the correlation R_n follows the same concepts as that for the polar NRZ line code. The magnitude $|P(f)|$ of the Fourier transform of the rectangular split-phase pulse $p(t)$, the impulse response of the LTI system here, is given by Equation 6.22. From Equation 6.17, the normalized power spectral density $S_{Split\ Phase}(f)$ of the split-phase line code is given by Equation 6.34.

$$S_{Split\ Phase}(f) = \frac{\left|P_{Split\ Phase}(f)\right|^2}{T_b}(1)$$
$$S_{Split\ Phase}(f) = A^2 T_b \operatorname{sinc}^2(2\pi(T_b/4)f)\sin^2(2\pi(T_b/4)f) \tag{6.34}$$

The normalized PSD of the split-phase line code is desirable in that it has no component at $f = 0$ Hz (DC). However, the $\sin^2$ term here implies that the split-phase line code has a wider transmission bandwidth, with the first null occurring at $f = 2/T_b = 2r_b$ Hz, than the AMI NRZ line code in Equation 6.33. Finally, the split-phase line code is fully transparent, because a long string of binary 1s or binary 0s does not imply the loss of timing and synchronization information.

The split-phase line code is used in the 10 Mb/sec Ethernet standard. A general mBnN code maps m bits into n bits and the split-phase line code is a 1B2B code.

Return-to-Zero Line Codes

A variant of any nonreturn-to-zero (NRZ) line code is the use of a return-to-zero (RZ) pulse. The coefficients R_n of the autocorrelation $R_x(\tau)$ do not change, and the normalized PSD of an equivalent RZ line code substitutes the magnitude $|P(f)|$ of the Fourier transform of the RZ pulse $p(t)$, given by Equation 6.21, into the normalized PSD of the appropriate NRZ line code. From the basis of Equation 6.25 then, Equation 6.35 is the normalized power spectral density $S_{polar\ RZ}(f)$ of the polar RZ line code.

$$S_{polar\ RZ}(f) = \frac{A^2 T_b}{4} \operatorname{sinc}^2(2\pi (T_b / 4) f) \qquad (6.35)$$

From the basis of Equation 6.28, Equation 6.36 is the normalized PSD of the unipolar RZ line code.

$$S_{unipolar\ RZ}(f) = \frac{A^2 T_b}{4} \operatorname{sinc}^2(2\pi (T_b / 4) f) \left[\frac{1}{4} + \frac{1}{4T_b} \sum_{n=-\infty}^{\infty} \delta(f - n/T_b) \right] \qquad (6.36)$$

Equation 6.28, unlike Equation 6.29, indicates that the normalized PSD of the unipolar rectangular RZ line code has components at multiples of the bit rate $r_b = 1/T_b$, because the sinc^2 term is not zero at the same frequency n/T_b Hz, $n \neq 0$. The unipolar rectangular RZ line code enables the extraction of timing and synchronization information, as described in Chapter 4.

Finally, from the basis of Equation 6.33, Equation 6.37 is the normalized PSD of the AMI RZ line code.

$$S_{AMI\ RZ}(f) = \frac{A^2 T_b}{4} \operatorname{sinc}^2(2\pi (T_b / 4) f) \sin^2\left(\pi f T_b\right) \qquad (6.37)$$

SystemVue Simulation of Line Codes

ON THE CD

The polar, unipolar, and AMI NRZ and RZ line codes, and the split-phase line code binary data generators are shown in Figure 6.22 (see Fig6-22.svu on the CD-ROM). The data source is a unipolar pseudonoise PN Sequence token from the Source Library, with parameters of a binary amplitude of +0.5 V, a data rate r_b of 1 kb/sec (a bit time T_b = 1 msec), and 0 V voltage and 0° phase offset. This PN Sequence token transmits a unipolar NRZ line code directly.

The Polynomial token from the Function Library converts the unipolar NRZ line code to the polar NRZ line code. The parameters of the Polynomial token are that the x^0 (offset) coefficient is –1, the x^1 (linear) coefficient is 2, and all other coefficients are zero. The polar and unipolar NRZ line codes are each converted to the polar and unipolar RZ line codes by the Multiplier tokens, with another input that is the Pulse Train token from the Source Library. The parameters of the Pulse Train token are an amplitude A = 1 V, a frequency f_o = 1 kHz (the same as the data rate $r_b = 1/T_b$), a pulse width τ = 500 μsec ($T_b/2$), and 0 V voltage and 0? phase offset. The polar and unipolar NRZ and RZ line codes generated are shown in Figure 6.23.

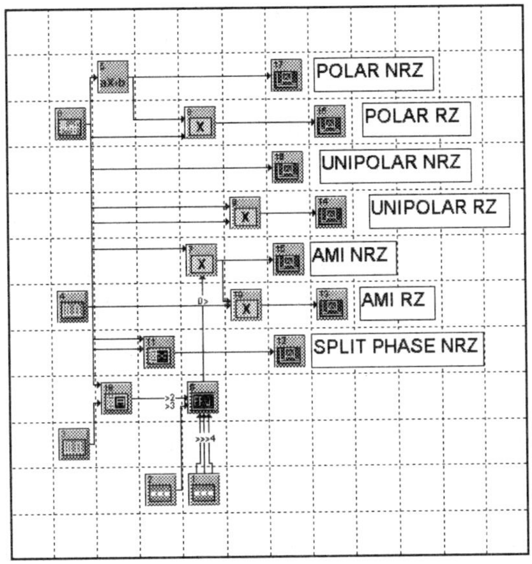

FIGURE 6.22 Polar, unipolar, and AMI NRZ and RZ line codes, and the split-phase NRZ line code binary data generators.

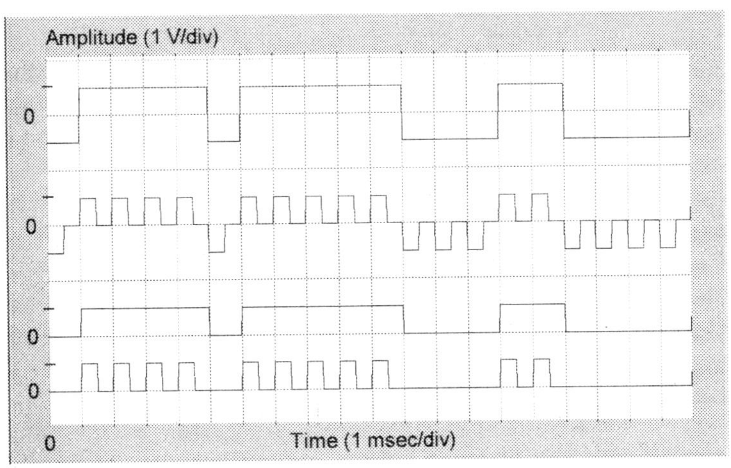

FIGURE 6.23 Polar NRZ (top), polar RZ, unipolar NRZ, and unipolar RZ (bottom) line codes from the binary data generators.

The AMI NRZ line code requires a 1-bit memory, because the pulse transmitted that represents an input binary data $b_k = 1$ alternates between $a_k = +1$ and $a_k = -1$. The J-NOT-K Flip-Flop token from the Logic Library implements the 1-bit memory. The binary data source PN Sequence token is logically ANDed by the

AND token from the Logic Library with the Pulse Train token from the Source Library to provide the Clock input of the J-NOT-K Flip-Flop token. The parameters of the Pulse Train token are an amplitude $A = 1$ V, a frequency $f_o = 1$ kHz, a pulse width $\tau = 20$ μsec (the System Time, T_{system}), and 0 V voltage and 0° phase offset.

The inactive high Set and Clear control signals and the NOT-J input of the J-NOT-K Flip-Flop token is provided by a Custom token from the Source Library, and one output with an algebraic simulation equation that is $p(0) = 1$ V. The K input is provided by a Custom token, and one output with an algebraic simulation equation that is $p(0) = 0$ V. With these control signals, the J-NOT-K Flip-Flop token toggles the output state on each positive edge of the Clock control signal. Because the binary data source PN Sequence token is ANDed with a clock synchronized to the bit rate r_b, only input binary data $b_k = 1$ toggles the J-NOT-K Flip-Flop token.

The AMI NRZ line code is converted to the AMI RZ line code by the Multiplier tokens, with another input that is the Pulse Train token from the Source Library, with a pulse width of $T_b/2$. The split-phase line code is obtained as the output of the XOR token from the Logic Library, with parameters of a logic true output of −1 V and a logic false output of +1 V (reverse logic). The inputs of the XOR token are the binary data source PN Sequence token and the Pulse Train token with a pulse width of $T_b/2$. The AMI NRZ, the AMI RZ, the unipolar NRZ (for reference), and the split-phase NRZ line codes generated are shown in Figure 6.24.

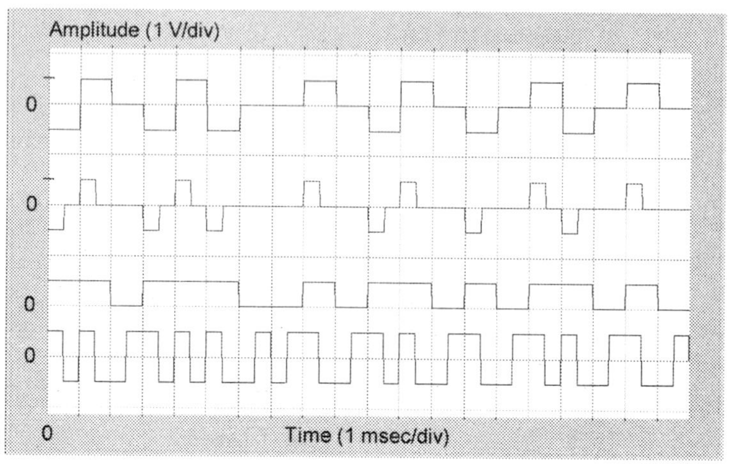

FIGURE 6.24 AMI NRZ (top), AMI RZ, unipolar NRZ, and split-phase NRZ (bottom) line codes from the binary data generators.

The System Time: Sample Rate is set to 50 kHz ($T_{system} = 20$ μsec), well above the nominal 1 kb/sec data rate. The System Time: Number of Samples is set to 262,144 (2^{18}) points, and results in a spectral resolution of 0.19 Hz. Here, the arbitrary amplitude A in the normalized power spectral density (PSD) is 1.

Figure 6.25 is the normalized PSD of the polar NRZ, unipolar NRZ, and AMI NRZ line codes. From Equations 6.25, 6.29, and 6.33, each of the PSDs show approximate spectral nulls at multiples of the data rate $r_b = 1/T_b = 1$ kHz. The PSD of the unipolar NRZ line code shows the spectral impulse at $f = 0$ Hz (DC), while the PSD of the AMI NRZ line code has a magnitude of 0 at $f = 0$ Hz.

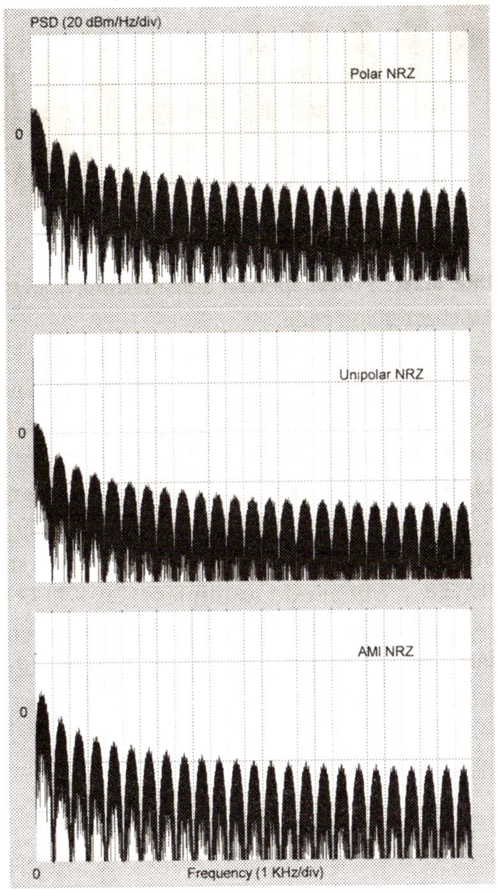

FIGURE 6.25 Power spectral densities of a 1 kb/sec polar NRZ, unipolar NRZ, and AMI NRZ line codes.

Figure 6.26 is the PSD of the split-phase NRZ or Manchester line code. From Equation 6.34, the PSD shows approximate spectral nulls at multiples of twice the data rate $2r_b = 2/T_b = 2$ kHz, and a magnitude of 0 at $f = 0$ Hz (DC).

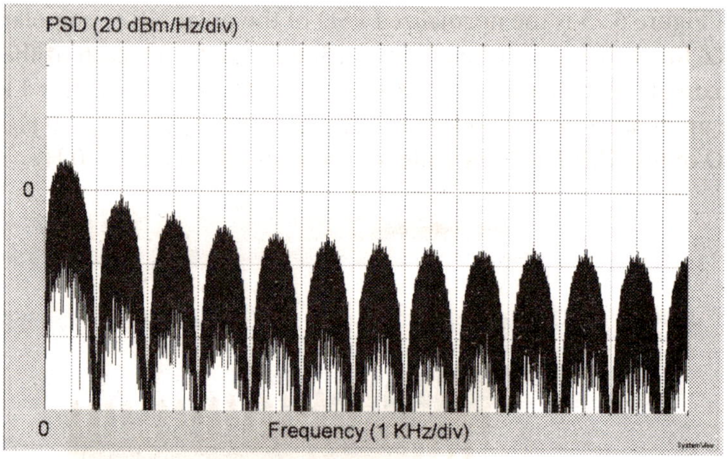

FIGURE 6.26 Power spectral densities of a 1 kb/sec split-phase NRZ or Manchester line code.

Figure 6.27 is the normalized PSD of the polar RZ, unipolar RZ, and AMI RZ line codes. From Equations 6.35, 6.36, and 6.37, each of the PSDs would show approximate spectral nulls at multiples of twice the data rate $2r_b = 2/T_b = 2$ kHz. However, the PSD of the AMI RZ line code also shows additional spectral nulls at multiples of the data rate $r_b = 1/T_b = 1$ kHz, because of the $\sin^2$ term in Equation 6.37.

In addition, the PSD of the unipolar RZ line code shows spectral impulses not only at $f = 0$ Hz (DC), but at odd multiples ($(2n + 1)$ kHz, $n = 1, 2, \ldots$) of the data rate $r_b = 1/T_b = 1$ kHz. The spectral impulses at the even multiples ($2n$ kHz, $n = 1, 2, \ldots$) of the data rate are suppressed, due to the spectral nulls at twice the data rate, because of the sinc^2 term in Equation 6.36.

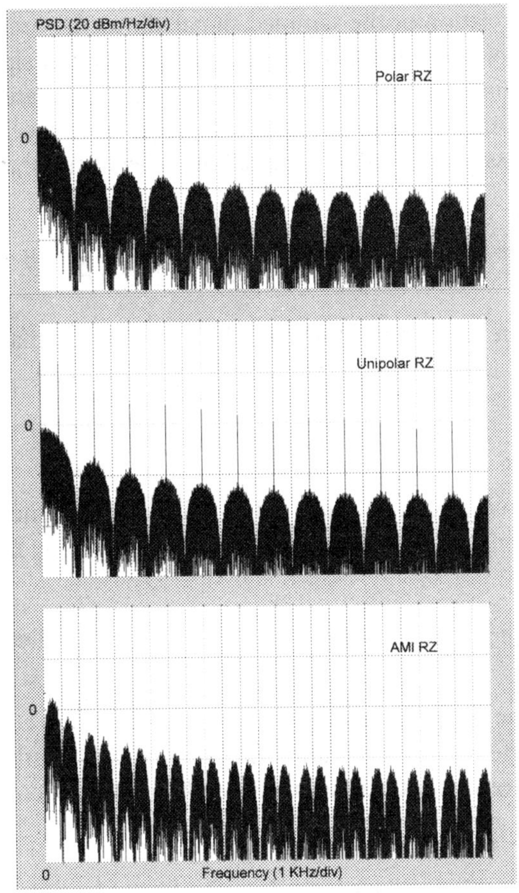

FIGURE 6.27 Power spectral densities of a 1 kb/sec polar RZ, unipolar RZ, and AMI RZ line codes.

DIFFERENTIAL PULSE CODE MODULATION

Differential pulse code modulation (DPCM) is an extension of baseband *delta modulation* (DM), which is described in Chapter 2. If an analog baseband signal is sampled at a rate higher than the Nyquist sampling rate ($f_s > 2\ f_{max}$ where f_{max} is the highest significant frequency content in the signal), the sampled signal has a high degree of correlation between adjacent samples. A high degree of correlation implies that the signal does not substantially vary from one sample to the next, and the difference between adjacent samples can be used to encode the transmitted signal at a reduced data rate [Haykin01].

The correlation of the sampled signal allows the prediction of future values of the signal from past behavior. The analog baseband signal $m(t)$ then is sampled with a continuous amplitude (not quantized) at the rate $f_s = f_{\text{DPCM}} = 1/T_s$ Hz. The next sample $m(t + T_s)$ can be predicted exactly from the Taylor series expansion for this signal (assuming that $m(t)$ has all of its derivatives), as given by Equation 6.38.

$$m(t+T_s) = m(t) + T_s \frac{\mathrm{d}\, m(t)}{\mathrm{d}t} + \frac{T_s^2}{2!}\frac{\mathrm{d}^2 m(t)}{\mathrm{d}t^2} + \frac{T_s^3}{3!}\frac{\mathrm{d}^3 m(t)}{\mathrm{d}t^3} + \ldots \tag{6.38}$$

The exact prediction of $m(t + T_s)$ can be approximated by using only the first two terms of Equation 6.38, as given by Equation 6.39.

$$m(t+T_s) \approx m(t) + T_s \frac{\mathrm{d}\, m(t)}{\mathrm{d}t} \tag{6.39}$$

Equation 6.39 can be rendered as a discrete sequence with $t = kT_s$ and the kth continuous (not quantized) sample as $m[k]$ is given by Equation 6.40.

$$m[k+1] \approx m[k] + T_s \left[\frac{m[k] - m[k-1]}{T_s}\right] = 2\,m[k] - m[k-1] \tag{6.40}$$

This is the *first order linear predictor* for the current continuous sample $m[k+1]$, which uses the two prior samples $m[k]$ and $m[k-1]$ [Lathi98]. The difference d between the current continuous sample $m[k+1]$ and the continuous first order linear prediction is given by Equation 6.41.

$$d[k+1] = m[k+1] - \left(2\,m[k] - m[k-1]\right) \tag{6.41}$$

DPCM transmits the *quantized difference* $d_q[k+1]$, as given by Equation 6.42.

$$d_q[k+1] = \left[(d[k+1])\right]_q = \left[m[k+1] - \left(2\,m[k] - m[k-1]\right)\right]_q \tag{6.42}$$

The function $[\ ]_q$ is the linear quantizer with n bits of resolution. The transmission of the quantized difference $d_q[k]$ is accomplished by any baseband or bandpass modulation method. The DPCM receiver reconstructs an estimate of the continuous sample $m_e[k+1]$ from the received quantized difference $r_q[k]$ and the prediction generated from past estimates, as given by Equation 6.43.

$$m_e[k+1] \approx r_q[k+1] + 2\,m_e[k] - m_e[k-1] \tag{6.43}$$

If the data transmission is error free, then the received quantized difference $r_q[k] = d_q[k]$. The reconstruction of the estimated analog baseband signal $m_e(t)$

from the estimated quantized sample $m_e[k] = m_e(kT_s)$ is accomplished by a low-pass filter (LPF).

ON THE CD

The first order linear predictor DPCM system is shown in Figure 6.28 (see Fig6-28.svu on the CD-ROM). For simplicity in the SystemVue simulation, the pulse code modulation (PCM) transmitter and receiver, as shown in Figure 6.19, is not implemented here. The analog baseband signal $m(t)$ is provided by the Sinusoid token from the Source Library, with parameters of an amplitude $A = 1$ V, a frequency $f_o = 1$ kHz, and 0° phase. The System Time: Sample Rate is set to 5 MHz ($T_{system} = 0.2$ μsec) and the DPCM sample rate $f_{\text{DPCM}} = 20$ kHz, well above the Nyquist sampling rate of 2×1 kHz $= 2$ kHz here.

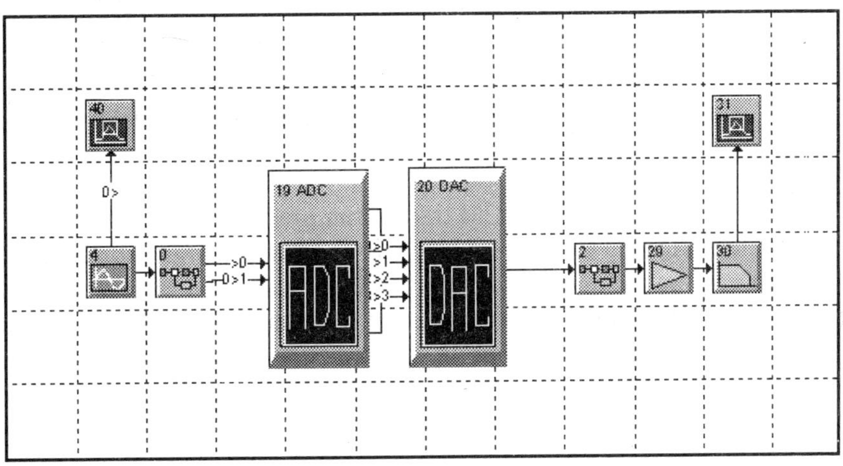

FIGURE 6.28 First order linear predictor DPCM system.

The output of the Sinusoid token is inputted to the first order linear predictor MetaSystem of the DPCM transmitter, as shown in Figure 6.29. The predictor consists of a feed forward system with three Sample and Hold tokens, three Delay tokens, and one Gain token from the Operator Library, an Adder token and a Pulse Train token from the Source Library, with parameters of an amplitude $A = 1$ V, a frequency $f_{\text{DPCM}} = 20$ kHz ($T_{\text{DPCM}} = 1/f_{\text{DPCM}} = 50$ μsec), a pulse width $\tau = 0.2$ μsec (the System Time, T_{system}), and 0 V voltage and 0° phase offset.

The DPCM transmitter predictor produces the continuous difference $d[k+1]$ between the current continuous sample $m[k+1]$ and the continuous first order linear prediction, as given by Equation 6.41. The parameters of the two Delay tokens of the predictor MetaSystem in Figure 6.29 are $T_{\text{DPCM}} - T_{system} = 50 - 0.2$ μsec $= 49.8$ μsec and $2\,(T_{\text{DPCM}} - T_{system}) = 98.6$ μsec.

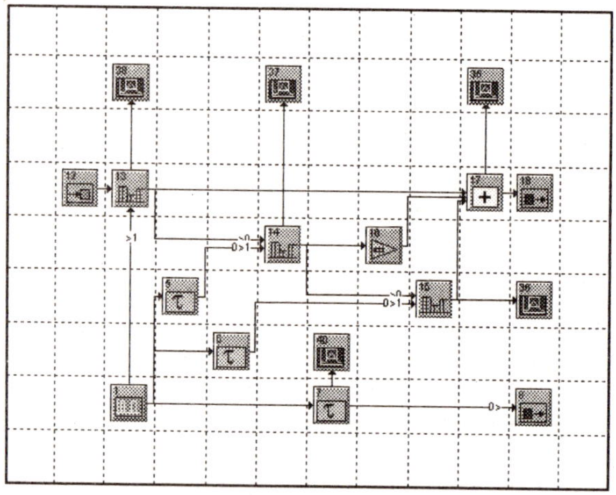

FIGURE 6.29 First order linear predictor MetaSystem of the transmitter of the DPCM system.

The output of the DPCM transmitter predictor MetaSystem is inputted to the Analog to Digital Converter (ADC) token from the Logic Library, which uniformly converts the continuous difference $d[k+1]$ to a binary number with parameters of 4 bits of resolution, two's complement signed integer parallel binary output, a maximum positive input voltage $V_{maxp} = 0.105$ V, a maximum negative input voltage $V_{maxn} = -0.12$ V, and a clock threshold of 0.5 V.

The ADC token sample clock input is provided by the Pulse Train and Delay tokens from the predictor MetaSystem in Figure 6.29. With the voltage range and number of bits specified, the actual uniform voltage step size Δ_{ADC} for the ADC token per bit of resolution here is 15 mV/bit, as given by Equation 6.5. The number of bits of resolution $n = 4$ is arbitrary, but illustrates that DPCM utilizes less than a significant number of bits to quantize and transmit the signal $m(t)$.

The value of Δ_{ADC} is selected by noting the range (≈ 0.1 V) of the continuous difference $d[k+1]$ after the initial transient response, as shown in Figure 6.30. The transient spike in the continuous different signal of Figure 6.30 is avoided by sampling at the midpoint, because the ADC token sample clock input is delayed by $T_{DPCM}/2 = 25$ μsec.

The 4 bits of output of the ADC token are connected in parallel to the 4 bits of input of the Digital to Analog Converter (DAC) token from the Logic Library, because an actual PCM transmitter and receiver is not implemented here. The paameters of the DAC token are the same as that for the ADC token, with 4 bits of resolution, $V_{maxp} = 0.105$ V, $V_{maxn} = -0.12$ V, and a logic threshold of 0.5 V.

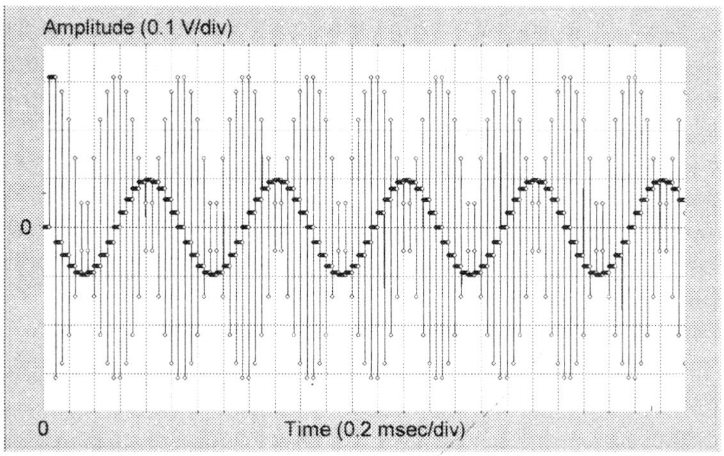

FIGURE 6.30 Continuous difference signal of the first order linear predictor.

The output of the DAC token is inputted to the first order linear predictor of the DPCM receiver, as shown in Figure 6.31. The receiver predictor consists of a feed forward system with three Sample and Hold tokens, three Delay tokens, one Gain token, and one Negate token from the Operator Library, an Adder token and a Pulse Train token from the Source Library, with parameters of an amplitude A = 1 V, a frequency f_{DPCM} = 20 kHz, a pulse width τ = 0.2 μsec (the System Time, T_{system}), and 0 V voltage and 0° phase offset.

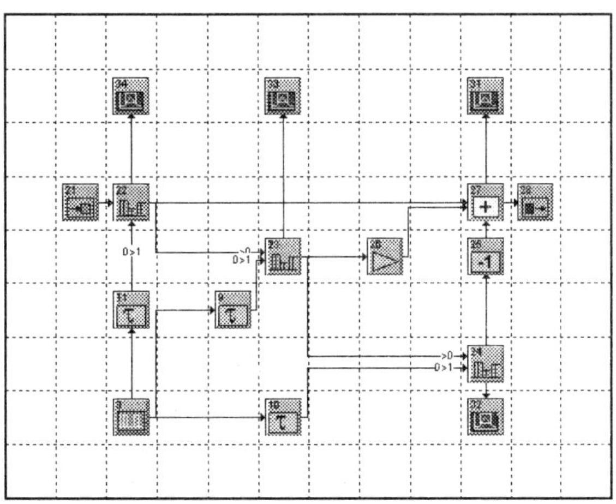

FIGURE 6.31 First order linear predictor MetaSystem of the receiver of the DPCM system.

The DPCM receiver predictor reconstructs an estimate of the continuous sample $m_e[k+1]$ from the received quantized difference $r_q[k]$ and the prediction generated from past estimates, as given by Equation 6.43. The parameters of the two Delay tokens of the predictor MetaSystem in Figure 6.31 are $T_d + T_{\text{DPCM}} - T_{system} = 25 + 50 - 0.2$ μsec = 74.8 μsec and $T_d + 2\,(T_{\text{DPCM}} - T_{system}) = 124.6$ μsec, where $T_d = 25$ μsec is the sampling delay. The parameter of the Delay token that provides the control input for the initial Sample and Hold token in Figure 6.31 is delayed by 25.2 μsec, to avoid the glitch in the received quantized difference $r_q[k]$ that occurs at the sampling point.

The output of the DPCM receiver predictor $m_e[k+1]$, as shown in Figure 6.32, is inputted to the Gain token from the Operator Library, with a parameter of a gain of 5 to restore the nominal amplitude of the original signal $m(t)$ (± 1 V). The DPCM receiver predictor output $m_e[k+1]$ shows a transient response and glitches that occur during the feed forward processing.

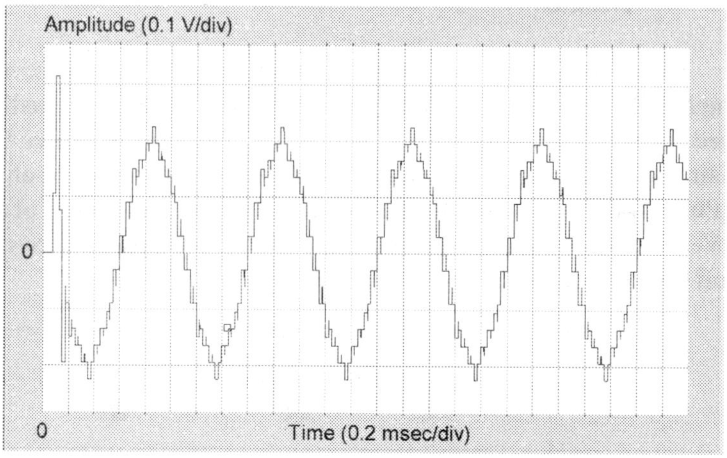

FIGURE 6.32 Quantized recieved estimate of the input signal of the first order linear predictor.

The output of the Gain token is inputted to a lowpass filter (LPF), which reconstructs the estimated analog baseband signal $m_e(t+T_s)$ from the estimated quantized sample $m_e[k+1] = m_e(kT_s+T_s)$. The Linear System Filter token from the Operator Library provides the LPF, which is selected to be a 9-pole Butterworth filter with a bandwidth of 1.5 kHz, which is appropriate for the 1 kHz sinusoidal signal $m(t)$ here. Figure 6.33 shows the original analog baseband signal $m(t)$ and the estimated signal $m_e(t)$ output by the DPCM system. Note the transient response and delay in the output $m_e(t)$.

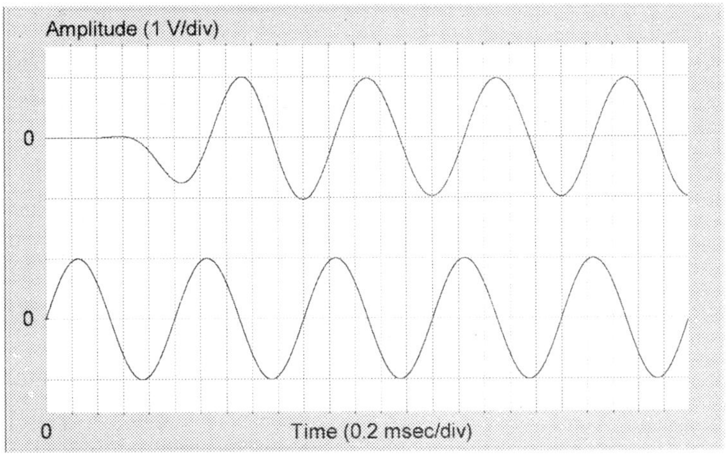

FIGURE 6.33 Original analog baseband signal (bottom) and the estimate of the input signal from the first order linear predictor DPCM system.

SAMPLING BANDPASS ANALOG SIGNALS

A bandpass analog signal has a spectrum that extends from a non-zero lower frequency limit f_L Hz to an upper frequency limit f_H Hz. The anticipated Nyquist sampling rate is 2 f_H Hz. However, a bandpass signal can also be sampled at a rate less than f_L Hz. The spectral aliases appear at frequencies lower and higher than the spectrum of the original signal. The spectral aliases do not overlap the original signal spectrum, and the original analog bandpass signal can be reconstructed without distortion by using a bandpass filter (BPF).

Sampling at less than the expected Nyquist rate for analog bandpass signals is known as bandpass sampling. This sampling technique can only be successful if the bandwidth of the analog signal $f_H - f_L$ Hz is less than f_L Hz, and the sampling rate f_s must be at least $2(f_H - f_L)$ Hz. This is the Nyquist sampling rate then for bandpass sampling.

ON THE CD

Figure 6.34 (see Fig6-34.svu on the CD-ROM) shows the SystemVue simulation of bandpass sampling. The source is the same combination of three Sinusoid tokens from the Source Library, as shown in Figure 6.1. The Adder token sums the output from the three Sinusoid tokens to form a composite baseband signal, as shown in Figure 6.2. The output of the Adder token is inputted to the DSB-AM (double sideband, amplitude modulation) token from the Communications Library, as described in Chapter 1.

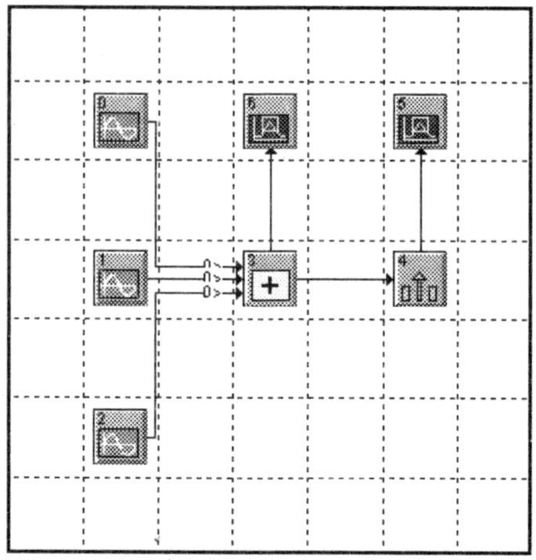

FIGURE 6.34 SystemVue simulation of bandpass sampling.

The parameters of the DSB-AM token are a carrier amplitude Ac = 10 V, a carrier frequency fc = 20 kHz, an amplitude modulation index $k_a = 0.75$, and 0° phase offset. The modulation index $k_a = 0.75$ ensures that the DSB-AM modulator does not distort, because $| k_a m(t) | < 1$ for all t, as given by Equation 1.15. The baseband analog information signal $| m(t) |_{max} \approx 1.15$, as shown in Figure 6.2.

The System Sampling Rate is set to 5 MHz and the System Time: Number of Samples specification to 4 194 304 (222) points, resulting in a spectral resolution of 1.19 Hz. To facilitate the simulation in SystemVue, the other Analysis token, except that for the output of the DSB-AM token, is deleted in the simulation model Fig6-34DT. As described in Chapter 1, the SystemVue Textbook Edition does not permit tokens to be deleted.

ON THE CD

The normalized ($R_L = 1\ \Omega$) power spectral density (PSD) of the DSB-AM token output is shown scaled and as a single-sided spectrum in Figure 6.35 in dBm/Hz. The DSB-AM spectrum consists of seven discrete components: six components at ± 500 Hz, ± 1.5 kHz, and ± 2.5 kHz above and below the carrier frequency f_c = 20 kHz, and the carrier component. For comparison, Figure 6.3 shows the baseband (unmodulated) spectrum of the sum of the three sinusoids.

The bandwidth of the bandpass analog signal is $f_H - f_L$ Hz = 22.5 – 17.5 kHz = 5 kHz, and the Nyquist bandpass sampling rate f_s must then be at least $2(f_H - f_L)$ Hz = 10 kHz. The System Sampling Rate now is set to 50 kHz for bandpass sampling. The System Time: Number of Samples specification is set to 32 768 (2^{15}) points, resulting in a comparable spectral resolution of 1.53 Hz. Because the System Time: Number of Samples has decreased from 4 194 304 to 32 768 (2^{19} to 2^{15}, less than 1% of the prior number of samples), the simulation time is also decreased significantly.

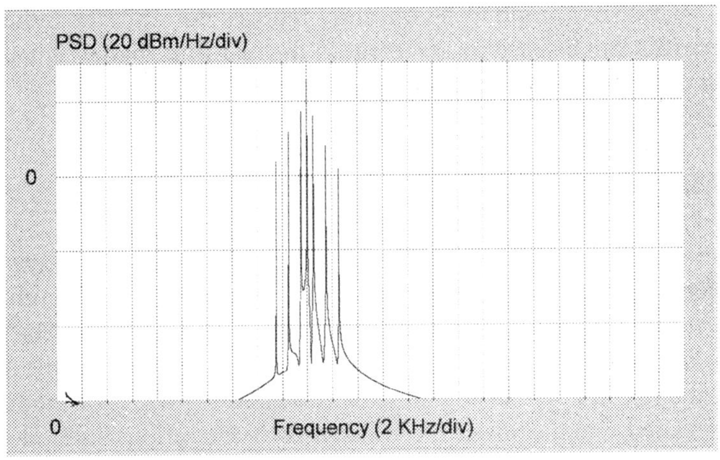

FIGURE 6.35 PSD of the DSB-AM signal.

The normalized PSD of the DSB-AM token output with bandpass sampling is shown scaled and as a single-sided spectrum in Figure 6.36 in dBm/Hz. A comparison of Figures 6.35 and 6.36 shows a nearly identical spectrum with the seven discrete components clearly evident. Finally, the maximum frequency shown in the scale of Figure 6.36, 50 kHz, is a value set only for direct comparison with Figure 6.35. The maximum discernable frequency in Figure 6.36 is $f_s/2$ = 50 kHz/2 = 25 kHz here.

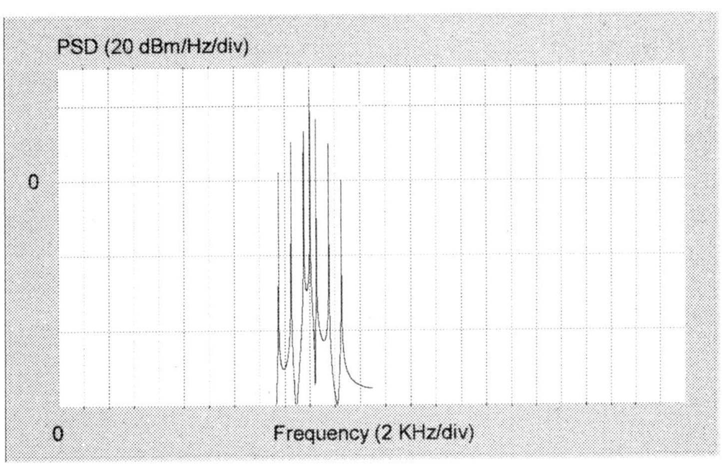

FIGURE 6.36 PSD of the DSB-AM signal with bandpass sampling.

Bandpass sampling significantly reduces not only the SystemVue simulation time, but can be used to advantage in the real-time digital signal processing of digital communication signals. [Gardner97].

SUMMARY

In this chapter, SystemVue was used to investigate and analyze the sampling and quantization of analog baseband signals. The performance of uniform and nonuniform quantization was assessed in a SystemVue simulation. The μ-law compandor for the processing of a voice signal was demonstrated with the input of a .wav file as a source in a SystemVue simulation. Baseband pulse code modulation (PCM) was simulated, and the power spectral density of the various line codes were obtained. A SystemVue simulation of differential PCM, as an extension of delta modulation in Chapter 2, used prediction to form a sequence of data to be transmitted. Finally, the sampling of an analog bandpass signal was shown to be facilitated by sampling at less than the customary Nyquist sampling rate with bandpass sampling.

REFERENCES

[Carlson02] Carlson, A. Bruce, et al., *Communication Systems*. McGraw-Hill, 2002.

[Gardner97] Gardner, Floyd, et al., *Simulation Techniques: Models of Communication Signals and Processes*. Wiley, 1997.

[Haykin01] Haykin, Simon, *Communication Systems*. Wiley, 2001.

[Lathi98] Lathi, B.P., Modern Digital and Analog Communication Systems, Oxford University Press, 1998.

[Stern04] Stern, Harold, et al., *Communication Systems Analysis and Design*. Prentice Hall, 2004.

[Xiong00] Xiong, Fuqin, *Digital Modulation Techniques*. Artech House, 2000.π

Appendix About the CD-ROM

In This Appendix

- System Requirements
- SystemVue Simulation Models

The CD-ROM to accompany *Digital Communication Systems Using SystemVue* contains the SystemVue Textbook Edition application software from Agilent EEsof (*http://eesof.tm.agilent.com*) that executes the simulation models in the text. The CD-ROM auto plays and installs the application software, the help files and the simulation models for each of the six chapters of this book, within individual chapter folders under the Examples folder. If the SystemVue Textbook Edition fails to install automatically, run the Setup application in the default directory of the CD-ROM. The simulation models are referenced by the figure number noted in the text, such as Fig1-1.svu.

The SystemVue Textbook Edition is a special application of the SystemVue Professional communication system simulator that, although somewhat restrictive, can execute the simulation models in the text. These restrictions include that the SystemVue simulation models here cannot be saved, no tokens can be added or deleted, and an existing token cannot be substituted for a different token in the same Standard Library or in the Optional Library.

However, the parameters of existing tokens in the SystemVue simulation models can be changed and there is no restriction on the SystemVue System Time. Advanced features of SystemVue that aren't used in the SystemVue simulation models here are also not available, such as the Automatic Program Generator, C Code Generation, and the Matlab M-link feature.

The SystemVue simulation models can execute without modification, except for those that use an external input text file for token parameter configuration or an external input and output file for Windows .wav files. For those simulation models, the Open External Source File text file parameter window (refer to Figure 1.12) should be modified to open the 16QAMV_32.txt, DQPSK.txt, and the QPSK.txt files in the Chapter 3 folder, and the QPSK.txt file in the Chapter 6 folder. The Open External Source File audio file parameter window (refer to Figure 1.13) should be modified to open the SVAudioIn.wav, SVAudioOut.wav, and the SVAudioOut2.wav files in the Chapter 1 folder, and the SVAudioIn.wav and SVAudioOut.wav files in the Chapter 6 folder. Browse the directory contents to find the locations substituted for these external files during the installation of the SystemVue Textbook Edition application software.

The SystemVue Textbook Edition does not permit you to delete tokens in a simulation model. However, an additional SystemVue simulation model is available where only the output of the bit error rate (BER) token is provided with a Final Value token. This additional model facilitates the analysis of the BER of a digital communication system by reducing the storage requirements of the simulation. These SystemVue simulation models are located in the chapter folders and have the same figure number reference, but are labeled "DT" (deleted Analysis tokens), as in Fig1-1DT.svu.

SYSTEM REQUIREMENTS

Digital communication system simulations require large amounts of RAM and adequate processor speeds. Minimum requirements are a Pentium III 1 GHz processor with at least 256 MB of RAM and the Windows 2000 or Windows XP operating system. Available hard drive storage requirements for the contents of the CD-ROM are approximately 50 MB, which includes the online SystemVue Help facility with manuals and application notes, and the SystemVue simulation models in the text.

SYSTEMVUE SIMULATION MODELS

The SystemVue simulation models are located in these folders:

Chapter 1: The SystemVue simulation models in this folder support Chapter 1, "Communication Simulation Techniques," which is an introduction to the simulation environment and the libraries and tokens. As a quick-start to the simulation environment, AM and FM analog communication systems are modeled in SystemVue.

Chapter 2: The SystemVue simulation models in this folder support Chapter 2, "Baseband Modulation and Demodulation." SystemVue simulation models are provided for rectangular, sinc, raised cosine, and multilevel pulse amplitude modulation. The optimum correlation receiver for baseband signals, partial response signaling, and delta modulation are also modeled in SystemVue.

Chapter 3: The SystemVue simulation models in this folder support Chapter 3, "Bandpass Modulation and Demodulation." SystemVue simulation models are provided for binary and multilevel amplitude, frequency, and phase shift keying. The optimum correlation receiver for bandpass signals, quadrature amplitude modulation, and differential phase shift keying are also modeled in SystemVue.

Chapter 4: The SystemVue simulation models in this folder support Chapter 4, "Synchronization and Equalization." Techniques for carrier frequency, carrier phase, and symbol synchronization are modeled in SystemVue. SystemVue simulation models are provided for channel equalization and channel models.

Chapter 5: The SystemVue simulation models in this folder support Chapter 5, "Multiplexing." SystemVue simulation models are provided for time, frequency, and code division multiplexing. Techniques for direct sequence and frequency hopping spread spectrum, and orthogonal frequency division multiplexing are modeled in SystemVue.

Chapter 6: The SystemVue simulation models in this folder support Chapter 6, "Sampling and Quantization." SystemVue simulation models are provided for sampling baseband and bandpass analog signals, companding, and pulse code modulation.

Bibliography

[Carlson02] Carlson, A. Bruce, et al., *Communication Systems.* McGraw-Hill, 2002.

[Dixon94] Dixon, Robert, *Spread Spectrum Systems with Commercial Applications.* Wiley, 1994.

[Gardner97] Gardner, Floyd, et al., *Simulation Techniques: Models of Communication Signals and Processes.* Wiley, 1997.

[Haykin01] Haykin, Simon, *Communication Systems.* Wiley, 2001.

[Holmes82] Holmes, John, *Coherent Spread Spectrum Systems.* Wiley, 1982.

[Lathi98] Lathi, B.P., *Modern Digital and Analog Communication Systems.* Oxford University Press, 1998.

[Lindsey91] Lindsey, William, et al., *Telecommunication Systems Engineering.* Dover, 1991.

[Papoulis02] Papoulis, Athanasios, *Probability, Random Variables and Stochastic Processes.* McGraw-Hill, 2002.

[Proakis01] Proakis, John, *Digital Communications.* McGraw-Hill, 2001.

[Proakis05] Proakis, John, et al., *Fundamentals of Communications Systems.* Prentice Hall, 2005.

[Shanmugan83] Shanmugan, K. Sam, *Digital and Analog Communication Systems.* Wiley, 1983.

[Simon95] Simon, Marvin, et al., *Digital Communications Techniques: Signal Design and Detection.* Prentice Hall, 1995.

[Simon01] Simon, Marvin, et al., *Spread Spectrum Communication Handbook.* McGraw-Hill, 2001.

[Sklar01] Sklar, Bernard, *Digital Communications.* Prentice Hall, 2001.

[Stern04] Stern, Harold, et al., *Communication Systems Analysis and Design.* Prentice Hall, 2004.

[Xiong00] Xiong, Fuqin, *Digital Modulation Techniques.* Artech House, 2000.

Index

C

D

E